Graduate Texts in Mathematics   257

T0155788

For other titles published in this series, go to
http://www.springer.com/series/136

Robin Hartshorne

Deformation Theory

 Springer

Robin Hartshorne
Department of Mathematics
University of California
Berkeley, CA 94720-3840
USA
robin@math.berkeley.edu

ISSN 0072-5285
ISBN 978-1-4614-2520-5   e-ISBN 978-1-4419-1596-2
DOI 10.1007/978-1-4419-1596-2
Springer New York Dordrecht Heidelberg London

Mathematics Subject Classification (2000): 14B07, 14B12, 14B10, 14B20, 13D10, 14D15, 14H60, 14D20

Printed on acid-free paper

Springer is part of Springer Science+Business Media (www.springer.com).

Contents

Preface

In the fall semester of 1979 I gave a course on deformation theory at Berkeley. My goal was to understand completely Grothendieck's local study of the Hilbert scheme using the cohomology of the normal bundle to characterize the Zariski tangent space and the obstructions to deformations. At the same time I started writing lecture notes for the course. However, the writing project soon foundered as the subject became more intricate, and the result was no more than five of a projected thirteen sections, corresponding roughly to sections 1, 2, 3, 5, 6 of the present book.

These handwritten notes circulated quietly for many years until David Eisenbud urged me to complete them and at the same time (without consulting me) mentioned to an editor at Springer, "You know Robin has these notes on deformation theory, which could easily become a book." When asked by Springer if I would write such a book, I immediately refused, since I was then planning another book on space curves. But on second thought, I decided this was, after all, a worthy project, and that by writing I might finally understand the subject myself.

So during 2004 I expanded the old notes into a rough draft, which I used to teach a course during the spring semester of 2005. Those notes, rewritten once more, with the addition of exercises, form the book you are now reading.

My goal in this book is to introduce the main ideas of deformation theory in algebraic geometry and to illustrate their use in a number of typical situations. I have made no effort to state results in the most general form, since I preferred to let the basic ideas shine forth unencumbered by technical details. Nor have I attempted to phrase results in the current "state of the art" language of stacks, since that requires a formidable apparatus of category theory. I hope that my elementary approach will be useful as a preparation for the new language in the same way that a thorough study of varieties is a good basis for understanding schemes and cohomology.

The prerequisite for reading this book is a basic familiarity with algebraic geometry as developed for example in [57].

Introduction

Deformation theory is the local study of deformations. Or, seen from another point of view, it is the infinitesimal study of a family in the neighborhood of a given element. A typical situation would be a flat morphism of schemes $f : X \to T$. For varying $t \in T$ we regard the fibers X_t as a family of schemes. Deformation theory is the infinitesimal study of the family in the neighborhood of a special fiber X_0.

Closely connected with deformation theory is the question of existence of varieties of moduli. Suppose we try to classify some set of objects, such as curves of genus g. Not only do we want to describe the set of isomorphism classes of curves as a set, but also we wish to describe families of curves. So we seek a universal family of curves, parametrized by a *variety of moduli* M, such that each isomorphism class of curves occurs exactly once in the family. Deformation theory would then help us infer properties of the variety of moduli M in the neighborhood of a point $0 \in M$ by studying deformations of the corresponding curve X_0. Even if the variety of moduli does not exist, deformation theory can be useful for the classification problem.

The purpose of this book is to establish the basic techniques of deformation theory, to see how they work in various standard situations, and to give some interesting examples and applications from the literature.

We will focus our attention on four standard situations.

Situation A. Subschemes of a fixed scheme X. The problem in this case is to deform the subschemes while keeping the ambient scheme fixed.

Situation B. Line bundles on a fixed scheme X.

Situation C. Vector bundles, or more generally coherent sheaves, on a fixed scheme X.

Situation D. Deformations of abstract schemes. This includes the local study of deformations of singularities, and the global study of deformations of nonsingular varieties.

R. Hartshorne, *Deformation Theory*, Graduate Texts in Mathematics 257,
DOI 10.1007/978-1-4419-1596-2_1, © Robin Hartshorne 2010

For each of these situations, we will consider a number of different questions. The ultimate goal is to have a global parameter space that classifies isomorphism classes of the objects in question. For example, in Situation A there is the Hilbert scheme, and in Situation D there is the variety of moduli of curves. In this book we will not prove the existence of these global parameter spaces. Our goal is rather to lay the foundations of the deformation theory that provides insight into the local structure of the global parameter space.

We start in Chapter 1 with deformations over the ring of dual numbers, which one can call first-order infinitesimal deformations. For Situations A, B, C, we can do this using the usual cohomology of coherent sheaves and the Ext groups. For Situation D we need something more, and for this purpose we introduce the cotangent complex and the T^i functors of Lichtenbaum and Schlessinger. Along the way, we see that nonsingular varieties play a special role, since their local deformations are all trivial. We show how they satisfy an infinitesimal lifting property, and that they are characterized by the vanishing of the T^1 functors. In any of our situations, when a good moduli space exists, the deformations over the dual numbers studied in this chapter will allow us to compute the Zariski tangent space to the moduli space.

In Chapter 2 we study higher-order deformations. The problem here is still infinitesimal: given a deformation over a local Artin ring A, can one extend it to a larger Artin ring A', and if so in how many ways? In general this is not always possible, and there is a corresponding obstruction theory. In the case of a good moduli space, the vanishing of the obstructions will imply that the moduli space is smooth. We show how this works for each of the four standard situations. In Situation A, we also describe several classes of subschemes for which there are no local obstructions, namely Cohen–Macaulay subschemes in codimension 2, Gorenstein subschemes in codimension 3, and locally complete intersection subschemes of any codimension. The obstruction theory allows us to give a bound on the dimension of the local rings of the parameter space, and we apply this to prove the classical result that the Hilbert scheme of curves of degree d in $\mathbb{P}^3$ has dimension at least $4d$ in every component. We then give as an application Mumford's example of a nonreduced component of the Hilbert scheme of nonsingular curves in $\mathbb{P}^3$.

Passing to the limit over larger and larger Artin rings gives rise to the notion of formal deformations, which we study in Chapter 3. Some situations are better than others. In the best possible case we get a formal deformation that accurately encodes all possible infinitesimal deformations, in which case we have a *pro-representable* functor of Artin rings. We give Schlessinger's criterion for pro-representability and show how it applies to each situation. If the functor is not pro-representable, there are the weaker notions of miniversal and versal families of deformations. As an example of the formal theory, we study the question of lifting varieties from characteristic p to characteristic 0 and give Serre's example of a nonliftable 3-fold.

To go from a formal family defined over a complete local ring to an algebraic family defined over a ring of finite type over the base, there is a theory of algebraizability due to Artin. We will mention this briefly, but without proofs, because to develop this theory fully would carry us too far afield.

In Chapter 4 we discuss global moduli questions. We introduce the language of functors and talk about a fine moduli space (corresponding to a representable functor) or a coarse moduli space. We describe various properties that are useful to test whether a functor is representable, though there is no satisfactory criterion (as there was in the case of the functors of Artin rings in Chapter 3) to determine whether a functor is representable. We illustrate these concepts in a number of cases: the Hilbert scheme for Situation A, the Picard scheme for Situation B, and the variety of moduli of stable vector bundles for Situation C. For Situation D, we describe in detail the moduli question for rational curves and elliptic curves. For curves of genus ≥ 2, we describe the *modular families* of Mumford, which help explain the functor of deformations of curves in the absence of a fine moduli space. As applications of the general theory we give Mori's theorem on the existence of rational curves in a nonsingular variety in characteristic p whose canonical divisor is not numerically effective. In a final section we study the question of smoothing singularities. We introduce the infinitesimal notion of *formally smoothable* scheme and use this to give examples of nonsmoothable singularities.

As the reader is probably aware, one of the big problems with global deformation questions is that the associated functor is not always representable by a scheme. This has led to various efforts to enlarge the category of schemes so that the functors will be representable, for example, by using the algebraic spaces of Artin and Knutson. More recently, the most promising way of dealing with global moduli questions seems to be with the theory of stacks, introduced by Deligne and Mumford. The reader may wonder why I say so little about stacks in this book. Two reasons are (a) it would take another whole book to do justice to the subject, and (b) I am not competent to write that book. I hope, however, that the present book will do a reasonable job of explaining deformation theory up to, but not including, the theory of stacks. And I believe that the material presented here, both by its successes and its failures, will provide good motivation for the study of stacks.

Perhaps I should also say what is not in this book. I do not include a proof of the existence of the Hilbert scheme, though I make frequent use of it in examples and proofs of other results. I do not discuss the geometric invariant theory of Mumford, and hence do not prove the existence of the coarse moduli schemes for curves and for stable vector bundles. I do not prove or make any use of Artin's approximation theorems. There are no simplicial complexes, no fibered categories, no differential graded algebras, and no derived categories. I preferred in each case to see how far one can go with elementary methods, even though some results could be sharpened and some proofs simplified by bringing in the big guns.

Finally, a remark on the generality of hypotheses. I will often state a result in a restricted situation to bring forth more clearly its essence. Later in the book, I may break one of the rules of mathematical exposition by applying it in a context wider than originally stated. I felt it necessary to make this compromise, since to state results in their most general formulation from the outset would make the book unreadable. For example, I usually assume that the ground field is algebraically closed, though that may not be necessary, so that there is one less thing to worry about. I am confident that the reader will have no difficulty disengaging the more general context in which the result may be true.

The book is divided into four chapters and twenty-nine sections. Cross-references to results in the main text are given by section and an internal number, e.g., (16.2), a theorem, or (29.10.3), an example. References to the exercises, which have additional results and examples, are preceded by Ex, e.g., (Ex. 5.8). References to the bibliography are in square brackets, e.g., [21].

I would like to thank all those people who have helped me in the preparation of this book, teachers, colleagues, and students: those who explained subtle points to me; those who answered my questions and provided references; those who asked questions prodding me to deeper understanding; and those who read parts of the manuscript and made valuable comments. If I were to begin to list your names, it would be a very long list and I would surely forget some, so I had better just say, thank you all, you know who you are.

I have done my best to state only true theorems and give only correct proofs, but in spite of all the help I have received, I am sure there are still some errors to challenge the careful reader. Please let me know when you find them.

1

First-Order Deformations

We start by introducing the Hilbert scheme, which will be a model for the other situations, and which will provide us with examples as we go along. Then in Section 2 we discuss deformations over the dual numbers for Situations A, B, and C. In Section 3 we introduce the cotangent complex and the T^i functors, which are needed to discuss deformations of abstract schemes (Situation D) in Section 5. In Section 4 we examine the special role of nonsingular varieties, using the infinitesimal lifting property and the T^i functors. We also show that the relative notion of a smooth morphism is characterized by the vanishing of the relative T^1 functors.

1. The Hilbert Scheme

As motivation for all the local study of deformations we are about to embark on, we will introduce the Hilbert scheme of Grothendieck, as a typical example of the goals of this work. The Hilbert scheme gives a particularly satisfactory answer to the problem of describing families of closed subschemes of a given scheme. In fact, when I first lectured on this subject and wrote some preliminary notes that have grown into this book, my goal was to understand completely the proof of the following theorem.

Theorem 1.1. *Let Y be a closed subscheme of the projective space $X = \mathbb{P}^n_k$ over a field k. Then*

(a) *There exists a projective scheme H, called the* Hilbert scheme, *parametrizing closed subschemes of X with the same Hilbert polynomial P as Y, and there exists a universal subscheme $W \subseteq X \times H$, flat over H, such that the fibers of W over closed points $h \in H$ are all closed subschemes of X with the same Hilbert polynomial P. Furthermore, H is universal in the sense that if T is any other scheme, if $W' \subseteq X \times T$ is a closed subscheme, flat over T, all of whose fibers are subschemes of X with the same Hilbert*

R. Hartshorne, *Deformation Theory*, Graduate Texts in Mathematics 257, DOI 10.1007/978-1-4419-1596-2_2, © Robin Hartshorne 2010

polynomial P, then there exists a unique morphism $\varphi : T \to H$ such that $W' = W \times_H T$ as subschemes of $X \times T$.

(b) *The Zariski tangent space to H at the point $y \in H$ corresponding to Y is given by $H^0(Y, \mathcal{N})$, where $\mathcal{N}$ is the normal sheaf of Y in X.*

(c) *If Y is a locally complete intersection, and if $H^1(Y, \mathcal{N}) = 0$, then H is nonsingular at the point y, of dimension equal to $h^0(Y, \mathcal{N}) = \dim_k H^0(Y, \mathcal{N})$.*

(d) *In any case, if Y is a locally complete intersection, the dimension of H at y is at least $h^0(Y, \mathcal{N}) - h^1(Y, \mathcal{N})$.*

Parts (a), (b), (c) of this theorem are due to Grothendieck [45]. For part (d) there are recent proofs due to Laudal [92] and Mori [109]. I do not know whether there is an earlier reference.

Since the main purpose of this book is to study the local theory, we will not prove the existence (a) of the Hilbert scheme. The proof of existence uses techniques quite different from those we consider here, and is not necessary for the comprehension of anything in this book. The reader who wishes to see a proof can consult any of many sources [45, exposé 221], [115], [151], [161], [152]. Parts (b), (c), (d) of the theorem will be proved in §2, §9, and §11, respectively.

Parts (b), (c), (d) of this theorem illustrate the benefit derived from Grothendieck's insistence on the systematic use of nilpotent elements. Let $D = k[t]/t^2$ be the ring of dual numbers. Taking D as our parameter scheme, we see from the universal property (a) that flat families $Y' \subseteq X \times D$ with closed fiber Y are in one-to-one correspondence with morphisms of schemes $\operatorname{Spec} D \to H$ that send the unique point to y. This set $\operatorname{Hom}_y(D, H)$ in turn can be interpreted as the Zariski tangent space to H at y [57, II, Ex. 2.8]. Thus to prove (b) of the theorem, we have only to classify schemes $Y' \subseteq X \times D$, flat over D, whose closed fiber is Y, which we will do in §2.

Part (c) of the theorem is related to *obstruction theory*. Given an infinitesimal deformation defined over an Artin ring A, to extend the deformation over a larger Artin ring there is usually some obstruction, whose vanishing is necessary and sufficient for the existence of an extended deformation. For closed subschemes with no local obstructions, such as locally complete intersection subschemes, the obstructions lie in $H^1(Y, \mathcal{N})$. If that group is zero, there are no obstructions, and the corresponding moduli space is nonsingular. The dimension estimate (d) comes out of obstruction theory.

Exercises.

1.1. Curves in $\mathbb{P}^2$.   Here we will verify the existence of the Hilbert scheme for curves in $\mathbb{P}^2$. Over an algebraically closed field k, we define a *curve* in $\mathbb{P}^2_k$ to be the closed subscheme defined by a homogeneous polynomial $f(x, y, z)$ of degree d in the coordinate ring $S = k[x, y, z]$. We can write f as $a_0 x^d + \cdots + a_n z^d$, $a_i \in k$, with $n = \binom{d+2}{2} - 1$ since f has that many terms. Consider $(a_0, \ldots, a_n)$ as a point in $\mathbb{P}^n_k$.

(a) Show that curves of degree d in $\mathbb{P}^2$ are in a one-to-one correspondence with points of $\mathbb{P}^n$ by this correspondence.

(b) Define $\mathcal{C} \subseteq \mathbb{P}^2 \times \mathbb{P}^n$ by the equation $f = a_0 x^d + \cdots + a_n z^d$ above, where the x, y, z are coordinates on $\mathbb{P}^2$ and $a_0, \ldots, a_n$ are coordinates on $\mathbb{P}^n$. Show that the correspondence of (a) is given by $a \in \mathbb{P}^n$ goes to the fiber $\mathcal{C}_a \subseteq \mathbb{P}^2$ over the point a. Therefore we call $\mathcal{C}$ a *tautological* family.

(c) For any finitely generated k-algebra A, we define a *family* of curves of degree d in $\mathbb{P}^2$ over A to be a closed subscheme $X \subseteq \mathbb{P}^2_A$, flat over A, whose fibers above closed points of Spec A are curves in $\mathbb{P}^2$. Show that the ideal $I_X \subseteq A[x, y, z]$ is generated by a single homogeneous polynomial f of degree d in $A[x, y, z]$.

(d) Conversely, if $f \in A[x, y, z]$ is homogeneous of degree d, what is the condition on f for the zero-scheme X defined by f to be flat over A? (Do not assume A reduced.)

(e) Show that the family $\mathcal{C}$ is *universal* in the sense that for any family $X \subseteq \mathbb{P}^2_A$ as in c), there is a unique morphism Spec $A \to \mathbb{P}^n$ such that $X = \mathcal{C} \times_{\mathbb{P}^n}$ Spec A.

(f) For any curve $X \subseteq \mathbb{P}^2_k$ of degree d, show that $h^0(\mathcal{N}_X) = n$ and $h^1(\mathcal{N}_X) = 0$. (Do not assume X nonsingular.)

1.2. Curves on quadric surfaces in $\mathbb{P}^3$. Consider the family $\mathcal{C}$ of all nonsingular curves C that lie on some nonsingular quadric surface Q in $\mathbb{P}^3$ and have bidegree (a, b) with $a, b > 0$.

(a) By considering the linear system of curves C on a fixed Q, and then varying Q, show that if the total degree d is equal to $a + b \geq 5$, then the dimension of the family $\mathcal{C}$ is $ab + a + b + 9$.

(b) If $a, b \geq 3$, show that $H^0(C, \mathcal{N}_C)$ has the same dimension $ab + a + b + 9$, using the exact sequence of normal bundles

$$0 \to \mathcal{N}_{C/Q} \to \mathcal{N}_C \to \mathcal{N}_Q|_C \to 0.$$

Show that $\mathcal{N}_{C/Q} \cong \mathcal{O}_C(C^2)$ is nonspecial, i.e., its H^1 is zero, so you can compute its H^0 by Riemann–Roch. Then note that $\mathcal{N}_Q|_C \cong \mathcal{O}_C(2)$, and compute its H^0 using the exact sequence

$$0 \to \mathcal{O}_Q(2 - C) \to \mathcal{O}_Q(2) \to \mathcal{O}_C(2) \to 0$$

and the vanishing theorems for H^1 of line bundles on Q given in [57, III, Ex. 5.6].

(c) Conclude that for $a, b \geq 3$ the family $\mathcal{C}$ gives (an open subset of) an irreducible component of the Hilbert scheme of dimension $ab + a + b + 9$, which is smooth at each of its points.

(d) What goes wrong with this argument if $a = 2$ and $b \geq 4$? Cf. (Ex. 6.4).

1.3. Complete intersection curves in $\mathbb{P}^3$. A curve C in $\mathbb{P}^3_k$ is a *complete intersection* if its homogeneous ideal $I \subseteq k[x, y, z, w]$ is generated by two homogeneous polynomials. Let C be a complete intersection curve defined by polynomials of degrees $a, b \geq 1$.

(a) The complete intersection curve C has degree $d = ab$ and arithmetic genus $g = \frac{1}{2}ab(a + b - 4) + 1$. The dualizing sheaf ω_C is isomorphic to $\mathcal{O}_C(a + b - 4)$. For any $a, b \geq 1$, a general such complete intersection curve is nonsingular. The family of all such curves is irreducible and of dimension $2\binom{a+3}{3} - 2$ if $a = b$ or $\binom{a+3}{3} + \binom{b+3}{3} - \binom{b-a+3}{3} - 2$ if $a < b$.

(b) The normal sheaf is $\mathcal{N}_C \cong \mathcal{O}_C(a) \oplus \mathcal{O}_C(b)$. Using the resolution

$$0 \to \mathcal{O}_{\mathbb{P}}(-a-b) \to \mathcal{O}_{\mathbb{P}}(-a) \oplus \mathcal{O}_{\mathbb{P}}(-b) \to \mathcal{I}_C \to 0,$$

verify that $H^0(\mathcal{N}_C)$ has dimension equal to the dimension of the family above, so that the family of complete intersection curves defined by polynomials of degrees a and b is a nonsingular open subset of an irreducible component of the Hilbert scheme.

1.4. The limit of a flat family of complete intersection curves in $\mathbb{P}^3$ need not be a complete intersection curve. In other words, the open set of the Hilbert scheme formed by complete intersection curves may not be closed. For an example, fix a $\lambda \in k$, $\lambda \neq 0, 1$, and consider the family of complete intersection curves over $k[t, t^{-1}]$ defined by the equations

$$\begin{cases} tyz - wx = 0, \\ yw - t(x-z)(x-\lambda z) = 0. \end{cases}$$

(a) Show that for any $t \neq 0$, these equations define a nonsingular cure $\mathcal{C}_t$ of degree 4 and genus 1.
(b) Now extend this family to a flat family over all of $k[t]$, and show that the special fiber C_0 over $t = 0$ is the union of a nonsingular plane cubic curve with a line not in that plane, but meeting the cubic curve at one point. Show also that C_0 is not a complete intersection. Since C_0 is a singular curve belonging to a flat family whose general member is nonsingular, we say that C_0 is a *smoothable* singular curve.

Note. What is happening in this example is that the curves C_t, as t approaches zero, are being pushed away from the point $P : (x, y, z, w) = (0, 0, 0, 1)$ of the curve toward the plane $w = 0$. In the end the irreducible curve C_t breaks into two pieces: the plane cubic curve plus a line through P.

1.5. Show that the Hilbert scheme of degree 4 and genus 1 curves is still nonsingular of dimension 16 at the point corresponding to the curve C_0 of (Ex. 1.4).

(a) First show that if a curve Y is the union of two nonsingular curves C and D in $\mathbb{P}^3$, meeting transversally at a single point P, then there are exact sequences of normal sheaves

$$0 \to \mathcal{N}_Y \to \mathcal{N}_Y|_D \oplus \mathcal{N}_Y|_C \to \mathcal{N}_Y \otimes k_P \to 0$$

and

$$0 \to \mathcal{N}_C \to \mathcal{N}_Y|_C \to k_P \to 0,$$
$$0 \to \mathcal{N}_D \to \mathcal{N}_Y|_D \to k_P \to 0.$$

(b) Apply these sequences to the union of a plane cubic curve and a line C_0 as above, to show that $h^0(\mathcal{N}_{C_0}) = 16$. Since C_0 is contained in the closure of the complete intersection curves, which form a family of dimension 16, this shows that the Hilbert scheme is smooth at C_0. For another proof of this fact, see (Ex. 8.3).

1.6. Twisted cubic curves. The twisted cubic curve in $\mathbb{P}^3$ is defined parametrically by $(x_0, x_1, x_2, x_3) = (u^3, tu^2, t^2u, t^3)$ for $(t, u) \in \mathbb{P}^1$. More generally we call any curve obtained from this one by a linear change of coordinates in $\mathbb{P}^3$ a *twisted cubic curve*.

(a) Show that any nonsingular curve of degree 3 and genus 0 in $\mathbb{P}^3$ is a twisted cubic curve. Show that these form a family of dimension 12, and that $H^0(C, \mathcal{N}_C) = 12$ for any such curve. Thus the twisted cubic curves form a nonsingular open subset of an irreducible component of the Hilbert scheme of curves with Hilbert polynomial $3z + 1$.

(b) Consider a subscheme $Y \subseteq \mathbb{P}^3$ that is a disjoint union of a plane cubic curve and a point. Show that these schemes form another nonsingular open subset of the Hilbert scheme of curves with Hilbert polynomial $3z + 1$. This component has dimension 15.

(c) There is a flat family of twisted cubic curves whose limit is a curve Y_0, supported on a plane nodal cubic curve, and having an embedded point at the node [57, III, 9.8.4]. Show that this curve is in the closure of both irreducible components mentioned above, hence corresponds to a singular point on the Hilbert scheme.

(d) Now show that $h^0(\mathcal{N}_{Y_0/\mathbb{P}^3}) = 16$, confirming that Y_0 is a singular point of the Hilbert scheme. *Hint:* Show that the homogeneous ideal of Y_0, $I = (z^2, yz, xz, y^2w - x^2(x + w))$, has a resolution over the polynomial ring $R = k[x, y, z, w]$ as follows:

$$R(-3)^3 \oplus R(-4) \to R(-2)^3 \oplus R(-3) \to I \to 0.$$

Tensor with $B = R/I$, then dualize and sheafify to get a resolution

$$0 \to \mathcal{N}_{Y/\mathbb{P}^3} \to \mathcal{O}_{Y_0}(2)^3 \oplus \mathcal{O}_{Y_0}(3) \to \mathcal{O}_{Y_0}(3)^3 \oplus \mathcal{O}_{Y_0}(4).$$

Compute explicitly with the sections of $\mathcal{O}_{Y_0}(2)$ and $\mathcal{O}_{Y_0}(3)$, which all come from polynomials in R, to show that $h^0(\mathcal{N}_{Y/\mathbb{P}^3}) = 16$.

Note. The structure of this Hilbert scheme is studied in detail in the paper [134].

1.7. Let C be a nonsingular curve in $\mathbb{P}^n$ that is *nonspecial*, i.e., $H^1(\mathcal{O}_C(1)) = 0$. Show that the Hilbert scheme is nonsingular at the point corresponding to C. *Hint:* Use the Euler sequence for the tangent bundle on $\mathbb{P}^n$, restricted to C, and use the exact sequence relating the tangent bundle of C, the tangent bundle of $\mathbb{P}^n$, and the normal bundle of C.

2. Structures over the Dual Numbers

The very first deformation question to study is structures over the dual numbers $D = k[t]/t^2$. That is, one gives a structure (e.g., a scheme, or a scheme with a subscheme, or a scheme with a sheaf on it) over k and one seeks to classify extensions of this structure over the dual numbers. These are also called first-order deformations.

To ensure that our structure is evenly spread out over the base, we will always assume that the extended structure is *flat* over D. Flatness is the technical condition that corresponds to the intuitive idea of a deformation.

In this section we will apply this study to Situations A, B, and C.

Recall that a module M is *flat* over a ring A if the functor $N \mapsto N \otimes_A M$ is exact on the category of A-modules. A morphism of schemes $f : X \to Y$ is *flat* if for every point $x \in X$, the local ring $\mathcal{O}_{x,X}$ is flat over the ring $\mathcal{O}_{f(x),Y}$. A sheaf of $\mathcal{O}_X$-modules $\mathcal{F}$ is *flat* over Y if for every $x \in X$, its stalk $\mathcal{F}_x$ is flat over $\mathcal{O}_{f(x),Y}$.

Lemma 2.1. *A module M over a noetherian ring A is flat if and only if for every prime ideal $\mathfrak{p} \subseteq A$, $\mathrm{Tor}_1^A(M, A/\mathfrak{p}) = 0$.*

Proof. The exactness of the functor $N \mapsto N \otimes_A M$ is equivalent to $\mathrm{Tor}_1(M, N) = 0$ for all A-modules N. Since Tor commutes with direct limits, it is sufficient to require $\mathrm{Tor}_1(M, N) = 0$ for all finitely generated A-modules N. Now over a noetherian ring A, a finitely generated module N has a filtration whose quotients are of the form $A/\mathfrak{p}_i$ for various prime ideals $\mathfrak{p}_i \subseteq A$ [103, p. 51]. Thus, using the exact sequence of Tor, we see that $\mathrm{Tor}_1(M, A/\mathfrak{p}) = 0$ for all $\mathfrak{p}$ implies $\mathrm{Tor}_1(M, N) = 0$ for all N; hence M is flat.

In the sequel, we will often make use of the following result, which is a special case of the "local criterion of flatness."

Proposition 2.2. *Let $A' \to A$ be a surjective homomorphism of noetherian rings whose kernel J has square zero. Then an A'-module M' is flat over A' if and only if*

(1) *$M = M' \otimes_{A'} A$ is flat over A, and*
(2) *the natural map $M \otimes_A J \to M'$ is injective.*

Proof. Note that since J has square zero, it is an A-module and we can identify $M' \otimes_{A'} J$ with $M \otimes_A J$.

If M' is flat over A', then (1) follows by base extension, and (2) follows by tensoring M' with the exact sequence

$$0 \to J \to A' \to A \to 0.$$

Suppose conversely that M' satisfies conditions (1) and (2). By the lemma, it is sufficient to show that $\mathrm{Tor}_1^{A'}(M', A'/\mathfrak{p}') = 0$ for every prime ideal $\mathfrak{p}' \subseteq A'$. Since J is nilpotent, it is contained in $\mathfrak{p}'$. Letting $\mathfrak{p}$ be the prime ideal $\mathfrak{p}'/J$ of A, we can write a diagram of exact sequences

$$
\begin{array}{ccccccccc}
& & & & 0 & & 0 & & \\
& & & & \downarrow & & \downarrow & & \\
0 & \to & J & \to & \mathfrak{p}' & \to & \mathfrak{p} & \to & 0 \\
& & \| & & \downarrow & & \downarrow & & \\
0 & \to & J & \to & A' & \to & A & \to & 0 \\
& & & & \downarrow & & \downarrow & & \\
& & & & A'/\mathfrak{p}' & = & A/\mathfrak{p} & & \\
& & & & \downarrow & & \downarrow & & \\
& & & & 0 & & 0 & &
\end{array}
$$

Tensoring with M' we obtain

$$
\begin{array}{ccc}
& 0 & 0 \\
& \downarrow & \downarrow \\
& \mathrm{Tor}_1^{A'}(M', A'/\mathfrak{p}') \to & \mathrm{Tor}_1^A(M, A/\mathfrak{p}) \\
& \downarrow & \downarrow \\
M \otimes_A J \to & M' \otimes_{A'} \mathfrak{p}' \to & M \otimes_A \mathfrak{p} \to 0 \\
\| & \downarrow & \downarrow \\
M \otimes_A J \to & M' \to & M \to 0 \\
& \downarrow & \downarrow \\
& M' \otimes_{A'} A'/\mathfrak{p}' = & M \otimes_A A/\mathfrak{p} \\
& \downarrow & \downarrow \\
& 0 & 0
\end{array}
$$

By hypothesis (2), the second (and therefore also the first) horizontal sequence is exact on the left. It follows from the snake lemma that the Tors at the top are isomorphic. The second is zero by hypothesis (1), so the first is also, as required.

Now we consider our first deformation problem, Situation A. Let X be a scheme over k and let Y be a closed subscheme of X. We define a *deformation of Y over D in X* to be a closed subscheme $Y' \subseteq X' = X \times D$, flat over D, such that $Y' \times_D k = Y$. We wish to classify all deformations of Y over D.

We consider the affine case first. Then X corresponds to a k-algebra B, and Y is defined by an ideal $I \subseteq B$. We are seeking ideals $I' \subseteq B' = B[t]/t^2$ with B'/I' flat over D and such that the image of I' in $B = B'/tB'$ is just I. Note that $(B'/I') \otimes_D k = B/I$. Since B is automatically flat over k, by (2.2) the flatness of B'/I' over D is equivalent to the exactness of the sequence

$$0 \to B/I \xrightarrow{t} B'/I' \to B/I \to 0.$$

Suppose I' is such an ideal, and consider the diagram

$$
\begin{array}{ccccccccc}
& & 0 & & 0 & & 0 & & \\
& & \downarrow & & \downarrow & & \downarrow & & \\
0 & \to & I & \xrightarrow{t} & I' & \to & I & \to & 0 \\
& & \downarrow & & \downarrow & & \downarrow & & \\
0 & \to & B & \xrightarrow{t} & B' & \to & B & \to & 0 \\
& & \downarrow & & \downarrow & & \downarrow & & \\
0 & \to & B/I & \xrightarrow{t} & B'/I' & \to & B/I & \to & 0 \\
& & \downarrow & & \downarrow & & \downarrow & & \\
& & 0 & & 0 & & 0 & &
\end{array}
$$

where the exactness of the bottom row implies the exactness of the top row.

Proposition 2.3. *In the situation above, to give $I' \subseteq B'$ such that B'/I' is flat over D and the image of I' in B is I is equivalent to giving an element $\varphi \in \operatorname{Hom}_B(I, B/I)$. In particular, $\varphi = 0$ corresponds to the trivial deformation given by $I' = I \oplus tI$ inside $B' \cong B \oplus tB$.*

Proof. We will make use of the splitting $B' = B \oplus tB$ as B-modules, or, equivalently, of the section $\sigma : B \to B'$ given by $\sigma(b) = b + 0 \cdot t$, which makes B' into a B-module.

Take any element $x \in I$. Lift it to an element of I', which, using the splitting of B', can be written $x + ty$ for some $y \in B$. Two liftings differ by something of the form tz with $z \in I$. Thus y is not uniquely determined, but its image $\bar{y} \in B/I$ is. Now sending x to $\bar{y}$ defines a mapping $\varphi : I \to B/I$. It is clear from the construction that it is a B-module homomorphism.

Conversely, suppose $\varphi \in \operatorname{Hom}_B(I, B/I)$ is given. Define

$$I' = \{x + ty \mid x \in I, \ y \in B, \text{ and the image of } y \text{ in } B/I \text{ is equal to } \varphi(x)\}.$$

Then one checks easily that I' is an ideal of B', that the image of I' in B is I, and that there is an exact sequence

$$0 \to I \xrightarrow{t} I' \to I \to 0.$$

Therefore there is a diagram as before, where this time the exactness of the top row implies the exactness of the bottom row, and hence that B'/I' is flat over D.

These two constructions are inverse to each other, so we obtain a natural one-to-one correspondence between the set of such I' and the set $\operatorname{Hom}_B(I, B/I)$, whereby the trivial deformation $I' = I \oplus tI$ corresponds to the zero element.

Now we wish to globalize this argument to the case of a scheme X over k and a given closed subscheme Y. There are two ways to do this. One is to cover X with open affine subsets and use the above result. The construction is compatible with localization, and the correspondence is natural, so we get a one-to-one correspondence between the flat deformations $Y' \subseteq X' = X \times D$ and elements of the set $\operatorname{Hom}_X(\mathcal{I}, \mathcal{O}_Y)$, where $\mathcal{I}$ is the ideal sheaf of Y in X.

The other method is to repeat the above proof in the global case, simply dealing with sheaves of ideals and rings, on the topological space of X (which is equal to the topological space of X').

Before stating the conclusion, we will define the normal sheaf of Y in X. Note that the group $\operatorname{Hom}_X(\mathcal{I}, \mathcal{O}_Y)$ can be regarded as $H^0(X, \mathcal{H}om_X(\mathcal{I}, \mathcal{O}_Y))$. Furthermore, homomorphisms of $\mathcal{I}$ to $\mathcal{O}_Y$ factor through $\mathcal{I}/\mathcal{I}^2$, which is a sheaf on Y. So

$$\mathcal{H}om_X(\mathcal{I}, \mathcal{O}_Y) = \mathcal{H}om_Y(\mathcal{I}/\mathcal{I}^2, \mathcal{O}_Y),$$

and this latter sheaf is called the *normal sheaf* of Y in X, and is denoted by $\mathcal{N}_{Y/X}$. If X is nonsingular and Y is a locally complete intersection in X, then

$\mathcal{I}/\mathcal{I}^2$ is locally free, so $\mathcal{N}_{Y/X}$ is locally free also and can be called the *normal bundle* of Y in X. This terminology derives from the fact that if Y is also nonsingular, there is an exact sequence

$$0 \to \mathcal{T}_Y \to \mathcal{T}_X|_Y \to \mathcal{N}_{Y/X} \to 0,$$

where $\mathcal{T}_Y$ and $\mathcal{T}_X$ denote the tangent sheaves to Y and X, respectively. In this case, therefore, $\mathcal{N}_{Y/X}$ is the usual normal bundle.

Summing up our results gives the following.

Theorem 2.4. *Let X be a scheme over a field k, and let Y be a closed subscheme of X. Then the deformations of Y over D in X are in natural one-to-one correspondence with elements of $H^0(Y, \mathcal{N}_{Y/X})$, the zero element corresponding to the trivial deformation.*

Corollary 2.5. *If Y is a closed subscheme of the projective space $X = \mathbb{P}^n_k$, then the Zariski tangent space of the Hilbert scheme H at the point y corresponding to Y is isomorphic to $H^0(Y, \mathcal{N}_{Y/X})$.*

Proof. The Zariski tangent space to H at y can be interpreted as the set of morphisms from the dual numbers D to H sending the closed point to y [57, II, Ex. 2.8]. Because of the universal property of the Hilbert scheme (1.1(a)), this set is in one-to-one correspondence with the set of deformations of Y over the dual numbers, which by (2.4) is $H^0(Y, \mathcal{N}_{Y/X})$.

Next we consider Situation B. Let X be a scheme over k and let $\mathcal{L}$ be an invertible sheaf on X. We will study the set of isomorphism classes of invertible sheaves $\mathcal{L}'$ on $X' = X \times D$ such that $\mathcal{L}' \otimes \mathcal{O}_X \cong \mathcal{L}$. In this case flatness is automatic, because $\mathcal{L}'$ is locally free and X' is flat over D.

Proposition 2.6. *Let X be a scheme over k, and $\mathcal{L}$ an invertible sheaf on X. The set of isomorphism classes of invertible sheaves $\mathcal{L}'$ on $X \times D$ such that $\mathcal{L}' \otimes \mathcal{O}_X \cong \mathcal{L}$ is in natural one-to-one correspondence with elements of the group $H^1(X, \mathcal{O}_X)$.*

Proof. We use the fact that on any ringed space X, the isomorphism classes of invertible sheaves are classified by $H^1(X, \mathcal{O}_X^*)$, where $\mathcal{O}_X^*$ is the sheaf of multiplicative groups of units in $\mathcal{O}_X$ [57, III, Ex. 4.5]. The exact sequence

$$0 \to \mathcal{O}_X \xrightarrow{t} \mathcal{O}_{X'} \to \mathcal{O}_X \to 0$$

gives rise to an exact sequence of sheaves of abelian groups

$$0 \to \mathcal{O}_X \xrightarrow{\alpha} \mathcal{O}_{X'}^* \to \mathcal{O}_X^* \to 0,$$

where $\alpha(x) = 1 + tx$. Here $\mathcal{O}_X$ is an additive group, while $\mathcal{O}_{X'}^*$ and $\mathcal{O}_X^*$ are multiplicative groups, and α is a truncated exponential map. Because the map of rings $D \to k$ has a section $k \to D$, it follows that this latter sequence is a split exact sequence of sheaves of abelian groups. So taking cohomology we obtain a split exact sequence

$$0 \to H^1(X, \mathcal{O}_X) \to H^1(X', \mathcal{O}_{X'}^*) \to H^1(X, \mathcal{O}_X^*) \to 0.$$

This shows that the set of isomorphism classes of invertible sheaves on X' restricting to a given isomorphism class on X is a coset of the group $H^1(X, \mathcal{O}_X)$. Letting 0 correspond to the trivial extension $\mathcal{L}' = \mathcal{L} \times D$, we obtain the result.

Proceeding to Situation C, we will actually consider a slightly more general set-up. Let X be a scheme over k, and let $\mathcal{F}$ be a coherent sheaf on X. We define a *deformation* of $\mathcal{F}$ over D to be a coherent sheaf $\mathcal{F}'$ on $X' = X \times D$, flat over D, together with a homomorphism $\mathcal{F}' \to \mathcal{F}$ such that the induced map $\mathcal{F}' \otimes_D k \to \mathcal{F}$ is an isomorphism. We say that two such deformations $\mathcal{F}_1' \to \mathcal{F}$ and $\mathcal{F}_2' \to \mathcal{F}$ are *equivalent* if there is an isomorphism $\mathcal{F}_1' \xrightarrow{\sim} \mathcal{F}_2'$ compatible with the given maps to $\mathcal{F}$.

Theorem 2.7. *Let X be a scheme over k, and let $\mathcal{F}$ be a coherent sheaf on X. The (equivalence classes of) deformations of $\mathcal{F}$ over D are in natural one-to-one correspondence with the elements of the group $\mathrm{Ext}_X^1(\mathcal{F}, \mathcal{F})$, where the zero-element corresponds to the trivial deformation.*

Proof. By (2.2), the flatness of $\mathcal{F}'$ over D is equivalent to the exactness of the sequence

$$0 \to \mathcal{F} \xrightarrow{t} \mathcal{F}' \to \mathcal{F} \to 0$$

obtained by tensoring $\mathcal{F}'$ with $0 \to k \xrightarrow{t} D \to k \to 0$. Since the latter sequence splits, we have a splitting $\mathcal{O}_X \to \mathcal{O}_{X'}$, and thus we can regard this sequence of sheaves as an exact sequence of $\mathcal{O}_X$-modules. By Yoneda's interpretation of the Ext groups [24, Ex. A3.26], we obtain an element $\xi \in \mathrm{Ext}_X^1(\mathcal{F}, \mathcal{F})$. Conversely, an element in that Ext group gives $\mathcal{F}'$ as an extension of $\mathcal{F}$ by $\mathcal{F}$ as $\mathcal{O}_X$-modules. To give a structure of an $\mathcal{O}_{X'}$-module on $\mathcal{F}'$ we have to specify multiplication by t. But this can be done in one and only one way compatible with the sequence above and the requirement that $\mathcal{F}' \otimes_D k \cong \mathcal{F}$, namely projection from $\mathcal{F}'$ to $\mathcal{F}$ followed by the injection $t : \mathcal{F} \to \mathcal{F}'$. Note finally that $\mathcal{F}' \to \mathcal{F}$ and $\mathcal{F}'' \to \mathcal{F}$ are equivalent as deformations of $\mathcal{F}$ if and only if the corresponding elements ξ, ξ' are equal. Thus the deformations $\mathcal{F}'$ are in natural one-to-one correspondence with elements of the group $\mathrm{Ext}^1(\mathcal{F}, \mathcal{F})$.

Remark 2.7.1. Given $\mathcal{F}$ on X, we can also pose a different problem, like the one in (2.6), namely to classify isomorphism classes of coherent sheaves $\mathcal{F}'$ on X', flat over D, such that $\mathcal{F}' \otimes_D k$ is isomorphic to $\mathcal{F}$ (without specifying the isomorphism). This set need not be the same as the set of deformations of $\mathcal{F}$, but we can explain their relationship as follows. The group $\mathrm{Aut}\,\mathcal{F}$ of automorphisms of $\mathcal{F}$ acts on the set of deformations of $\mathcal{F}$ by letting $\alpha \in \mathrm{Aut}\,\mathcal{F}$

applied to $f : \mathcal{F}' \to \mathcal{F}$ be $\alpha f : \mathcal{F}' \to \mathcal{F}$. Now let $f : \mathcal{F}' \to \mathcal{F}$ and $g : \mathcal{F}'' \to \mathcal{F}$ be two deformations of $\mathcal{F}$. One sees easily that $\mathcal{F}'$ and $\mathcal{F}''$ are isomorphic as sheaves on X' if and only if there exists an $\alpha \in \operatorname{Aut} \mathcal{F}$ such that αf and g are equivalent as deformations of $\mathcal{F}$. Thus the set of $\mathcal{F}''$s up to isomorphism as sheaves on X' is the orbit space of $\operatorname{Ext}^1_X(\mathcal{F}, \mathcal{F})$ under the action of $\operatorname{Aut} \mathcal{F}$. This kind of subtle distinction will play an important role in questions of pro-representability (Chapter 3).

Corollary 2.8. *If $\mathcal{E}$ is a vector bundle over X, then the deformations of $\mathcal{E}$ over D are in natural one-to-one correspondence with the elements of $H^1(X, \mathcal{E}nd\,\mathcal{E})$, where $\mathcal{E}nd\,\mathcal{E} = \mathcal{H}om(\mathcal{E}, \mathcal{E})$ is the sheaf of endomorphisms of $\mathcal{E}$. The trivial deformation corresponds to the zero element.*

Proof. In this case, since $\mathcal{E}$ is locally free, $\operatorname{Ext}^1(\mathcal{E}, \mathcal{E}) = \operatorname{Ext}^1(\mathcal{O}_X, \mathcal{E}nd\,\mathcal{E}) = H^1(X, \mathcal{E}nd\,\mathcal{E})$.

Remark 2.8.1. If $\mathcal{E}$ is a line bundle, i.e., an invertible sheaf $\mathcal{L}$ on X, then $\mathcal{E}nd\,\mathcal{E} \cong \mathcal{O}_X$, and the deformations of $\mathcal{L}$ are classified by $H^1(\mathcal{O}_X)$. We get the same answer as in (2.6) because $\operatorname{Aut} \mathcal{L} = H^0(\mathcal{O}_X^*)$ and for any $\mathcal{L}'$ invertible on X', $\operatorname{Aut} \mathcal{L}' = H^0(\mathcal{O}_{X'}^*)$. Now $H^0(\mathcal{O}_{X'}^*) \to H^0(\mathcal{O}_X^*)$ is surjective because of the split exact sequence mentioned in the proof of (2.6), and from this it follows that two deformations $\mathcal{L}_1' \to \mathcal{L}$ and $\mathcal{L}_2' \to \mathcal{L}$ are equivalent as deformations of $\mathcal{L}$ if and only if $\mathcal{L}_1'$ and $\mathcal{L}_2'$ are isomorphic as invertible sheaves on X'.

Remark 2.8.2. *Use of the word "natural."* In each of the main results of this section, we have said that a certain set was in *natural* one-to-one correspondence with the set of elements of a certain group. We have not said exactly what we mean by this word natural. So for the time being, you may understand it something like this: If I say there is a natural mapping from one set to another, that means I have a particular construction in mind for that mapping, and if you see my construction, you will agree that it is natural. It does not involve any unnatural choices. Use of the word natural carries with it the expectation (but not the promise) that the same construction carried out in parallel situations will give compatible results. It should be compatible with localization, base-change, etc. However, natural does not mean unique. It is quite possible that someone else could find another mapping between these two sets, different from this one, but also natural from a different point of view.

In contrast to the natural correspondences of this section, we will see later situations in which there are nonnatural one-to-one correspondences. Having fixed one deformation, any other will define an element of a certain group, thus giving a one-to-one correspondence between the set of all deformations and the elements of the group, with the fixed deformation corresponding to the zero element. So there is a one-to-one correspondence, but it depends on the choice of a fixed deformation, and there may be no such choice that is natural, i.e., no one we can single out as a "trivial" deformation. In this case we say that the set is a principal homogeneous space or torsor under the action of the group—cf. §6 for examples.

References for this section. The notion of flatness is due to Serre [153], who showed that there is a one-to-one correspondence between coherent algebraic sheaves on a projective variety over $\mathbb{C}$ and the coherent analytic sheaves on the associated complex analytic space. He observed that the algebraic and analytic local rings have the same completion, and that this makes them a "flat couple." The observation that localization and completion both enjoy this property, and that flat modules are those that are acyclic for the Tor functors, explained and simplified a number of situations by combining them into one concept. Then in the hands of Grothendieck, flatness became a central tool for managing families of structures of all kinds in algebraic geometry. The local criterion of flatness is developed in [47, IV, §5]. Our statement is [loc. cit., 5.5]. A note before [loc. cit. 5.2] says "La proposition suivante a été dégagée au moment du Séminaire par Serre; elle permet des simplifications substantielles dans le présent numéro."

The infinitesimal study of the Hilbert scheme is in Grothendieck's Bourbaki seminar [45, exposé 221].

Exercises.

2.1. If X is a scheme with $H^1(\mathcal{O}_X) = 0$, then by (2.6) there are no nontrivial extensions of an invertible sheaf to a deformation of X over the dual numbers. This suggests that perhaps there are no global nontrivial families either. Indeed this is true with the following hypotheses. Let X be an integral projective scheme over k with $H^1(X, \mathcal{O}_X) = 0$. Let T be a connected scheme with a closed point t_0. Let $\mathcal{L}$ be an invertible sheaf on $X \times T$, and let $\mathcal{L}_0 = \mathcal{L} \otimes \mathcal{O}_{X_0}$ be the restriction of $\mathcal{L}$ to the fiber $X_0 = X \times k(t_0)$ over t_0. Show then that there is an invertible sheaf $\mathcal{M}$ on T such that $\mathcal{L} \cong p_1^*\mathcal{L}_0 \otimes p_2^*\mathcal{M}$. In particular, all the fibers of $\mathcal{L}$ over points of T are isomorphic. (*Hint:* Use [57, III, Ex. 12.6].)

2.2. The Jacobian of an elliptic curve. Let C be an elliptic curve over k, that is, a nonsingular projective curve of genus 1 with a fixed point P_0. Then any invertible sheaf $\mathcal{L}$ of degree 0 on C is isomorphic to $\mathcal{O}_C(P - P_0)$ for a uniquely determined point $P \in C$. Thus the curve C itself acts as a parameter space for the group $\mathrm{Pic}^0(C)$ of invertible sheaves of degree 0, and as such is called the *Jacobian variety* J of C. Describe explicitly the functorial properties of J as a classifying space and thus justify the identification of the one-dimensional space $H^1(\mathcal{O}_C)$ with the Zariski tangent space to J at any point (cf. [57, III, §4]).

2.3. Vector bundles on $\mathbb{P}^1$. One knows that every vector bundle on $\mathbb{P}^1$ is a direct sum of line bundles $\mathcal{O}(a_i)$ for various $a_i \in \mathbb{Z}$ [57, V, Ex. 2.6]. Thus the set of isomorphism classes of vector bundles of given rank and degree is a discrete set. Nevertheless, there are nontrivial deformations of bundles on $\mathbb{P}^1$. Let $\mathcal{E}_0 = \mathcal{O}(-1) \oplus \mathcal{O}(1)$ and show that $H^1(\mathbb{P}^1, \mathcal{E}nd\,\mathcal{E}_0)$ has dimension one. A nontrivial family containing $\mathcal{E}_0$ is given by the extensions

$$0 \to \mathcal{O}(-1) \to \mathcal{E}_t \to \mathcal{O}(1) \to 0$$

for $t \in \mathrm{Ext}^1(\mathcal{O}(1), \mathcal{O}(-1)) = H^1(\mathcal{O}(-2))$. Show that for $t \neq 0$, $\mathcal{E}_t \cong \mathcal{O} \oplus \mathcal{O}$, while for $t = 0$ we get $\mathcal{E}_0$.

2.4. Rank 2 bundles on an elliptic curve. Let C be an elliptic curve. Let $\mathcal{E}$ be a rank 2 vector bundle obtained as a nonsplit extension

$$0 \to \mathcal{O}_C \to \mathcal{E} \to \mathcal{O}_C(P) \to 0$$

for some point $P \in C$.

(a) Show that $\mathcal{E}$ is *normalized* in the sense that $H^0(\mathcal{E}) \neq 0$, but for any invertible sheaf $\mathcal{L}$ with $\deg \mathcal{L} < 0$, $H^0(\mathcal{E} \otimes \mathcal{L}) = 0$. Show also that $\mathcal{E}$ is uniquely determined by P, up to isomorphism.
(b) Show that $h^0(\mathcal{E}) = 1$ and $h^1(\mathcal{E}nd\,\mathcal{E}) = 1$.
(c) Show that any normalized rank 2 vector bundle of degree 1 on C is isomorphic to an $\mathcal{E}$ as above, for a uniquely determined point $P \in C$. Thus the family of all such bundles is parametrized by the curve C, consistent with the calculation $h^1(\mathcal{E}nd\,\mathcal{E}) = 1$.

2.5. A line bundle and its associated divisor. Let X be an integral projective scheme. Let $\mathcal{L}$ be an invertible sheaf on X, let $s \in H^0(\mathcal{L})$ be a global section, and let $Y = (s)_0$ be the associated divisor on X. We wish to compare deformations of $\mathcal{L}$ as an invertible sheaf on X with deformations of Y as a closed subscheme of X.

(a) Show that the normal bundle of Y in X is isomorphic to $\mathcal{L}_Y = \mathcal{L} \otimes \mathcal{O}_Y$. Then use the exact sequence

$$0 \to \mathcal{O}_X \to \mathcal{L} \to \mathcal{L}_Y \to 0$$

to obtain a long exact sequence of cohomology

$$0 \to H^0(\mathcal{O}_X) \xrightarrow{s} H^0(\mathcal{L}) \xrightarrow{\alpha} H^0(\mathcal{L}_Y) \xrightarrow{\beta} H^1(\mathcal{O}_X) \xrightarrow{\gamma} H^1(\mathcal{L}) \to \cdots .$$

We interpret this as follows. The image of α corresponds to deformations of Y within the linear system $|Y|$. The map β gives the deformation of $\mathcal{L}$ associated to a deformation of Y. If the map γ is nonzero, then some deformations of $\mathcal{L}$ may not come from a deformation of Y, because the section s does not lift to the deformation of $\mathcal{L}$.
(b) For an example of this latter situation, let X be a nonsingular projective curve of genus $g \geq 2$, let $P \in X$ be a point, and let $\mathcal{L} = \mathcal{O}_X(P)$. If Q is another point, we can consider the family of invertible sheaves $\mathcal{L}_Q = \mathcal{O}_X(2P - Q)$. For $Q = P$ we recover $\mathcal{L}$. For $Q \neq P$, the sheaf $\mathcal{L}_Q$ has no global sections (assuming $2P$ is not in the linear system g_2^1 if X is hyperelliptic). In this case the sheaf deforms, but the section does not.
(c) The exact sequence in (a) shows that if $H^1(\mathcal{L}) = 0$, then for any lifting $\mathcal{L}'$ of $\mathcal{L}$ over the dual numbers, the section s lifts to a section of $\mathcal{L}'$. A corresponding global result also holds: Changing notation, let $\mathcal{L}$ be an invertible sheaf on $X \times T$ for some scheme T, let $\mathcal{L}_0$ be the restriction to the fiber over a point $t_0 \in T$, and assume that $H^1(X, \mathcal{L}_0) = 0$. Show that $p_{2*}\mathcal{L}$ is locally free on T, so that every section of $\mathcal{L}_0$ on X extends to a section of $\mathcal{L}$ over some neighborhood of $t_0 \in T$. (*Hint:* Use the theorem of cohomology and base change [57, III, 12.11].)

2.6. Rank 2 vector bundles on $\mathbb{P}^3$. Let $\mathcal{E}$ be a rank 2 vector bundle on $\mathbb{P}^3$, let s be a section of $H^0(\mathcal{E})$ that does not vanish on any divisor, and let $Y = (s)_0$ be the curve of zeros of s. Then there is an exact sequence

$$0 \to \mathcal{O} \xrightarrow{s} \mathcal{E} \to \mathcal{I}_Y(a) \to 0,$$

where $a = c_1(\mathcal{E})$ is the first Chern class of $\mathcal{E}$. We wish to compare deformations of $\mathcal{E}$ with deformations of the closed subscheme Y in $\mathbb{P}^3$.

(a) Show that the normal bundle of Y in $\mathbb{P}^3$ is $\mathcal{E}_Y = \mathcal{E} \otimes \mathcal{O}_Y$. (Note that since $\mathcal{E}$ has rank 2, its dual $\mathcal{E}^\vee$ is isomorphic to $\mathcal{E}(-a)$.)

(b) Show that there are exact sequences

$$0 \to \mathcal{E}^\vee \to \mathcal{E}nd\,\mathcal{E} \to \mathcal{E} \otimes \mathcal{I}_Y \to 0$$

and

$$0 \to \mathcal{E} \otimes \mathcal{I}_Y \to \mathcal{E} \to \mathcal{E}_Y \to 0$$

from which one can obtain exact sequences of cohomology

$$\to H^1(\mathcal{E}^\vee) \to H^1(\mathcal{E}nd\,\mathcal{E}) \to H^1(\mathcal{E} \otimes \mathcal{I}_Y) \to H^2(\mathcal{E}^\vee) \to \cdots$$
$$\|$$
$$\to H^0(\mathcal{E}) \to H^0(\mathcal{E}_Y) \to H^1(\mathcal{E} \otimes \mathcal{I}_Y) \to H^1(\mathcal{E}) \to \cdots .$$

Here $H^1(\mathcal{E}nd\,\mathcal{E})$ represents deformations of $\mathcal{E}$, and $H^0(\mathcal{E}_Y)$ represents deformations of Y in $\mathbb{P}^3$. In general a deformation of one may not correspond to a deformation of the other.

(c) Now consider a particular case, the so-called *null-correlation bundle* on $\mathbb{P}^3$. It belongs to a sequence

$$0 \to \mathcal{O} \to \mathcal{E} \to \mathcal{I}_Y(2) \to 0,$$

where Y is a disjoint union of two lines in $\mathbb{P}^3$. For existence of such bundles, show that $\operatorname{Ext}^1(\mathcal{I}_Y(2), \mathcal{O}) \cong \operatorname{Ext}^2(\mathcal{O}_Y(2), \mathcal{O}) \cong H^0(\mathcal{O}_Y)$, so that an extension as above may be determined by choosing two scalars, one for each of the two lines in Y.

(d) For the bundles in (c) verify that $h^0(\mathcal{E}nd\,\mathcal{E}) = 1$, $h^1(\mathcal{E}nd\,\mathcal{E}) = 5$; $h^0(\mathcal{E}) = 5$, $h^0(\mathcal{E}_Y) = 8$, $h^1(\mathcal{E} \otimes \mathcal{I}_Y) = 4$ and $h^1(\mathcal{E}) = h^2(\mathcal{E}^\vee) = 0$. So in this case, any deformation of $\mathcal{E}$ corresponds to a deformation of Y and vice versa. In fact, there is a 5-dimensional global family of such bundles, parametrized by $\mathbb{P}^5$ minus the four-dimensional Grassmann variety $G(1,3)$ of lines in $\mathbb{P}^3$ [58, 8.4.1], consistent with the calculation that $h^1(\mathcal{E}nd\,\mathcal{E}) = 5$.

3. The T^i Functors

In this section we will present the construction and main properties of the T^i functors introduced by Lichtenbaum and Schlessinger [96]. For any ring homomorphism $A \to B$ and any B-module M they define functors $T^i(B/A, M)$, for $i = 0, 1, 2$. With A and B fixed these form a cohomological functor in M, giving a nine-term exact sequence associated to a short exact sequence of modules $0 \to M' \to M \to M'' \to 0$. On the other hand, if $A \to B \to C$ are three rings and homomorphisms, and if M is a C-module, then there is a nine-term exact sequence of T^i functors associated with the three ring homomorphisms $A \to B$, $A \to C$, and $B \to C$. The principal application of these functors for us is the study of deformations of rings and schemes (Situation D). We will see that deformations of a ring are classified by a certain T^1 group (§5), and that obstructions lie in a certain T^2 group (§10). We will also see in §4 that the vanishing of the T^1 functor characterizes smooth morphisms and the vanishing of the T^2 functor characterizes locally complete intersection morphisms.

Construction 3.1. Let $A \to B$ be a homomorphism of rings and let M be a B-module. Here we will construct the groups $T^i(B/A, M)$ for $i = 0, 1, 2$. The rings are assumed to be commutative with identity, but we do not impose any finiteness conditions yet.

First choose a polynomial ring $R = A[x]$ in a set of variables $x = \{x_i\}$ (possibly infinite) such that B can be written as a quotient of R as an A-algebra. Let I be the ideal defining B, so that there is an exact sequence

$$0 \to I \to R \to B \to 0.$$

Second choose a free R-module F and a surjection $j : F \to I \to 0$ and let Q be the kernel:

$$0 \to Q \to F \xrightarrow{j} I \to 0.$$

Having chosen R and F as above, the construction proceeds with no further choices. Let F_0 be the submodule of F generated by all "Koszul relations" of the form $j(a)b - j(b)a$ for $a, b \in F$. Note that $j(F_0) = 0$ so $F_0 \subseteq Q$.

We define a complex of B-modules, called the *cotangent complex*,

$$L_2 \xrightarrow{d_2} L_1 \xrightarrow{d_1} L_0$$

as follows. Take $L_2 = Q/F_0$. Why is L_2 a B-module? A priori it is an R-module. But if $x \in I$ and $a \in Q$, we can write $x = j(x')$ for some $x' \in F$ and then $xa = j(x')a \equiv j(a)x' \pmod{F_0}$. But $j(a) = 0$, since $a \in Q$, so we see that $xa = 0$. Therefore L_2 is a B-module.

Take $L_1 = F \otimes_R B = F/IF$, and let $d_2 : L_2 \to L_1$ be the map induced from the inclusion $Q \to F$.

Take $L_0 = \Omega_{R/A} \otimes_R B$, where $\Omega_{R/A}$ is the module of relative differentials. To define d_1 just map L_1 to I/I^2, then apply the derivation $d : R \to \Omega_{R/A}$, which induces a B-module homomorphism $I/I^2 \to L_0$.

Clearly $d_1 d_2 = 0$, so we have defined a complex of B-modules. Note also that L_1 and L_0 are *free* B-modules: L_1 is free because it is defined from the free R-module F; L_0 is free because R is a polynomial ring over A and so $\Omega_{R/A}$ is a free R-module.

For any B-module M we now define the modules

$$T^i(B/A, M) = h^i(\mathrm{Hom}_B(L_\bullet, M))$$

as the cohomology modules of the complex of homomorphisms of the complex $L_\bullet$ into M.

To show that these modules are well-defined (up to isomorphism), we must verify that they are independent of the choices made in the construction.

Lemma 3.2. *The modules $T^i(B/A, M)$ constructed above are independent of the choice of F (keeping R fixed).*

Proof. If F and F' are two choices of free R-modules mapping onto I, then $F \oplus F'$ is a third choice, so by symmetry it is sufficient to compare F with $F \oplus F'$. Since F' is free, the map $j' : F' \to I$ factors through F, i.e., $j' = jp$ for some map $p : F' \to F$. Changing bases in $F \oplus F'$, replacing each generator e' of F' by $e' - p(e')$, we may assume that the map $F \oplus F' \to I$ is just j on the first factor and 0 on the second factor. Thus we have the diagram

$$
\begin{array}{ccccccccc}
0 & \to & Q \oplus F' & \to & F \oplus F' & \overset{(j,0)}{\to} & I & \to & 0 \\
 & & \downarrow & & \downarrow pr_1 & & \downarrow id & & \\
0 & \to & Q & \to & F & \overset{j}{\to} & I & \to & 0
\end{array}
$$

showing that the kernel of $(j, 0) : F \oplus F' \to I$ is just $Q \oplus F'$. Then clearly $(F \oplus F')_0 = F_0 + IF'$. Denoting by $L'_\bullet$ the complex obtained from the new construction, we see that $L'_2 = L_2 \oplus F'/IF'$, $L'_1 = L_1 \oplus (F' \otimes_R B)$, and $L'_0 = L_0$. Since $F' \otimes_R B = F'/IF'$ is a free B-module, the complex $L'_\bullet$ is obtained by taking the direct sum of $L_\bullet$ with the free acyclic complex $F' \otimes_R B \to F' \otimes_R B$. Hence when we take Hom of these complexes into M and then cohomology, the result is the same.

Lemma 3.3. *The modules $T^i(B/A, M)$ are independent of the choice of R.*

Proof. Let $R = A[x]$ and $R' = A[y]$ be two choices of polynomial rings with surjections to B. As in the previous proof, it will be sufficient to compare R with $R'' = A[x, y]$. Furthermore, the map $A[y] \to B$ can be factored through $A[x]$ by a homomorphism $p : A[y] \to A[x]$. Then, changing variables in $A[x, y]$, replacing each y_i by $y_i - p(y_i)$, we may assume that all the y_i go to zero in the ring homomorphism $A[x, y] \to B$. Then we have the diagram

$$
\begin{array}{ccccccccc}
0 & \to & IR'' + yR'' & \to & R'' & \to & B & \to & 0 \\
 & & \uparrow & & \uparrow\downarrow p & & \uparrow id & & \\
0 & \to & I & \to & R & \to & B & \to & 0
\end{array}
$$

showing that the kernel of $R'' \to B$ is generated by I and all the y-variables.

Since we have already shown that the construction is independent of the choice of F, we may use any F's we like in the present proof. Take any free

R-module F mapping surjectively to I. Take F' a free R''-module on the same number of generators as F, and take G' a free R''-module on the index set of the y variables. Then we have

$$\begin{array}{ccccccccc}
0 & \to & Q' & \to & F' \oplus G' & \to & IR'' + yR'' & \to & 0 \\
& & \uparrow & & \uparrow & & \uparrow & & \\
0 & \to & Q & \to & F & \to & I & \to & 0
\end{array}$$

Observe that since the y_i are independent variables, and G' has a basis e_i going to y_i, the kernel Q' in the upper row must be generated by (1) things in Q, (2) things of the form $y_i a - j(a)e_i$ with $a \in F$, and (3) things of the form $y_i e_j - y_j e_i$. Clearly the elements of types (2) and (3) are in $(F' \oplus G')_0$. Therefore $Q'/(F' \oplus G')_0$ is a B-module generated by the image of Q, so $L_2 = L_2'$.

On the other hand, $L_1' = L_1 \oplus (G' \otimes_{R'} B)$, and $L_0' = L_0 \oplus (\Omega_{A[y]/A} \otimes B)$. Thus L_1' has an extra free B-module generated by the e_i, L_0' has an extra free B-module generated by the dy_i, and the map d_1 takes e_i to dy_i. As in the previous proof we see that $L_\bullet'$ is obtained from $L_\bullet$ by adding a free acyclic complex, and hence the modules $T^i(B/A, M)$ are the same.

Remark 3.3.1. Even though the complex $L_\bullet$ is not unique, the proofs of (3.2) and (3.3) show that it gives a well-defined element of the derived category of the category of B-modules.

Theorem 3.4. *Let $A \to B$ be a homomorphism of rings. Then for $i = 0, 1, 2$, $T^i(B/A, \cdot)$ is a covariant, additive functor from the category of B-modules to itself. If*

$$0 \to M' \to M \to M'' \to 0$$

is a short exact sequence of B-modules, then there is a long exact sequence

$$0 \to T^0(B/A, M') \to T^0(B/A, M) \to T^0(B/A, M'') \to$$
$$\to T^1(B/A, M') \to T^1(B/A, M) \to T^1(B/A, M'') \to$$
$$\to T^2(B/A, M') \to T^2(B/A, M) \to T^2(B/A, M'').$$

In the language of [57, III, §1], the T^i's form a truncated δ-functor.

Proof. We have seen that the $T^i(B/A, M)$ are well-defined. By construction they are covariant additive functors. Given a short exact sequence of modules as above, since the terms L_1 and L_0 of the complex $L_\bullet$ are free, we get a sequence of complexes

$$0 \to \operatorname{Hom}_B(L_\bullet, M') \to \operatorname{Hom}_B(L_\bullet, M) \to \operatorname{Hom}_B(L_\bullet, M'') \to 0$$

that is exact except possibly for the map

$$\operatorname{Hom}_B(L_2, M) \to \operatorname{Hom}_B(L_2, M''),$$

which may not be surjective. This sequence of complexes gives the long exact sequence of cohomology above. Note that since the complex $L_\bullet$ is unique up to adding free acyclic complexes, the coboundary maps of the long exact sequence are also functorial.

Theorem 3.5. *Let $A \to B \to C$ be rings and homomorphisms, and let M be a C-module. Then there is an exact sequence of C-modules*

$$0 \to T^0(C/B, M) \to T^0(C/A, M) \to T^0(B/A, M) \to$$
$$\to T^1(C/B, M) \to T^1(C/A, M) \to T^1(B/A, M) \to$$
$$\to T^2(C/B, M) \to T^2(C/A, M) \to T^2(B/A, M).$$

Proof. To prove this theorem, we will show that for suitable choices in the construction (3.1), the resulting complexes form a sequence

$$0 \to L_\bullet(B/A) \otimes_B C \to L_\bullet(C/A) \to L_\bullet(C/B) \to 0$$

that is split exact on the degree 0 and 1 terms, and right exact on the degree 2 terms. Given this, taking $\mathrm{Hom}(\cdot, M)$ will give a sequence of complexes that is exact on the degree 0 and 1 terms, and left exact on the degree 2 terms. Taking cohomology will give the nine-term exact sequence above.

First choose a surjection $A[x] \to B \to 0$ with kernel I, and a surjection $F \to I \to 0$ with kernel Q, where F is a free $A[x]$-module, to calculate the functors $T^i(B/A, M)$.

Next choose a surjection $B[y] \to C \to 0$ with kernel J, and a surjection $G \to J \to 0$ of a free $B[y]$-module G with kernel P, to calculate $T^i(C/B, M)$.

To calculate the functors T^i for C/A, take a polynomial ring $A[x, y]$ in the x-variables and the y-variables. Then $A[x, y] \to B[y] \to C$ gives a surjection of $A[x, y] \to C$. If K is its kernel then there is an exact sequence

$$0 \to I[y] \to K \to J \to 0,$$

where $I[y]$ denotes polynomials in y with coefficients in I. Take F' and G' to be free $A[x, y]$-modules on the same index sets as F and G respectively. Choose a lifting of the map $G \to J$ to a map $G' \to K$. Then adding the natural map $F' \to K$ we get a surjection $F' \oplus G' \to K$. Let S be its kernel:

$$0 \to S \to F' \oplus G' \to K \to 0.$$

Now we are ready to calculate. Out of the choices thus made there are induced maps of complexes

$$L_\bullet(B/A) \otimes_B C \to L_\bullet(C/A) \to L_\bullet(C/B).$$

On the degree 0 level we have

$$\Omega_{A[x]/A} \otimes C \to \Omega_{A[x,y]/A} \otimes C \to \Omega_{B[y]/B} \otimes C.$$

These are free C-modules with bases $\{dx_i\}$ on the left, $\{dy_i\}$ on the right, and $\{dx_i, dy_i\}$ in the middle. So this sequence is clearly split exact.

On the degree 1 level we have

$$F \otimes C \to (F' \oplus G') \otimes C \to G \otimes C,$$

which is split exact by construction.

On the degree 2 level we have

$$(Q/F_0) \otimes_B C \to S/(F' \oplus G')_0 \to P/G_0.$$

The right-hand map is surjective because the map $S \to P$ is surjective. Clearly the composition of the two maps is 0. We make no claim of injectivity for the left-hand map. So to complete our proof it remains only to show exactness in the middle.

Let $s = f' + g'$ be an element of S, and assume that its image in P is contained in G_0. We must show that s can be written as a sum of something in $(F' \oplus G')_0$ and something in the image of $Q[y]$. In the map $S \to P$, the element f' goes to 0. Let g be the image of g'. Then $g \in G_0$, so g can be written as a linear combination of expressions $j(a)b - j(b)a$ with $a, b \in G$. Lift a, b to elements a', b' in G'. Then the expressions $j(a')b' - j(b')a'$ are in S. Let g'' be g' minus a linear combination of these expressions $j(a')b' - j(b')a'$. We get a new element $s' = f' + g''$ in S, differing from s by something in $(F' \oplus G')_0$, and where now g'' is in the kernel of the map $G' \to G$, which is IG'. So we can write g'' as a sum of elements xh with $x \in I$ and $h \in G'$. Let $x' \in F$ map to x by j. Then $xh = j(x')h \equiv j(h)x' (\mod F_0)$. Therefore $s' \equiv f' + \Sigma j(h)x' (\mod(F' \oplus G')_0)$, and this last expression is in $F' \cap S$, and therefore is in $Q[y]$.

Now we will give some special cases and remarks concerning these functors.

Proposition 3.6. *For any $A \to B$ and any M, $T^0(B/A, M) = \mathrm{Hom}_B(\Omega_{B/A}, M) = \mathrm{Der}_A(B, M)$. In particular, $T^0(B/A, B) = \mathrm{Hom}_B(\Omega_{B/A}, B)$ is* the tangent module $T_{B/A}$ of B over A.

Proof. Write B as a quotient of a polynomial ring R, with kernel I. Then there is an exact sequence [57, II, 8.4A]

$$I/I^2 \xrightarrow{d} \Omega_{R/A} \otimes_R B \to \Omega_{B/A} \to 0.$$

Since $F \to I$ is surjective, there is an induced surjective map $L_1 \to I/I^2 \to 0$. Thus the sequence

$$L_1 \to L_0 \to \Omega_{B/A} \to 0$$

is exact. Taking $\mathrm{Hom}(\cdot, M)$, which is left exact, we see that $T^0(B/A, M) = \mathrm{Hom}_B(\Omega_{B/A}, M)$.

Proposition 3.7. *If B is a polynomial ring over A, then $T^i(B/A, M) = 0$ for $i = 1, 2$ and for all M.*

Proof. In this case we can take $R = B$ in the construction. Then $I = 0$, $F = 0$, so $L_2 = L_1 = 0$, and the complex $L_\bullet$ is reduced to the L_0 term. Therefore $T^i = 0$ for $i = 1, 2$ and any M.

Remark 3.7.1. We will see later that the vanishing of the T^1 functor characterizes smooth morphisms (4.11).

Proposition 3.8. *If $A \to B$ is a surjective ring homomorphism with kernel I, then $T^0(B/A, M) = 0$ for all M, and $T^1(B/A, M) = \operatorname{Hom}_B(I/I^2, M)$. In particular, $T^1(B/A, B) = \operatorname{Hom}_B(I/I^2, B)$ is the normal module $N_{B/A}$ of $\operatorname{Spec} B$ in $\operatorname{Spec} A$.*

Proof. In this case we can take $R = A$, so that $L_0 = 0$. Thus $T^0 = 0$ for any M. Furthermore, the exact sequence

$$0 \to Q \to F \to I \to 0,$$

tensored with B, gives an exact sequence

$$Q \otimes_A B \to F \otimes_A B \to I/I^2 \to 0.$$

There is also a surjective map $Q \otimes_A B \to Q/F_0$, since the latter is a B-module, so we have an exact sequence

$$L_2 \to L_1 \to I/I^2 \to 0.$$

Taking $\operatorname{Hom}(\cdot, M)$ shows that $T^1(B/A, M) = \operatorname{Hom}_B(I/I^2, M)$.

A useful special case is the following.

Corollary 3.9. *If A is a local ring and B is a quotient A/I, where I is generated by a regular sequence $a_1, \ldots, a_r$, then $T^2(B/A, M) = 0$ for all M.*

Proof. Indeed, in this case, since the Koszul complex of a regular sequence is exact [104, 16.5], we find $Q = F_0$ in the construction of the T^i-functors. Thus $L_2 = 0$ and $T^2(B/A, M) = 0$ for all M.

Remark 3.9.1. We will see later that the vanishing of the T^2 functor characterizes relative local complete intersection morphisms (4.13).

Another useful special case is given by the following proposition.

Proposition 3.10. *Suppose $A = k[x_1, \ldots, x_n]$ and $B = A/I$. Then for any M there is an exact sequence*

$$0 \to T^0(B/k, M) \to \operatorname{Hom}(\Omega_{A/k}, M) \to \operatorname{Hom}(I/I^2, M) \to T^1(B/k, M) \to 0$$

and an isomorphism

$$T^2(B/A, M) \xrightarrow{\sim} T^2(B/k, M).$$

Proof. Write the long exact sequence of T^i-functors for the composition $k \rightarrow A \rightarrow B$ and use (3.6), (3.7), and (3.8). The same works for any base ring k, not necessarily a field.

Remark 3.10.1. Throughout this section we have not made any finiteness assumptions on the rings and modules. However, it is easy to see that if A is a noetherian ring, B a finitely generated A-algebra, and M a finitely generated B-module, then the B-modules $T^i(B/A, M)$ are also finitely generated. Indeed, we can take R to be a polynomial ring in finitely many variables over A, which is therefore noetherian. Then I is finitely generated and we can take F to be a finitely generated R-module. Thus the complex $L_\bullet$ consists of finitely generated B-modules, whence the result.

Notation. In the sequel we will often denote the modules $T^i(B/A, B)$ and $T^i(B/k, B)$ by $T^i_{B/A}$ and $T^i_{B/k}$, or even T^i_B, if there is no confusion as to the base. Furthermore $T^0_{B/A}$ will be written $T_{B/A}$, the tangent module of B over A. Similarly for the sheaves $\mathcal{T}^i(X/Y, \mathcal{O}_X)$ and $\mathcal{T}^i(X/k, \mathcal{O}_X)$ (see (Ex. 3.5)), we will write $\mathcal{T}^i_{X/Y}$ and $\mathcal{T}^i_{X/k}$, or even $\mathcal{T}^i_X$. The sheaf $\mathcal{T}^0_X$ will be written $\mathcal{T}_X$, the tangent sheaf of X.

References for this section. The development of the T^i functors presented here is due to Lichtenbaum and Schlessinger [96]. A more general cohomology theory for commutative rings, extending the definition to functors T^i for all $i \geq 0$ has been developed independently by André and Quillen. Quillen states [138] that the T^1 and T^2 functors of Lichtenbaum and Schlessinger are the same as those defined more generally, though I have not seen a direct proof of this fact. Later Illusie [73] globalized those theories by constructing the cotangent complex of a morphism of schemes. This has been extended to stacks in [93]. Independently, Laudal [92] gave another globalization of André's cohomology of algebras. For a computational approach with many examples, see [160].

Exercises.

3.1. Let $B = k[x, y]/(xy)$. Show that $T^1(B/k, M) = M \otimes k$ and $T^2(B/k, M) = 0$ for any B-module M.

3.2. More generally, if $B = k[x, y]/(f)$, then $T^1(B/k, M) = M/(f_x, f_y)M$ for any M, where f_x and f_y are the partial derivatives of f with respect to x and y.

3.3. Let $B = k[x, y]/(x^2, xy, y^2)$. Show that $T^0(B/k, B) = k^4$, $T^1(B/k, B) = k^4$, and $T^2(B/k, B) = k$.

3.4. Let B be a finitely generated integral domain over an algebraically closed field k, and let M be a torsion-free B-module. Show that $T^1(B/k, M) = \operatorname{Ext}^1_B(\Omega^1_{B/k}, M)$. *Hint:* Compare the exact sequence of (3.10) with an exact sequence arising from the cotangent sequence [57, II, 8.4A], and use the fact that $\operatorname{Spec} B$ has a dense open subset that is nonsingular.

3.5. Localization. Show that the construction of the T^i functors is compatible with localization, and thus define sheaves $T^i(X/Y, \mathcal{F})$ for any morphism of schemes $f : X \to Y$ and any sheaf $\mathcal{F}$ of $\mathcal{O}_X$-modules, such that for any open affine $V \subseteq Y$ and any open affine $U \subseteq f^{-1}(V)$, where $\mathcal{F} = \tilde{M}$, the sections of $T^i(X/Y, \mathcal{F})$ over U give $T^i(U/V, M)$. Show that the T^i sheaves satisfy a long exact sequence for change of $\mathcal{F}$ analogous to (3.4) and a long exact sequence analogous to (3.5) for change of schemes. Note that since the complexes $L_\bullet$ used to define the T^i functors are not unique, one cannot in general define an analogous complex of sheaves $\mathcal{L}_\bullet$ on X by this method.

3.6. Global construction of $\boldsymbol{T^i}$ sheaves. Let $f : X \to Y$ be a projective morphism, so that X can be realized as a closed subscheme of the projective space $\mathbb{P}^n_Y$ over Y for some n. Show that in this case one can define a global complex $\mathcal{L}_\bullet$ of sheaves on X such that for any $\mathcal{O}_X$-module $\mathcal{F}$, the sheaf T^i functors can be computed as $h^i(\mathcal{H}om_X(\mathcal{L}_\bullet, \mathcal{F}))$.

3.7. Base change I. Assume A noetherian, B a finitely generated A-algebra, and M a B-module. Let $A \to A'$ be a flat morphism, and let $B' = B \otimes_A A'$ and $M' = M \otimes_B B'$ be obtained by base extension. Show that $T^i(B/A, M) \otimes_A A' \cong T^i(B'/A', M')$ for each i.

3.8. Base change II. Again with A noetherian, B finitely generated, and $A \to A'$ a base extension, this time assume that B is flat over A. Let $B' = B \otimes_A A'$, and let M' be a B'-module. Show that $T^i(B/A, M') = T^i(B'/A', M')$ for each i.

4. The Infinitesimal Lifting Property

In this section we first review the properties of nonsingular varieties. Then we show that nonsingularity can be characterized by an "infinitesimal lifting property" that is closely related to deformation theory. We also show that nonsingular varieties and smooth morphisms are characterized by the vanishing of the T^1 functors, and that local complete intersections are characterized by the vanishing of the T^2 functors. As a matter of terminology, we will use the word "nonsingular" only for varieties over an algebraically closed field. Otherwise we talk of a "smooth morphism," or a scheme "smooth" over a base scheme. If the base scheme is an algebraically closed field, the two notions coincide.

Let us consider a scheme X of finite type over an algebraically closed ground field k. After the affine space $\mathbb{A}^n_k$ and the projective space $\mathbb{P}^n_k$, the nicest kind of scheme is a nonsingular one. The property of being nonsingular can be defined extrinsically on open affine pieces by the Jacobian criterion [57, I, §5]. Let Y be a closed subscheme of $\mathbb{A}^n$, with $\dim Y = r$. Let $f_1, \ldots, f_s \in k[x_1, \ldots, x_n]$ be a set of generators for the ideal I_Y of Y. Then Y is *nonsingular* at a closed point $P \in Y$ if the rank of the Jacobian matrix $\|(\partial f_i/\partial x_j)(P)\|$ is equal to $n - r$. We say that Y is nonsingular if it is nonsingular at every closed point. A scheme X is *nonsingular* if it can be covered by open affine subsets that are nonsingular.

This definition is awkward, because it is not obvious that the property of being nonsingular is independent of the affine embedding used in the definition. For this reason it is useful to have an intrinsic criterion for nonsingularity.

Proposition 4.1. *A scheme X of finite type over an algebraically closed field k is nonsingular if and only if the local ring $\mathcal{O}_{P,X}$ is a regular local ring for every point $P \in X$ [57, I, 5.1; II, 8.14A].*

Using differentials we have another characterization of nonsingular varieties.

Proposition 4.2. *Let X be a scheme over k algebraically closed. Then X is nonsingular if and only if the sheaf of differentials $\Omega^1_{X/k}$ is locally free of rank $n = \dim X$ at every point of X [57, II, 8.15].*

This result is closely related to the original definition using the Jacobian criterion. The generalization of the Jacobian criterion describes when a closed subscheme Y of a nonsingular scheme X over k is nonsingular.

Proposition 4.3. *Let Y be an irreducible closed subscheme of a nonsingular scheme X over k algebraically closed, defined by a sheaf of ideals $\mathcal{I}$. Then Y is nonsingular if and only if*

(1) *$\Omega_{Y/k}$ is locally free, and*
(2) *the sequence of differentials [57, II, 8.12]*

$$0 \to \mathcal{I}/\mathcal{I}^2 \to \Omega^1_{X/k} \otimes \mathcal{O}_Y \to \Omega^1_{Y/k} \to 0$$

is exact on the left.

Furthermore, in this case $\mathcal{I}$ is locally generated by $n - r = \dim X - \dim Y$ elements, and $\mathcal{I}/\mathcal{I}^2$ is locally free on Y of rank $n - r$ [57, II, 8.17].

In this section we will see that nonsingular schemes have a special property related to deformation theory, called the infinitesimal lifting property. The general question is this. Suppose we are given a morphism $f : Y \to X$ of schemes and an *infinitesimal thickening* $Y \subseteq Y'$. This means that Y is a closed subscheme of another scheme Y', and that the ideal $\mathcal{I}$ defining Y inside Y' is nilpotent. Then the question is, does there exist a *lifting* $g : Y' \to X$, i.e., a morphism such that g restricted to Y is f? Of course, there is no reason for this to hold in general, but we will see that if Y and X are affine, and X is nonsingular, then it does hold, and this property of X, for all such morphisms $f : Y \to X$, characterizes nonsingular schemes.

Proposition 4.4 (Infinitesimal Lifting Property). *Let X be a nonsingular affine scheme of finite type over k, let $f : Y \to X$ be a morphism from an affine scheme Y over k, and let $Y \subseteq Y'$ be an infinitesimal thickening of Y. Then the morphism f lifts to a morphism $g : Y' \to X$ such that $g|_Y = f$.*

Proof (cf. [57, II, Ex. 8.6]). First we note that Y' is also affine [57, III, Ex. 3.1], so we can rephrase the problem in algebraic terms. Let $X = \operatorname{Spec} A$, let $Y = \operatorname{Spec} B$, and let $Y' = \operatorname{Spec} B'$. Then f corresponds to a ring homomorphism, which (by abuse of notation) we call $f : A \to B$. On the other hand, B is a quotient of B' by an ideal I with $I^n = 0$ for some n. The problem is to find a homomorphism $g : A \to B'$ lifting f, i.e., such that g followed by the projection $B' \to B$ is f.

If we filter I by its powers and consider the sequence $B' = B'/I^n \to B'/I^{n-1} \to \cdots \to B'/I^2 \to B'/I$, it will be sufficient to lift one step at a time. Thus (changing notation) we reduce to the case $I^2 = 0$.

Since X is of finite type over k, we can write A as a quotient of a polynomial ring $P = k[x_1, \ldots, x_n]$ by an ideal J. Composing the projection $P \to A$ with f we get a homomorphism $P \to B$, which we can lift to a homomorphism $h : P \to B'$, since one can send the variables x_i to any liftings of their images in B (this corresponds to the fact that the polynomial ring is a free object in the category of k-algebras):

$$
\begin{array}{ccccccccc}
0 & \to & J & \to & P & \to & A & \to & 0 \\
 & & & & \downarrow h & & \downarrow f & & \\
0 & \to & I & \to & B' & \to & B & \to & 0
\end{array}
$$

Now h induces a map $h : J \to I$, and since $I^2 = 0$, this gives a map $\bar{h} : J/J^2 \to I$.

Next we note that the homomorphism $P \to A$ gives an embedding of X in an affine n-space $\mathbb{A}^n_k$. By (4.3), we obtain an exact sequence

$$
0 \to J/J^2 \to \Omega^1_{P/k} \otimes_P A \to \Omega^1_{A/k} \to 0,
$$

and note that these modules correspond to locally free sheaves on X, hence are projective A-modules. Via the maps h, f, we get a P-module structure on B', and A-module structures on B, I. Applying the functor $\operatorname{Hom}_A(\cdot, I)$ to the above sequence gives another exact sequence

$$
0 \to \operatorname{Hom}_A(\Omega^1_{A/k}, I) \to \operatorname{Hom}_P(\Omega^1_{P/k}, I) \to \operatorname{Hom}_A(J/J^2, I) \to 0.
$$

Let $\theta \in \operatorname{Hom}_P(\Omega^1_{P/k}, I)$ be an element whose image is $\bar{h} \in \operatorname{Hom}_A(J/J^2, I)$. We can regard θ as a k-derivation of P to the module I. Then we define a new map $h' : P \to B'$ by $h' = h - \theta$. I claim that h' is a ring homomorphism lifting f and with $h'(J) = 0$. The first statement is a consequence of the lemma (4.5) below. To see that $h'(J) = 0$, let $y \in J$. Then $h'(y) = h(y) - \theta(y)$. We need only consider $y \bmod J^2$, and then $h(y) = \theta(y)$ by choice of θ, so $h'(y) = 0$. Now since $h'(J) = 0$, h' descends to give the desired homomorphism $g : A \to B'$ lifting f.

Lemma 4.5. *Let $B' \to B$ be a surjective homomorphism of k-algebras with kernel I of square zero. Let $R \to B$ be a homomorphism of k-algebras.*

(a) *If $f, g : R \to B'$ are two liftings of the map $R \to B$ to B', then $\theta = g - f$ is a k-derivation of R to I.*

(b) *Conversely, if $f : R \to B'$ is one lifting, and $\theta : R \to I$ a derivation, then $g = f + \theta$ is another homomorphism of R to B' lifting the given map $R \to B$.*

In other words, if it is nonempty, the set of liftings $R \to B$ to k-algebra homomorphisms of R to B' is a principal homogeneous space under the action by addition of the group $\mathrm{Der}_k(R, I) = \mathrm{Hom}_R(\Omega_{R/k}, I)$. (Note that since $I^2 = 0$, I has a natural structure of a B-module and hence also of an R-module.)

Proof. (a) Let $f, g : R \to B'$ and let $\theta = g - f$. As a k-linear map, θ followed by the projection $B' \to B$ is zero, so θ sends R to I. Let $x, y \in R$. Then

$$
\begin{aligned}
\theta(xy) &= g(xy) - f(xy) \\
&= g(x)g(y) - f(x)f(y) \\
&= g(x)(g(y) - f(y)) + f(y)(g(x) - f(x)) \\
&= g(x)\theta(y) + f(y)\theta(x) \\
&= x\theta(y) + y\theta(x),
\end{aligned}
$$

the last step being because $g(x)$ and $f(y)$ act in I just like x, y. Thus θ is a k-derivation of R to I.

(b) Conversely, given f and θ as above, let $g = f + \theta$. Then

$$
\begin{aligned}
g(xy) &= f(xy) + \theta(xy) \\
&= f(x)f(y) + x\theta(y) + y\theta(x) \\
&= (f(x) + \theta(x))(f(y) + \theta(y)) \\
&= g(x)g(y),
\end{aligned}
$$

where we note that $\theta(x)\theta(y) = 0$, since $I^2 = 0$. Thus g is a homomorphism of $R \to B'$ lifting R to B.

For a converse to (4.4), we need only a special case of the infinitesimal lifting property.

Proposition 4.6. *Let X be a scheme of finite type over k algebraically closed. Suppose that for every morphism $f : Y \to X$ of a punctual scheme Y (meaning Y is the Spec of a local Artin ring), finite over k, and for every infinitesimal thickening $Y \subseteq Y'$ with ideal sheaf of square zero, there is a lifting $g : Y' \to X$. Then X is nonsingular.*

Proof. It is sufficient (4.1) to show that the local ring $\mathcal{O}_{P,X}$ is a regular local ring for every closed point $P \in X$. So again we reduce to an algebraic question, namely, let $A, \mathfrak{m}$ be a local k-algebra, essentially of finite type over k, and with residue field k. Assume that for every homomorphism $f : A \to B$, where B

is a local artinian k-algebra and for every thickening $0 \to I \to B' \to B \to 0$ with $I^2 = 0$, there is a lifting $g : A \to B'$. Then A is a regular local ring.

Let $a_1, \ldots, a_n$ be a minimal set of generators for the maximal ideal $\mathfrak{m}$ of A. Then there is a surjective homomorphism f of the formal power series ring $P = k[[x_1, \ldots, x_n]]$ to $\hat{A}$, the completion of A, sending x_i to a_i, and creating an isomorphism of $P/\mathfrak{n}^2$ to $A/\mathfrak{m}^2$, where $\mathfrak{n} = (x_1, \ldots, x_n)$ is the maximal ideal of P.

Consider the surjections $P/\mathfrak{n}^{i+1} \to P/\mathfrak{n}^i$, each defined by an ideal of square zero. Starting with the map $A \to A/\mathfrak{m}^2 \cong P/\mathfrak{n}^2$, we can lift step by step to get maps of $A \to P/\mathfrak{n}^i$ for each i, and hence a map g into the inverse limit, which is P. Passing to $\hat{A}$, we have maps $P \xrightarrow{f} \hat{A} \xrightarrow{g} P$ with the property that $g \circ f$ is an isomorphism on $P/\mathfrak{n}^2$. It follows that $g \circ f$ is an automorphism of P (Ex. 4.1). Hence $g \circ f$ has no kernel, so f is injective. But f was surjective by construction, so f is an isomorphism, and $\hat{A}$ is regular. From this it follows that A is regular, as required.

Corollary 4.7. *Let A be a local ring, essentially of finite type over an algebraically closed field k, with residue field k. Then A is a regular local ring if and only if it has the infinitesimal lifting property for local Artin rings $B' \to B$ finite over k.*

Proof. Just localize (4.4) and (4.6).

The following result shows that infinitesimal deformations of nonsingular affine schemes are trivial.

Corollary 4.8. *Let X be a nonsingular affine scheme over k. Let A be a local Artin ring over k, and let X' be a scheme, flat over $\operatorname{Spec} A$, such that $X' \times_A k$ (where by abuse of notation we mean $X' \times_{\operatorname{Spec} A} \operatorname{Spec} k$) is isomorphic to X. Then X' is isomorphic to the trivial deformation $X \times_k A$ of X over A.*

Proof. We apply (4.4) to the identity map of X to X and the infinitesimal thickening $i : X \hookrightarrow X'$ defined by the isomorphism $X' \times_A k \cong X$. Therefore there is a lifting $p : X' \to X$ such that $p \circ i = \operatorname{id}_X$. The maps of X' to X and to $\operatorname{Spec} A$ define a map to the product: $X' \to X \times_k A$. Both of these schemes are flat over A, and this map restricts to the identity on X, so it is an isomorphism (Ex. 4.2).

Remark 4.8.1. The infinitesimal lifting property for nonsingular varieties over an algebraically closed field that we have explained here can be generalized to the relative case of a morphism of schemes, giving a characterization of smooth morphisms (Ex. 4.7, Ex. 4.8). In fact, Grothendieck takes the infinitesimal lifting property as one of the equivalent definitions of smooth morphisms. See [48, IV, §17].

Next we investigate the relation between nonsingularity and the T^i functors.

Theorem 4.9. *Let* $X = \operatorname{Spec} B$ *be an affine scheme over* k *algebraically closed. Then* X *is nonsingular if and only if* $T^1(B/k, M) = 0$ *for all* B-*modules* M. *Furthermore, if* X *is nonsingular, then also* $T^2(B/k, M) = 0$ *for all* M.

Proof. Write B as a quotient of a polynomial ring $A = k[x_1, \ldots, x_n]$ over k. Then $\operatorname{Spec} A$ is nonsingular, and we can use the criterion of (4.3), which shows that X is nonsingular if and only if the conormal sequence

$$0 \to I/I^2 \to \Omega_{A/k} \otimes_A B \to \Omega_{B/k} \to 0$$

is exact and $\Omega_{B/k}$ is locally free, i.e., a projective B-module. Since $\Omega_{A/k}$ is a free A-module, the sequence will split, so we see that X is nonsingular if and only if this sequence is split exact. By (3.10), $T^1(B/k, M) = 0$ for all M if and only if the map

$$\operatorname{Hom}(\Omega_{A/k}, M) \to \operatorname{Hom}(I/I^2, M)$$

is surjective for all M, and this is equivalent to the splitting of the sequence above (just consider the case $M = I/I^2$). Thus X is nonsingular if and only if $T^1(B/k, M) = 0$ for all M.

For the vanishing of $T^2(B/k, M)$, suppose X is nonsingular. By (3.10) again, $T^2(B/k, M) = T^2(B/A, M)$. Localizing at any point $x \in X$, by (4.3) the ideal I_x is generated by $n - r = \dim A - \dim B$ elements in the regular local ring A_x. Hence these generators form a regular sequence, and (3.9) shows that $T^2(B_x/A_x, M) = 0$ for all B_x-modules M. Thus $T^2(B/A, M) = 0$ by localization (Ex. 3.5).

Corollary 4.10. *Let* B *be a local* k-*algebra with residue field* k *algebraically closed. Then* B *is a regular local ring if and only if* $T^1(B/k, M) = 0$ *for all* B-*modules* M, *and in this case* $T^2(B/k, M) = 0$ *for all* M.

Proof. By localization, using (4.1) and (4.9).

From this theorem we can deduce a relative version. We say that a morphism of finite type $f : X \to Y$ of noetherian schemes is *smooth* if f is flat, and for every point $y \in Y$, the *geometric fiber* $X_y \otimes_{k(y)} \overline{k(y)}$ is nonsingular over $\overline{k(y)}$, where $\overline{k(y)}$ is the algebraic closure of $k(y)$ (cf. [57, III, 10.2]).

Theorem 4.11. *A morphism of finite type* $f : X \to Y$ *of noetherian schemes is smooth if and only if it is flat, and* $T^1(X/Y, \mathcal{F}) = 0$ *for all coherent sheaves* $\mathcal{F}$ *on* X. *Furthermore, if* f *is smooth, then also* $T^2(X/Y, \mathcal{F}) = 0$ *for all* $\mathcal{F}$.

Proof. The question is local, so we may assume that $X = \operatorname{Spec} B$ and $Y = \operatorname{Spec} A$ are affine and that f is given by a ring homomorphism $A \to B$.

First suppose B is flat over A and $T^1(B/A, M) = 0$ for all B-modules M. Let $y \in Y$ be a point, corresponding to a prime ideal $\mathfrak{p} \subseteq A$, and let $k = k(y)$

be its residue field. Let $A' = A/\mathfrak{p}$ and $B' = B \otimes_A A' = B/\mathfrak{p}B$. Then for any B'-module M we obtain $T^1(B'/A', M) = T^1(B/A, M) = 0$ by base change II (Ex. 3.8). Write B' as a quotient of a polynomial ring $R = A'[x_1, \ldots, x_n]$, with kernel I. Then in particular $T^1(B'/A', I/I^2) = 0$.

Now consider the flat base extension from A' to $\bar{k}$, where $\bar{k}$ is the algebraic closure of $k(y)$, which is the quotient field of A'. By base change I (Ex. 3.7), $T^1(B' \otimes \bar{k}/\bar{k}, (I/I^2) \otimes \bar{k}) = 0$. But since the base change is flat, it follows that $B' \otimes \bar{k}$ is the quotient of the polynomial ring $R \otimes \bar{k} = \bar{k}[x_1, \ldots, x_n]$ with kernel $\bar{I} = I \otimes \bar{k}$, and $(I/I^2) \otimes \bar{k} = \bar{I}/\bar{I}^2$. Then from the proof of (4.9) it follows that $\operatorname{Spec} B' \otimes \bar{k}$ is nonsingular over $\bar{k}$. Thus the geometric fibers of the morphism f are nonsingular, and f is smooth.

For the converse, suppose that B is smooth over A. First we will show that $T^1(B/A, B/\mathfrak{m}) = 0$ for every maximal ideal $\mathfrak{m} \subseteq B$. Let $\mathfrak{m}$ correspond to the point $x \in \operatorname{Spec} B$, let $f(x) = y$, and let k be the residue field of y. Then $B/\mathfrak{m}$ is a module over the ring $B \otimes_A k$, so by base change II (Ex. 3.8), we obtain $T^1(B/A, B/\mathfrak{m}) = T^1(B \otimes_A k/k, B/\mathfrak{m})$. Then by base change I (Ex. 3.7), this latter module, tensored with $\bar{k}$, the algebraic closure of k, is equal to $T^1(B \otimes \bar{k}/\bar{k}, (B/\mathfrak{m}) \otimes \bar{k})$, and this one is zero, since the geometric fibers are nonsingular. Since $k \to \bar{k}$ is a faithfully flat extension, it follows that $T^1(B \otimes k/k, B/\mathfrak{m}) = 0$ and hence $T^1(B/A, B/\mathfrak{m}) = 0$.

We observe that the functor $T^1(B/A, \cdot)$ is an additive functor from finitely generated B-modules to finitely generated B-modules, and is *semi-exact* in the sense that to each short exact sequence of modules it gives a sequence of three modules that is exact in the middle. It follows from the lemma of Dévissage below (4.12) that $T^1(B/A, M) = 0$ for all finitely generated B-modules, and hence for all B-modules, since the T^i functors commute with direct limits. The same argument shows also that $T^2(B/A, M) = 0$ for all M.

Lemma 4.12 (Dévissage). *Let B be a noetherian ring, and let F be a semi-exact additive functor from finitely generated B-modules to finitely generated B-modules. Assume that $F(B/\mathfrak{m}) = 0$ for every maximal ideal $\mathfrak{m}$ of B. Then $F(M) = 0$ for all finitely generated B-modules.*

Proof. Any finitely generated B-module M has a composition series whose quotients are $B/\mathfrak{p}_i$ for various prime ideals $\mathfrak{p}_i$. By semi-exactness, it is sufficient to show that F vanishes on each of these. Thus we may assume $M = B/\mathfrak{p}$.

We proceed by induction on the dimension of the support of M. If $\dim \operatorname{Supp} M = 0$, then M is just $B/\mathfrak{m}$ for some maximal ideal, and $F(M) = 0$ by hypothesis. For the general case, let $M = B/\mathfrak{p}$ have some dimension r. For any maximal ideal $\mathfrak{m} \supseteq \mathfrak{p}$, choose an element $t \in \mathfrak{m} - \mathfrak{p}$. Then t is a non-zero-divisor for M and we can write

$$0 \to M \xrightarrow{t} M \to M' \to 0,$$

where M' is a module with support of dimension $< r$. Hence by the induction hypothesis, $F(M') = 0$ and we get a surjection $F(M) \xrightarrow{t} F(M) \to 0$. It follows from Nakayama's lemma that $F(M)$ localized at $\mathfrak{m}$ is zero. This holds for any $\mathfrak{m} \supseteq \mathfrak{p}$, i.e., any point of $\operatorname{Spec} B/\mathfrak{p}$, and so $F(M) = 0$.

If A is a regular local ring and $B = A/I$ is a quotient, we say that B is a *local complete intersection in A* if the ideal I can be generated by $\dim A - \dim B$ elements.

Theorem 4.13. *Let A be a regular local k-algebra with residue field k algebraically closed, and let $B = A/I$ be a quotient of A. Then B is a local complete intersection in A if and only if $T^2(B/k, M) = 0$ for all B-modules M.*

Proof. Since A is regular, we have $T^1(A/k, M) = 0$ for $i = 1, 2$ and all M by (4.10). Then from the exact sequence (3.5) we obtain $T^2(B/k, M) = T^2(B/A, M)$ for all M. If B is a local complete intersection in A, then the vanishing of T^2 follows from (3.9).

Conversely, suppose that $T^2(B/k, M) = 0$ for all M. As above, this implies $T^2(B/A, M) = 0$ for all M. To compute this group, in (3.1) we can take $R = A$, $I = I$, and let F map to a minimal set of generators $(a_1, \ldots, a_s)$ of I, with kernel Q. Then the hypothesis $T^2(B/A, M) = 0$ for all M implies that

$$\operatorname{Hom}(F/IF, M) \to \operatorname{Hom}(Q/F_0, M)$$

is surjective for all M, and this in turn (taking $M = Q/F_0$) implies that the mapping $d_2 : Q/F_0 \to F/IF$ has a splitting, i.e., a map $p : F/IF \to Q/F_0$ such that $p \circ d_2 = \operatorname{id}_{Q/F_0}$. Since we chose a minimal set of generators for I, it follows that $Q \subseteq \mathfrak{m}F$, where $\mathfrak{m}$ is the maximal ideal of A. Thus the identity map $p \circ d_2$ sends Q/F_0 into $\mathfrak{m}(Q/F_0)$, and so by Nakayama's lemma, $Q/F_0 = 0$. But Q/F_0 is just the first homology group of the Koszul complex $K_\bullet(a_1, \ldots, a_s)$ over A, and the vanishing of this group is equivalent to $a_1, \ldots, a_s$ being a regular sequence [104, 16.5]. Thus B is a local complete intersection in A.

Remark 4.13.1. Since the condition $T^2(B/k, M) = 0$ for all B-modules M depends only on B, and not on A, it follows that if B is a local complete intersection in one regular local ring, then it will be a local complete intersection in any regular local ring of which it is a quotient. Thus we can say simply that B is a *local complete intersection ring* without mentioning A.

Example 4.13.2. The node of (Ex. 3.1) is a local complete intersection and correspondingly has $T^2 = 0$ for all M. The thick point of (Ex. 3.3) is not a local complete intersection and has $T^2(B/k, B) \neq 0$.

Remark 4.13.3. If we define a *relative local complete intersection* morphism $f : X \to Y$ to be one that is flat and whose geometric fibers are local complete intersection schemes, then an argument similar to the proof of (4.13) shows that f is a relative local complete intersection morphism if and only if $T^2(X/Y, \mathcal{F}) = 0$ for all coherent sheaves $\mathcal{F}$ on X.

References for this section. The definition of smooth morphisms and their characterization by the infinitesimal lifting property are due to Grothendieck [47]. The characterizations of smooth and local complete intersection morphisms using the T^i functors are due to Lichtenbaum and Schlessinger [96].

Exercises.

4.1. Let A be a local k-algebra with residue field k. Let $f : A \to A$ be a k-algebra homomorphism inducing an isomorphism $A/\mathfrak{m}^2 \to A/\mathfrak{m}^2$, where $\mathfrak{m}$ is the maximal ideal of A. Show that f itself is an isomorphism, i.e., an automorphism of A.

4.2. Let A be a local artinian k-algebra, let X_1 and X_2 be schemes of finite type, flat over A, and let $f : X_1 \to X_2$ be an A-morphism that induces an isomorphism of closed fibers $f \otimes_A k : X_1 \times_A k \to X_2 \times_A k$. Show that f itself is an isomorphism.

4.3. If X is any scheme of finite type over k, the sheaves $T^i(X/k, \mathcal{F})$ for $i = 1, 2$ and any coherent $\mathcal{O}_X$-module $\mathcal{F}$ have support in the singular locus of X.

4.4. Let B and B' be local rings, essentially of finite type over k, having isolated singularities at the closed points, and assume that B and B' are *analytically isomorphic*, i.e., the completions $\hat{B}$ and $\hat{B}'$ are isomorphic. Show that the modules $T^i_{B/k}$ and $T^i_{B'/k}$ for $i = 1, 2$ are isomorphic as modules over the isomorphic completions of B and B'. In particular, they have the same lengths.

4.5. Let Y be a closed subscheme of a nonsingular scheme X over k with ideal sheaf of $\mathcal{I}$. Then for any coherent $\mathcal{O}_Y$-module $\mathcal{F}$ there is an exact sequence of sheaves

$$0 \to T^0(Y/k, \mathcal{F}) \to \mathcal{H}om_X(\Omega_{X/k}, \mathcal{F}) \to \mathcal{H}om_Y(\mathcal{I}/\mathcal{I}^2, \mathcal{F}) \to T^1(Y/k, \mathcal{F}) \to 0.$$

4.6. Let A be a regular local k-algebra with residue field k, let $B = A/I$ for some ideal I, and assume that $I \subseteq \mathfrak{m}^2$, where $\mathfrak{m}$ is the maximal ideal of A. Show that $T^1(B/k, k)$ is a k-vector space of dimension equal to the minimal number of generators of I. Conclude that a local ring B is regular if and only if $T^1(B/k, k) = 0$.

4.7. Prove the relative version of the infinitesimal lifting property: Let X be smooth over a scheme S, let $f : Y \to X$ be a morphism of an affine scheme over S to X, and let $Y \subseteq Y'$ be an infinitesimal thickening of S-schemes. Then f lifts to a morphism $g : Y' \to X$ such that $g/y = f$. *Hint:* Imitate the proof of (4.4), using (4.11) to characterize smoothness.

4.8. Prove the converse to (Ex. 4.7): assuming X/S flat, and that the infinitesimal lifting property holds for punctual schemes Y over S, as in (4.6), show that X/S is smooth. *Hint:* For any point $s \in S$, consider the base extension $\mathrm{Spec}\,\overline{k(s)} \to S$.

4.9. Affine elliptic curves. Let k be an algebraically closed field, fix $\lambda \in k$, $\lambda \neq 0, 1$, and consider the family of affine elliptic curves over $k[t]$ defined by the equation $y^2 = x(x - 1)\,(x - (\lambda + t))$.

(a) This family is not trivial over any neighborhood of $t = 0$ because over a field, the j-invariant is already determined by any open affine piece of an elliptic curve, and the j-invariant varies in this family—cf. [57, IV, §4].

(b) This family is still not trivial over the complete local ring $k[[t]]$ at the origin, because one can look at the j-invariant over the field of fractions of this ring.

(c) The projective completion of this family in $\mathbb{P}^2$ is not trivial even over the Artin ring $C = k[t]/t^n$ for any $n \geq 2$, because the computation of the j-invariant can be made to work over the ring C.

(d) However, the affine family over the Artin ring C is trivial for any n because of (4.8). Your task, should you choose to accept it, is to find an explicit isomorphism of this family over C with the trivial family $y^2 = x(x-1)(x-\lambda)$. Hints follow.

(e) Find $a, b, c, d \in C$ such that the transformation $x' = (ax+b)/(cx+d)$ sends $0, 1, \lambda$ to $0, 1, \lambda + t$.

(f) Substitute for x' in the equation

$$y'^2 = x'(x'-1)(x'-(\lambda+t))$$

and show that the result can be written as

$$y'^2 \frac{(cx+d)^3}{a(a-c)(a-c(\lambda+t))} = x(x-1)(x-\lambda).$$

(g) Now, using the fact that t is nilpotent, show that one can find $f(x,t)$ and $g(x,t)$ in $C[x]$ such that the substitutions

$$x' = x + tf(x,t),$$
$$y' = y(1 + tg(x,t)),$$

bring the equation into the form $y^2 = x(x-1)(x-\lambda)$.

(h) Show that the transformation $(x, y) \mapsto (x', y')$ is an automorphism of the ring $C[x, y]$, so the two families are isomorphic over C.

5. Deformations of Rings

In this section we use the T^i functors to study deformations of arbitrary schemes (Situation D). We will see that the deformations of an affine scheme $X = \operatorname{Spec} B$ over k are given by $T^1(B/k, B)$, and that the deformations of a nonsingular scheme X are given by $H^1(X, \mathcal{T}_X)$, where $\mathcal{T}_X$ is the tangent sheaf. As an application, we study deformations of cones.

Definition. If X is a scheme over k, and A an Artin ring over k, we define a *deformation of X over A* to be a scheme X', flat over A, together with a closed immersion $i : X \hookrightarrow X'$ such that the induced map $i \times_A k : X \to X' \times_A k$ is an isomorphism. Two such deformations X_1', i_1 and X_2', i_2 are *equivalent* if there is an isomorphism $f : X_1' \to X_2'$ over A compatible with i_1 and i_2, i.e., such that $i_2 = f \circ i_1$.

Remark 5.0.1. Why don't we simply define a deformation of X to be a scheme X' flat over A for which $X' \times_A k$ is isomorphic to X, without specifying the isomorphism? The reason is that the set of these is less well behaved functorially than the definition we have given, and in any case, this latter set can be recovered from the former by dividing by the action of the group of automorphisms of X over k. This matter of automorphisms makes most deformation problems more complicated than Situation A, where an isomorphism of closed subschemes is just equality, and there are no automorphisms. (See §18, where automorphisms play an important role.)

We start by considering deformations of affine schemes. Let B be a k-algebra. A deformation of $\operatorname{Spec} B$ over the dual numbers D is then a D-algebra B', flat over D, together with a homomorphism $B' \to B$ inducing an isomorphism $B' \otimes_D k \to B$. Because of (2.2) the flatness of B' is equivalent to the exactness of the sequence

$$0 \to B \xrightarrow{t} B' \to B \to 0. \qquad (*)$$

Here we think of B' and B on the right as rings, and B on the left as an ideal of square 0, which is a B-module. Furthermore, B' is a D-algebra and B is a k-algebra. On the other hand, we can forget the D-algebra structure of B' and regard it simply as a k-algebra via the inclusion $k \subseteq D$. Then, as in (2.7), we see that the D-algebra structure of B' can be recovered in a unique way compatible with the exact sequence $(*)$. We need only specify multiplication by t, and this must be done by passing from B' to B on the right, followed by the inclusion $B \to B'$ on the left.

Thus we see that equivalence classes of deformations of B over D are in one-to-one correspondence with equivalence classes of exact sequences $(*)$, where B' and B are regarded only as k-algebras. We say in that case that B' is an extension as k-algebras of the k-algebra B by the B-module B.

This discussion suggests that we consider a more general situation. Let A be a ring, let B be an A-algebra, and let M be a B-module. We define an *extension* of B by M as A-algebras to be an exact sequence

$$0 \to M \to B' \to B \to 0,$$

where $B' \to B$ is a homomorphism of A-algebras, and M is an ideal in B' with $M^2 = 0$. Two such extensions B', B'' are *equivalent* if there is an isomorphism $B' \to B''$ compatible in the exact sequences with the identity maps on B and M. The *trivial* extension is given by $B' = B \oplus M$ made into a ring by the rule $(b, m) \cdot (b_1, m_1) = (bb_1, bm_1 + b_1 m)$.

Theorem 5.1. *Let A be a ring, B an A-algebra, and M a B-module. Then equivalence classes of extensions of B by M as A-algebras are in natural one-to-one correspondence with elements of the group $T^1(B/A, M)$. The trivial extension corresponds to the zero element.*

Proof. Let $A[x] \to B$ be a surjective map of a polynomial ring over A to B, let $\{e_i\}$ be a set of generators of the B-module M, and let $y = \{y_i\}$ be a set of indeterminates with the same index set as $\{e_i\}$. We consider the polynomial ring $A[x, y]$, and note that if B' is any extension of B by M, then one can find a surjective ring homomorphism $f : A[x, y] \to B'$, not unique, that makes a commutative diagram

$$
\begin{array}{ccccccccc}
0 & \to & (y) & \to & A[x, y] & \to & A[x] & \to & 0 \\
& & \downarrow & & \downarrow f & & \downarrow & & \\
0 & \to & M & \to & B' & \to & B & \to & 0 \\
& & \downarrow & & \downarrow & & \downarrow & & \\
& & 0 & & 0 & & 0 & &
\end{array}
$$

where the two outer vertical arrows are determined by the construction. Here (y) denotes the ideal in $A[x, y]$ generated by the y_i, and the map $(y) \to M$ sends y_i to e_i.

Now we proceed in two steps. First we classify quotients $f : A[x, y] \to B'$ that form a diagram as above. Then we ask, for a given extension B', how many different ways are there to express B' as a quotient of $A[x, y]$? Dividing out by this ambiguity will give us a description of the set of extensions B'.

For the first step, we complete the above diagram by adding a top row consisting of the kernels of the vertical arrows:

$$
0 \to Q \to I' \to I \to 0.
$$

Giving B' as a quotient of $A[x, y]$ is equivalent to giving the ideal I' in $A[x, y]$. Since we have a splitting of the middle row given by the ring inclusion $A[x] \to A[x, y]$, the argument used in the proof of (2.3) shows that the set of such diagrams is in natural one-to-one correspondence with the group $\text{Hom}_{A[x]}(I, M) = \text{Hom}_B(I/I^2, M)$.

For the second step, we use (4.5), whose proof works over any ring A in place of k, taking $R = A[x]$, to see that the set of possible choices for $f : A[x, y] \to B'$ forms a principal homogeneous space under the action of $\text{Der}_A(A[x], M)$. Note that because of the inclusion $A[x] \to A[x, y]$, any map $f : A[x, y] \to B'$ determines a map $g : A[x] \to B'$ and is uniquely determined by it.

Now write the long exact sequence of T^i functors (3.5) for the three rings $A \to A[x] \to B$ and the module M. The part that interests us is

$$
T^0(A[x]/A, M) \to T^1(B/A[x], M) \to T^1(B/A, M) \to T^1(A[x]/A, M).
$$

The first term here, by (3.6), is $\text{Hom}_{A[x]}(\Omega_{A[x]/A}, M)$, which is just the module of derivations $\text{Der}_A(A[x], M)$. The second term, by (3.8), is $\text{Hom}_B(I/I^2, M)$. The fourth term, by (3.7), is 0. Therefore $T^1(B/A, M)$ appears as the cokernel of a natural map

$$
\text{Der}_A(A[x], M) \to \text{Hom}_B(I/I^2, M).
$$

Under the interpretations above we see that this cokernel is the set of diagrams $A[x, y] \to B'$ as above, modulo the ambiguity of choice of the map $A[x, y] \to B'$, and so $T^1(B/A, M)$ is in one-to-one correspondence with the set of extensions B', as required.

Corollary 5.2. *Let k be a field and let B be a k-algebra. Then the set of deformations of B over the dual numbers is in natural one-to-one correspondence with the group $T^1(B/k, B)$.*

Proof. This follows from the theorem and the discussion at the beginning of this section, which showed that such deformations are in one-to-one correspondence with the k-algebra extensions of B by B.

Next we consider deformations of a nonsingular variety. We use Čech cohomology of an open covering, knowing that deformations of affine nonsingular varieties are trivial.

Theorem 5.3. *Let X be a nonsingular variety over k. Then the deformations of X over the dual numbers are in natural one-to-one correspondence with the elements of the group $H^1(X, \mathcal{T}_X)$, where $\mathcal{T}_X = \mathcal{H}om_X(\Omega_{X/k}, \mathcal{O}_X)$ is the tangent sheaf of X.*

Proof (cf. [57, III, 9.13.2]). Let X' be a deformation of X, and let $\mathcal{U} = (U_i)$ be an open affine covering of X. Over each U_i the induced deformation U'_i is trivial by (4.8), or by (4.9) combined with (5.2), so we can choose an isomorphism $\varphi_i : U_i \times_k D \xrightarrow{\sim} U'_i$. Then on $U_{ij} = U_i \cap U_j$ we get an automorphism $\psi_{ij} = \varphi_j^{-1} \varphi_i$ of $U_{ij} \times_k D$, which corresponds to an element $\theta_{ij} \in H^0(U_{ij}, \mathcal{T}_X)$ by (Ex. 5.2). By construction, on U_{ijk} we have $\theta_{ij} + \theta_{jk} + \theta_{ki} = 0$, since composition of automorphisms corresponds to addition of derivations. Therefore (θ_{ij}) is a Čech 1-cocycle for the covering $\mathcal{U}$ and the sheaf $\mathcal{T}_X$. If we replace the original chosen isomorphisms $\varphi_i : U_i \times_k D \xrightarrow{\sim} U'_i$ by some others φ'_i, then $\varphi'^{-1}_i \varphi_i$ will be an automorphism of $U_i \times_k D$ coming from a section $\alpha_i \in H^0(U_i, \mathcal{T}_X)$, and the new $\theta'_{ij} = \theta_{ij} + \alpha_i - \alpha_j$. So the new 1-cocycle θ'_{ij} differs from θ_{ij} by a coboundary, and we obtain a well-defined element θ in the Čech cohomology group $\check{H}^1(\mathcal{U}, \mathcal{T}_X)$. Since $\mathcal{U}$ is an open affine covering and $\mathcal{T}_X$ is a coherent sheaf, this is equal to the usual cohomology group $H^1(X, \mathcal{T}_X)$. Clearly θ is independent of the covering chosen.

Reversing this process, an element $\theta \in H^1(X, \mathcal{T}_X)$ is represented on $\mathcal{U}$ by a 1-cocycle θ_{ij}, and these θ_{ij} define automorphisms of the trivial deformations $U_{ij} \times_k D$ that can be glued together to make a global deformation X' of X. So we see that the deformations of X over D are given by $H^1(X, \mathcal{T}_X)$.

Example 5.3.1. If $X = \mathbb{P}^n_k$ for $n \geq 1$, then $H^1(\mathcal{T}_X) = 0$, so every deformation of X over the dual numbers is trivial. Thus X is an example of a *rigid* scheme, by which we mean a scheme all of whose deformations over the dual numbers are trivial. We have already seen that any affine nonsingular scheme is rigid (4.8). This result also follows from (5.3), since an affine scheme has no higher cohomology. We will see examples of singular rigid schemes in (5.5.1) and the exercises of this section.

One needs to exercise some caution around this notion of rigid scheme, lest intuition lead one into error. One might think, for example, that a rigid scheme has no nontrivial deformations. We will show indeed (Ex. 10.3) that every deformation over an Artin ring is trivial. However, there can be non-trivial global deformations of a rigid affine scheme (Ex. 4.9). For an even more striking example, consider a family C_t of nonsingular plane cubic curves degenerating to C_0, a union of two lines L, M, with M counted twice. Now pass to the affine plane by removing M. Then we have a family of affine elliptic curves whose limit is the affine line $\mathbb{A}^1$, which is rigid.

Another intuition might say that a singular rigid scheme cannot be embedded in a flat family whose general member is smooth, i.e., is not smoothable. This is true, but the proof is not obvious (29.6).

For projective varieties, the situation is somewhat better. If X_0 is a rigid projective scheme, then one can show that nearby fibers in a flat family of projective schemes are isomorphic to X_0 (Ex. 24.7c). If X_0 is an affine rigid scheme with an isolated singularity, the best one can hope for is that nearby fibers in a flat family have analytically isomorphic singularities (Ex. 18.8).

Example 5.3.2. Let C be a nonsingular projective curve of genus g. Then by Serre duality $H^1(\mathcal{T}_C)$ is dual to $H^0(\Omega_C^{\otimes 2})$, which has degree $4g - 4$. For $g \geq 2$ this is nonspecial, so by Riemann–Roch, $H^1(\mathcal{T}_C)$ has dimension $3g - 3$.

Now, as an application, we will study deformations of cones. Let Y be a nonsingular subvariety of $P = \mathbb{P}_k^n$, and let $X = \operatorname{Spec} B$ be the affine cone over Y inside $\mathbb{A}^{n+1} = \operatorname{Spec} R$, where $R = k[x_0, \ldots, x_n]$ is the homogeneous coordinate ring of P. We wish to study the deformations of X, i.e., the module $T_{B/k}^1 = T^1(B/k, B)$, in terms of properties of Y. To relate the two we will compare each of them to the open subset $U = X - \{x\}$, where $x = (x_0, \ldots, x_n)$ is the vertex of the cone.

Theorem 5.4. *In the situation above, if* $\operatorname{depth}_x B \geq 2$, *then there is an exact sequence*

$$0 \to T_{B/k}^1 \to H^1(U, \mathcal{T}_U) \to H^1(U, T_R|_U).$$

If furthermore, $\operatorname{depth}_x B \geq 3$, *then* $T_{B/k}^1 \cong H^1(U, \mathcal{T}_U)$ *and there is an injection*

$$0 \to T_{B/k}^1 \to \bigoplus_{\nu \in \mathbb{Z}} H^1(Y, \mathcal{T}_Y(\nu)).$$

Proof. Since U is nonsingular, we have an exact conormal sequence of sheaves

$$0 \to \mathcal{T}_U \to T_R|_U \to \mathcal{N}_{U/R} \to 0.$$

We consider the following diagram, where the second row is the cohomology sequence of this exact sequence of sheaves, and the first row comes from (3.10) applied to $\operatorname{Spec} B$ in $\operatorname{Spec} R$. The vertical arrows are restriction maps from X to U. Note that each of the modules $T_{B/k}^0, T_R \otimes B, N_{B/R}$ is the dual of some B-module. Because of the hypothesis $\operatorname{depth}_x B \geq 2$, these modules are reflexive, and hence of depth ≥ 2 [63, 1.8, 1.9]. Hence by the exact sequence of local cohomology with support at x and the cohomological interpretation of depth [57, III, Ex. 3.4], the restriction maps are isomorphisms:

$$
\begin{array}{ccccccccc}
0 \to & T^0_{B/k} & \to & T_R \otimes B & \to & N_{B/R} & \to & T^1_{B/k} & \to & 0 \\
& \downarrow\cong & & \downarrow\cong & & \downarrow\cong & & & & \\
0 \to & H^0(U, \mathcal{T}_U) & \to & H^0(U, T_R|_U) & \to & H^0(U, \mathcal{N}_{U/R}) & \to & H^1(U, \mathcal{T}_U) & \to & H^1(U, T_R|_U) \to \cdots
\end{array}
$$

From this we obtain the first exact sequence of the theorem.

Now suppose that $\mathrm{depth}_x B \geq 3$. Since $T_R|_U$ is a free $\mathcal{O}_U$-sheaf, and $H^1(U, \mathcal{O}_U) \cong H^2_x(B) = 0$ by the depth condition, we obtain $T^1_{B/k} \cong H^1(U, \mathcal{T}_U)$.

To compare this to Y, we use the exact sequence of relative tangent sheaves (Ex. 5.3)

$$0 \to \mathcal{O}_U \to \mathcal{T}_U \to \pi^* \mathcal{T}_Y \to 0.$$

Since $\mathrm{depth}_x B \geq 3$, we have $H^1(U, \mathcal{O}_U) = 0$. On the other hand, $U \to Y$ is an affine morphism with fibers that are punctured affine lines $\mathbb{A}^1 - \{0\}$. So $H^1(U, \pi^* \mathcal{T}_Y) \cong \bigoplus_{\nu \in \mathbb{Z}} H^1(Y, \mathcal{T}_Y(\nu))$. Thus $H^1(U, \mathcal{T}_U)$ injects into this latter group.

Remark 5.4.1. The depth conditions on B can be expressed in terms of Y. Thus $\mathrm{depth}_x B \geq 2$ is equivalent to saying that Y is projectively normal, which in turn is equivalent to $H^0(\mathcal{O}_P(\nu)) \to H^0(\mathcal{O}_Y(\nu))$ being surjective for all ν. And $\mathrm{depth}_x B \geq 3$ if and only if in addition, $H^1(\mathcal{O}_Y(\nu)) = 0$ for all ν.

Corollary 5.5. *If Y is a nonsingular projectively normal subvariety of $P = \mathbb{P}^n_k$, and if $H^1(\mathcal{O}_Y(\nu)) = H^1(\mathcal{T}_Y(\nu)) = 0$ for all $\nu \in \mathbb{Z}$, then the affine cone X over Y is a rigid scheme.*

Proof. Indeed, taking into account (5.4.1), the theorem implies $T^1_{B/k} = 0$.

Example 5.5.1. Let Y be the Veronese surface in $\mathbb{P}^5$, which is the 2-uple embedding of $\mathbb{P}^2$ in $\mathbb{P}^5$. It is easy to see that Y is projectively normal and that $H^1(\mathcal{O}_Y(\nu)) = 0$ for all ν. There is just one twist of the tangent sheaf $\mathcal{T}_{\mathbb{P}^2}$ that has a nonzero H^1, namely $H^1(\mathbb{P}^2, \mathcal{T}_{\mathbb{P}^2}(-3)) = k$. However, since we are dealing with the 2-uple embedding, $H^1(Y, \mathcal{T}_Y(\nu)) = H^1(\mathbb{P}^2, \mathcal{T}_{\mathbb{P}^2}(2\nu))$ for each ν, and this will be 0 for all ν. Thus the cone over the Veronese surface is rigid.

References for this section. The applications of the T^1 functors to deformations of rings appear in [96]. The deformations of cones are treated in two papers by Schlessinger [146], [147]. See also [8]. For a further study of deformations of cones, see §29.

Exercises.

5.1. Show that

(a) A node $k[x, y]/(xy)$ has a 1-dimensional space of deformations over the dual numbers.

(b) A cusp $k[x, y]/(y^2 - x^3)$ has a 2-dimensional space of deformations.

(c) An ordinary double point of a surface $k[x, y, z]/(xy - z^2)$ has a 1-dimensional space of deformations.

5.2. Automorphisms. Examining the proof of (5.1) more carefully, show that automorphisms of extensions B' of B by M are given by $T^0(B/A, M)$. Hence if B' is a deformation of B/k over D as in (5.2), the automorphisms of B' are given by $T^0(B/k, B)$, which is the tangent module of B/k.

5.3. (a) Let $P = \mathbb{P}^n_k$, let $R = k[x_0, \ldots, x_n]$ be its homogeneous coordinate ring, and let $V = \operatorname{Spec} R - \{x\}$, where x is the closed point $(x_0, \ldots, x_n)$. Using the projection $\pi : V \to P$, and comparing the Euler sequences on P and $\mathbb{P}^{n+1}$, which contains V, show that there is an exact sequence

$$0 \to \mathcal{O}_V \to \mathcal{T}_V \to \pi^* \mathcal{T}_P \to 0.$$

(b) Now let Y be a nonsingular closed subscheme of P, let X be the affine cone over Y in $\operatorname{Spec} R$, and let $U = X - \{x\}$. Show similarly that there is an exact sequence

$$0 \to \mathcal{O}_U \to \mathcal{T}_U \to \pi^* \mathcal{T}_Y \to 0.$$

5.4. Using the criterion of (5.4), show that the cone X in $\mathbb{A}^6$ over the Segre embedding of $\mathbb{P}^1 \times \mathbb{P}^2$ in $\mathbb{P}^5$ is rigid.

5.5. Let $X \subseteq \mathbb{A}^4$ be the union of two planes meeting at a point. This can be regarded as the cone over two skew lines in $\mathbb{P}^3$. Use the method of proof of (5.4) to show that X is rigid. *Hints:* Be careful, because in this case, $\operatorname{depth}_x B$ is only 1! However, two special features of this example save the day. One is that U is a disjoint union of two punctured affine planes, so that the conormal sequence for U is split exact. The other is that $H^1_x(B) = k$, and one can show by a direct analysis of the situation that the composed map $H^0(U, \mathcal{T}_U) \to H^1_x(T_R \otimes B)$ is surjective. A surprising consequence of this is that $N_{B/R}$ has depth 2, even though B only has depth 1.

5.6. Abstract versus embedded deformations.   The question is, when Y is a closed subscheme of X, can every abstract deformation of Y be realized as an embedded deformation of Y in X? Cf. (Ex. 10.1) for higher-order deformations, and §20 for further study of this question.

(a) If Y is affine and X is nonsingular, then every abstract deformation of Y over the dual numbers can be realized as an embedded deformation.

(b) On the other hand, if C is a nonsingular projective plane curve of degree $d \geq 5$, then there are abstract deformations of C over the dual numbers that cannot be realized as embedded deformations in $\mathbb{P}^2$.

5.7. Deformations of nonaffine schemes.   Let X be a scheme over k, and let X' be a deformation of X over the dual numbers. For each open affine subset $U_i \subseteq X$, the restriction of X' to U_i is a deformation of U_i, so determines an element α_i in $T^1(U, \mathcal{O}_U)$. These glue to make an element $\alpha \in H^0(X, \mathcal{T}^1_X)$, where $\mathcal{T}^1_X$ is the sheaf $\mathcal{T}^1(X/k, \mathcal{O}_X)$. If α is zero, we say that X' is locally trivial. Show that the locally trivial deformations are classified by $H^1(X, \mathcal{T}_X)$. On the other hand, given an $\alpha \in H^0(\mathcal{T}^1_X)$, show how to construct an element δ_α in $H^2(\mathcal{T}_X)$ with the property that $\delta_\alpha = 0$ if and only if α comes from a global deformation of X. Thus, if we denote by $\mathrm{Def}(X/k, D)$ the set of global deformations of X over D, there is an exact sequence

$$0 \to H^1(X, \mathcal{T}_X) \to \mathrm{Def}(X/k, D) \to H^0(X, \mathcal{T}_X^1) \overset{\delta}{\to} H^2(X, \mathcal{T}_X).$$

(Perhaps some astute reader will recognize this as the exact sequence of terms of low degree of a suitable spectral sequence.)

5.8. Hilb^8($\mathbb{P}^4$) is not irreducible. Consider the Hilbert scheme of zero-dimensional closed subschemes of $\mathbb{P}_k^4$ of length 8. There is one component of dimension 32 that has a nonsingular open subset corresponding to sets of eight distinct points. We will exhibit another component containing a nonsingular open subset of dimension 25.

(a) Let $R = k[x, y, z, w]$, let $\mathfrak{m}$ be a maximal ideal, and let $I = V + \mathfrak{m}^3$, where V is a 7-dimensional subvector space of $\mathfrak{m}^2/\mathfrak{m}^3$. Let $B = R/I$, and let Z be the associated closed subscheme of $\mathbb{A}^4 \subseteq \mathbb{P}^4$. Show that the set of all such Z, as the point of its support ranges over $\mathbb{P}^4$, forms an irreducible 25-dimensional subset of the Hilbert scheme $H = \mathrm{Hilb}^8(\mathbb{P}^4)$.

(b) Now look at the particular case

$$I = (x^2, xy, y^2, z^2, zw, w^2, xz - yw).$$

Show that $\dim_k \mathrm{Hom}(I/I^2, B) = 25$ as follows. First show that any homomorphism $\varphi : I/I^2 \to B$ has image contained in $\mathfrak{m}B$. Second, observe that homomorphisms φ with image in $\mathfrak{m}^2 B$ form a vector space of dimension 21. Third, show that if the image of φ in $\mathfrak{m}B/\mathfrak{m}^2 B$ is nonzero, then φ is completely determined, modulo those ψ mapping I/I^2 to $\mathfrak{m}^2 B$, by $\varphi(xz - yw)$, and for this one there is a four-dimensional vector space of choices.

(c) Conclude that the family of all Z's described in (a) forms an irreducible component of H of dimension 25 that is nonsingular at the point studied in (b). In particular, by reason of dimension, the zero-scheme described in (b) is not in the closure of the component corresponding to sets of eight distinct points. It is therefore a *nonsmoothable* subscheme of $\mathbb{P}^4$.

(d) Show that the image of the natural map $\mathrm{Hom}(\Omega_R, B) \to \mathrm{Hom}(I/I^2, B)$ has dimension 12, generated by the four homomorphisms that can be described as $\partial/\partial x$, $\partial/\partial y$, $\partial/\partial z$, $\partial/\partial w$, so that $T^1(B/k, B)$ has dimension 13. Thus this singularity is not rigid.

Note. This is a slight variant of an example discovered by Iarrobino and Emsalem [72].

5.9. Reduced locally complete intersection curves. (a) Let C be a reduced locally complete intersection curve in $P = \mathbb{P}^n$, with ideal sheaf $\mathcal{I}$. Then $\mathcal{I}/\mathcal{I}^2$ is locally free of rank $n - 1$ and there is an exact sequence

$$0 \to \mathcal{I}/\mathcal{I}^2 \to \Omega_P \otimes \mathcal{O}_C \to \Omega_C \to 0.$$

From this, deduce a map

$$\overset{n-1}{\bigwedge}(\mathcal{I}/\mathcal{I}^2) \otimes \Omega_C \to \omega_P \otimes C$$

and then by tensoring with $\bigwedge^{n-1}(\mathcal{I}/\mathcal{I}^2)^\vee$, a map

$$\varphi : \Omega_C \to \omega_C,$$

where ω_C is the dualizing sheaf of C (cf. [57, III.7.11]).

(b) Now suppose C is contained in a nonsingular surface X. Tensoring the exact sequence

$$0 \to \mathcal{T}_{C/k} \to \mathcal{T}_{X/k} \otimes \mathcal{O}_C \to \mathcal{N}_{C/X} \to \mathcal{T}^1_{C/k} \to 0$$

with ω_X, we get a sequence

$$0 \to \mathcal{T}_{C/k} \otimes \omega_X \to \Omega_{X/k} \otimes \mathcal{O}_C \xrightarrow{\psi} \omega_C \to \mathcal{T}^1_{C/k} \otimes \omega_X \to 0.$$

Show that the map ψ factors through the projection of $\Omega_X \otimes \mathcal{O}_C$ to Ω_C and the map φ above, so that the cokernel of φ is $\mathcal{T}^1_C \otimes \omega_X$. Show also that the kernel of φ is the torsion submodule of Ω_C, which is isomorphic to $\mathcal{T}^{1*}_C \otimes \omega_X^\vee$, where $*$ denotes the dual vector space with the appropriate module structure.

(c) Conclude that if C is a reduced locally complete intersection curve in $\mathbb{P}^n$ whose singularities all have embedding dimension 2, then φ sits in an exact sequence

$$0 \to \mathcal{R} \to \Omega_C \xrightarrow{\varphi} \omega_C \to \mathcal{S} \to 0,$$

where $\mathcal{S}$ is locally isomorphic to $\mathcal{T}^1_C$ and $\mathcal{R}$ is locally isomorphic to $\mathcal{T}^{1*}_C$.

(d) With C a curve as in (c), let $\Delta = $ length $\mathcal{T}^1_C$. Show that the tangent sheaf $\mathcal{T}_C$ has degree $2 - 2p_a + \Delta$, where p_a is the arithmetic genus of C. Conclude from Riemann–Roch that if $H^0(\mathcal{T}_C) = 0$ (which is the case, for example, if C is integral and $\Delta < 2p_a - 2$), then $H^1(\mathcal{T}_C)$, which gives the locally trivial deformations of C over the dual numbers, has dimension $3p_a - 3 - \Delta$. Taking into account (Ex. 5.7) show that the total space of (abstract) deformations of C over the dual numbers has dimension $3p_a - 3$.

5.10. Deformations of a double line.   Let L be a line on a nonsingular cubic surface X in $\mathbb{P}^3$, and let Y be the scheme associated to the divisor $2L$ on X. Then Y is a curve of degree 2, supported on the line L, and there is an exact sequence

$$0 \to \mathcal{O}_L(1) \to \mathcal{O}_Y \to \mathcal{O}_L \to 0.$$

The curve Y is obviously not smoothable in $\mathbb{P}^3$, because it has arithmetic genus $p_a = -2$, and there are no nonsingular curves of that degree and genus in $\mathbb{P}^3$. The only nonsingular curves of degree 2 are the conic, with $p_a = 0$, and a disjoint union of two lines, with $p_a = -1$.

(a) Show that the family of all such double lines in $\mathbb{P}^3$ has dimension 9; also show that $H^0(\mathcal{N}_{Y/\mathbb{P}^3}) = 9$, so these curves correspond to a nonsingular open subset of an irreducible component of the Hilbert scheme for Hilbert polynomial $2z - 1$.

(b) By looking at an affine piece, show that the sheaf $\mathcal{T}^1_Y$ is locally isomorphic to $\mathcal{O}_L$. Then, by looking at the sequence

$$0 \to \mathcal{T}_Y \to \mathcal{T}_X \otimes \mathcal{O}_Y \to \mathcal{N}_{Y/X} \to \mathcal{T}^1_Y \to 0$$

conclude that $\mathcal{T}^1_Y \cong \mathcal{O}_L(-2)$. It follows that even though there are many local deformations of Y, the sheaf $\mathcal{T}^1_Y$ has no global sections, so by (Ex. 5.7) every global abstract deformation of Y is locally trivial.

(c) Now show that $\mathcal{T}_Y$ belongs to an exact sequence

$$0 \to \mathcal{O}_L(-1) \oplus \mathcal{O}_L(2) \to \mathcal{T}_Y \to \mathcal{O}_L(1) \to 0.$$

Conclude that $H^1(\mathcal{T}_Y) = 0$, so as an abstract scheme, Y is rigid.

5.11.   Use the method of (5.5.1) to show that for any $n \geq 2$ and any $d \geq 2$ the cone over the Veronesean d-uple embedding of $\mathbb{P}^n$ in $\mathbb{P}^N$ is rigid. (Watch out for the case $n = 2$, $d = 3$!)

2

Higher-Order Deformations

In Chapter 1 we studied deformations of a structure over the dual numbers. These are the first-order infinitesimal deformations. In this chapter we discuss deformations over arbitrary Artin rings, deformations of higher order. A new feature of these is that having a deformation over one Artin ring, it is not always possible to extend it to a larger Artin ring: there may be obstructions. So in each particular case we investigate the obstructions, and when extensions do exist, try to enumerate them. For closed subschemes, invertible sheaves, and vector bundles, that is, Situations A, B, C, we do this in Sections 6, 7. For abstract schemes, Situation D, the results are in Section 10. In Sections 8, 9 we deal with three special cases, namely Cohen–Macaulay subschemes of codimension 2, locally complete intersection schemes, and Gorenstein subschemes in codimension 3. In each of these cases we can track the deformations explicitly and show that they are unobstructed. In Section 11 we show how an obstruction theory affects the local ring of the corresponding parameter space, and in Section 12 we apply this to prove a classical bound on the dimension of the Hilbert scheme of curves in $\mathbb{P}^3$. In Section 13 we describe one of Mumford's examples of "pathologies" in algebraic geometry, a family of nonsingular curves in $\mathbb{P}^3$ whose Hilbert scheme is generically nonreduced.

6. Subschemes and Invertible Sheaves

A general context in which to study higher-order deformations is the following. Suppose we are given a structure S over a field k. S could be a scheme, or a closed subscheme of a given scheme, or a vector bundle on a given scheme, etc. We look for deformations of S over the ring $A_n = k[t]/t^{n+1}$. These would be called nth-order deformations of S. Since it is hard to classify these all at once, we consider an easier problem. Suppose S_n is a given deformation of S over A_n. Then we seek to classify all deformations S_{n+1} over A_{n+1} whose restriction to A_n is the given deformation S_n. In this case we say that S_{n+1} is an extension of S_n over the ring A_{n+1}.

R. Hartshorne, *Deformation Theory*, Graduate Texts in Mathematics 257, DOI 10.1007/978-1-4419-1596-2_3, © Robin Hartshorne 2010

Typically the answer to such a problem comes in two parts: there is an obstruction to the existence of S_{n+1}; then if the obstruction is zero, the set of extensions S_{n+1} is classified by some group. However, it is not a natural correspondence as in the earlier sections. Rather it works like this: given one such extension S_{n+1}, any other S'_{n+1} determines an element of a group. We say that the set of extensions is a principal homogeneous space or torsor under the action of the group, defined as follows.

Definition. Let G be a group acting on a set S, i.e., there is a map $G \times S \to S$, written $\langle g, s \rangle \mapsto g(s)$, such that for any $g, h \in G$, $(gh)(s) = g(h(s))$. We say that S is a *principal homogeneous space* or *torsor* under the action of G if there exists an element $s_0 \in S$ such that the mapping $g \mapsto g(s_0)$ is a bijective mapping of G to S. Note that if there exists one such $s_0 \in S$, then the same is true for any other element $s_1 \in S$. Thus S is a torsor under the action of G if and only if it satisfies the following two conditions:

(1) For every $s \in S$ the induced mapping $g \mapsto g(s)$ is bijective from G to S, and

(2) S is nonempty.

If condition (1) is satisfied we say that S is a *pseudotorsor*.

Although it seems natural to discuss deformations over the rings $A_n = k[t]/t^{n+1}$ as described above, it will be useful for later purposes to work in a slightly more general context.

Notation 6.1. In this chapter we will consider deformation problems over a sequence

$$0 \to J \to C' \to C \to 0,$$

where C is a local Artin ring with residue field k, C' is another local Artin ring mapping to C, and J is an ideal with $\mathfrak{m}_{C'} J = 0$, so that J can be considered as a k-vector space. (The importance of this hypothesis is that in comparing deformations over C and C', the term on the left depends only on the initial data of the original object over k.) We suppose that some structure S_0 is given over k. Let S be a deformation of S_0 over C, and we seek to classify extensions of S to C'. This more general setting includes the earlier one, but also includes the nonequicharacteristic case, such as $\mathbb{Z}/p^{n+1} \to \mathbb{Z}/p^n$, where the rings are not k-algebras for any field k. This will be useful for lifting questions (§22).

Let us consider Situation A, deformations of closed subschemes. Suppose we are given Y_0, a closed subscheme of X_0 over a field k. As in (6.1), let X be a deformation of X_0 over C in the sense of §5, and let $Y \subseteq X$ be a closed subscheme such that $Y \times_C k = Y_0$. Further, let X' be an extension of X over C', that is, X' is flat over C' and there is a given closed immersion $X \hookrightarrow X'$ inducing an isomorphism $X \xrightarrow{\sim} X' \times_{C'} C$. We seek to classify extensions of Y over C', as closed subschemes, namely those $Y' \subseteq X'$, flat over C', such that $Y' \times_{C'} C = Y$.

Theorem 6.2. *In the above situation:*

(a) *The set of extensions of Y over C' in X' is a pseudotorsor under the action of the group $H^0(Y_0, \mathcal{N}_0 \otimes_k J)$, where $\mathcal{N}_0$ is the normal sheaf $\mathcal{N}_{Y_0/X_0}$.*

(b) *If extensions of Y over C' exist locally on X, then there is an obstruction $\alpha \in H^1(Y_0, \mathcal{N}_0 \otimes_k J)$ whose vanishing is necessary and sufficient for the global existence of Y'. If one extension Y' of Y exists, then the set of all such is a torsor under $H^0(Y_0, \mathcal{N}_0 \otimes_k J)$.*

Remark 6.2.1. In particular, part (b) applies if Y_0 in X_0 is a local complete intersection, is Cohen–Macaulay of codimension 2, or is Gorenstein of codimension 3; cf. (8.5), and §9 below. See (10.4) for the contribution of local obstructions when they are present.

Proof. We will prove this theorem in several stages. First we consider the affine case $X = \operatorname{Spec} A$, $X' = \operatorname{Spec} A'$, $Y = \operatorname{Spec} B$. Then we have a diagram

$$
\begin{array}{ccccccccc}
& & 0 & & 0 & & 0 & & \\
& & \downarrow & & \downarrow & & \downarrow & & \\
0 & \to & J \otimes_C I & \to & I' & \to & I & \to & 0 \\
& & \downarrow & & \downarrow & & \downarrow & & \\
0 & \to & J \otimes_C A & \to & A' & \to & A & \to & 0 \\
& & \downarrow & & \downarrow & & \downarrow & & \\
0 & \to & J \otimes_C B & \to & B' & \to & B & \to & 0 \\
& & \downarrow & & \downarrow & & \downarrow & & \\
& & 0 & & 0 & & 0 & &
\end{array}
$$

where all parts except I' and B' are given, and we seek to classify the possible B' (resp. I') to fill in the diagram. The exactness of the bottom two rows is equivalent to the flatness of A' and B' over C' by (2.2). The exactness of the first column results from flatness of B over C.

Compare this diagram with the diagram preceding (2.3). In that case we had a splitting $A \to A'$ and so were able to show that the possible diagrams were in natural one-to-one correspondence with $\operatorname{Hom}_B(I, A/I)$. In the present case we do not have a splitting. However, we can use a similar line of reasoning if I' and I'' are two choices of I' to fill in the diagram. Given $x \in I$, lift it to $x' \in I'$ and to $x'' \in I''$. Then $x'' - x' \in A'$ and its image in A is zero. Hence $x'' - x' \in J \otimes_C A$, and we denote its image in $J \otimes_C B$ by $\varphi(x)$. Note that the choices of x' and x'' are not unique. They are defined only up to something in $J \otimes_C I$, but this goes to 0 in $J \otimes_C B$, so φ is a well-defined additive map, in fact an A-linear homomorphism $\varphi \in \operatorname{Hom}_A(I, J \otimes_C B)$.

Conversely, given I' and given $\varphi \in \operatorname{Hom}_A(I, J \otimes_C B)$, we define another ideal I'' solving our problem as follows: I'' is the set of $x'' \in A'$ whose image in A is in I, say x, and such that for any lifting x' of x to I', the image of $x'' - x'$ in $J \otimes_C B$ is equal to $\varphi(x)$.

Note finally that if I', I'', I''' are three choices of I', and if φ_1 is defined by I', I'' as above, φ_2 defined by I'', I''', and φ_3 defined by I', I''', then $\varphi_3 = \varphi_1 +$

φ_2. Thus the operation $\langle I', \varphi \rangle \mapsto I''$ is an action of the group $\operatorname{Hom}_A(I, J \otimes_C B)$ on the set of ideals I' solving our problem, and what we have just shown is that the set of deformations of B over C' is a pseudotorsor for this group action. We have not yet discussed existence, so we cannot assert that it is a torsor.

Note that $\operatorname{Hom}_A(I, J \otimes_C B) = \operatorname{Hom}_A(I, J \otimes_k B_0)$, since J is a k-vector space, and this in turn is equal to $\operatorname{Hom}_{A_0}(I_0, J \otimes_k B_0)$, which is $H^0(Y_0, \mathcal{N}_0 \otimes_k J)$ in the affine case. (Here I_0, B_0 are the restrictions to X_0 of I, B.)

To pass from the affine case to the global case, we note that the action of $H^0(Y_0, \mathcal{N}_0 \otimes_k J)$ on the set of Y', which is defined locally, is a natural action, and that it glues together on the overlaps to give a global action of $H^0(Y_0, \mathcal{N}_0 \otimes_k J)$ on the set of solutions. The fact of being a pseudotorsor also globalizes, so we have proved (a) of the theorem.

To prove (b) suppose that local deformations of Y over C' exist. In other words, we assume that there exists an open affine covering $\mathcal{U} = (U_i)$ of X such that on each U_i there exists a deformation Y_i' of $Y \cap U_i$ in $U_i' \subseteq X'$. Choose one such Y_i' for each i. Then on $U_{ij} = U_i \cap U_j$ there are two extensions $Y_i' \cap U_{ij}'$ and $Y_j' \cap U_{ij}'$. By the previous part (a) already proved, these define an element $\alpha_{ij} \in H^0(U_{ij}, \mathcal{N}_0 \otimes_k J)$. On the intersection U_{ijk} of three open sets, there are three extensions Y_i', Y_j', Y_k', whose differences define elements $\alpha_{ij}, \alpha_{jk}, \alpha_{ik}$, and since by (a) the set of extensions is a torsor, we have $\alpha_{ik} = \alpha_{ij} + \alpha_{jk}$ in $H^0(U_{ijk}, \mathcal{N}_0 \otimes_k J)$. So we see that (α_{ij}) is a 1-cocycle for the covering $\mathcal{U}$ and the sheaf $\mathcal{N}_0 \otimes_k J$. Finally, note that this cocycle apparently depends on the choices of deformations Y_i' over U_i. If Y_i'' is another set of such choices, then Y_i' and Y_i'' define an element $\beta_i \in H^0(U_i, \mathcal{N}_0 \otimes_k J)$, and the new 1-cocycle (α_{ij}'') defined using the Y_i'' satisfies $\alpha_{ij}'' = \alpha_{ij} + \beta_j - \beta_i$. So the cohomology class $\alpha \in H^1(Y, \mathcal{N}_0 \otimes_k J)$ is well-defined. It depends only on the given Y over C.

This α is the obstruction to the existence of a global deformation Y' of Y over C'. Indeed, if Y' exists, we can take $Y_i' = Y' \cap U_i'$ for each i. Then $\alpha_{ij} = 0$, so $\alpha = 0$. Conversely, if $\alpha = 0$ in $H^1(Y, \mathcal{N}_0 \otimes_k J)$, then it is already 0 in the Čech group $\check{H}^1(\mathcal{U}, \mathcal{N}_0 \otimes_k J)$, so the cocycle α_{ij} must be a coboundary, $\alpha_{ij} = \beta_j - \beta_i$. Then using the β_i, we modify the choices Y_i' to new choices Y_i'', which then glue to form a global deformation Y'. This proves (b).

Example 6.2.2. Let X_0 be a nonsingular projective surface over k, and let $Y_0 \subseteq X_0$ be an exceptional curve of the first kind, that is, Y_0 is a nonsingular curve of genus zero, and $Y_0^2 = -1$ in X_0. Then $\mathcal{N}_{Y_0/X_0} \cong \mathcal{O}_{Y_0}(-1)$, so that $H^0(\mathcal{N}_{Y_0/X_0}) = H^1(\mathcal{N}_{Y_0/X_0}) = 0$. Hence there are no obstructions, and Y_0 extends uniquely to any deformation of X_0.

Since Y_0 can be blown down to a point P on another nonsingular surface X_0', this suggests that deformations of X_0 all arise from deformations of X_0' and changing the position of P in X_0'. See (Ex. 10.5) to verify this.

Of course other kinds of exceptional curves may not extend to a deformation (Ex. 6.7), but if they do extend, the extension will be unique.

Corollary 6.3. *Let Y be a closed subscheme of $X = \mathbb{P}_k^n$. Assume that there are no local obstructions to deformations of Y, and that $H^1(Y, \mathcal{N}_{Y/X}) = 0$. Then the Hilbert scheme H is nonsingular at the point y corresponding to Y.*

Proof. According to the infinitesimal lifting property (4.6), to show that H is nonsingular at Y, it is sufficient to show that for any local Artin ring C over k and a morphism $f : \operatorname{Spec} C \to H$ sending the closed point to y, and for any surjection of local Artin rings $C' \to C$, the morphism f lifts to a morphism $g : \operatorname{Spec} C' \to H$. If J is the kernel of $C' \to C$, then filtering J by the ideals in $\mathfrak{m}^i J$, and extending successively, we reduce to the case $\mathfrak{m} J = 0$, so as to conform with (6.1). Now by the universal property of the Hilbert scheme (1.1(a)), f corresponds to a deformation of Y in $\mathbb{P}_C^n$ over C and the problem is to extend this to Y' in $\mathbb{P}_{C'}^n$. By our hypotheses, the theorem applies to show the existence of Y' and hence of g. Hence H is nonsingular at y.

Remark 6.3.1. The hypothesis "no local obstructions to deformations of Y" applies if Y is a local complete intersection, or is Cohen–Macaulay in codimension 2, or is Gorenstein in codimension 3, as we will see in §§8, 9. In particular, it holds if Y is nonsingular (Ex. 6.1).

Now let us study Situation B, deformations of invertible sheaves. We suppose that we are given a scheme X flat over C and an extension X' of X over C', as in (6.1). Let $\mathcal{L}$ be a given invertible sheaf on X. We seek to classify invertible sheaves $\mathcal{L}'$ on X' such that $\mathcal{L}' \otimes \mathcal{O}_X \cong \mathcal{L}$.

Theorem 6.4. *In the above situation:*

(a) *There is an obstruction $\delta \in H^2(J \otimes_C \mathcal{O}_X)$ whose vanishing is a necessary and sufficient condition for the existence of $\mathcal{L}'$ on X'.*

(b) *If an $\mathcal{L}'$ exists, the group $H^1(J \otimes_C \mathcal{O}_X)$ acts transitively on the set of all isomorphism classes of such $\mathcal{L}'$ on X'.*

(c) *The set of isomorphism classes of such $\mathcal{L}'$ is a torsor under the action of $H^1(J \otimes_C \mathcal{O}_X)$ if and only if the natural map $H^0(\mathcal{O}_{X'}^*) \to H^0(\mathcal{O}_X^*)$ is surjective.*

(d) *A sufficient condition for the property of (c) to hold is that $H^0(\mathcal{O}_{X_0}) = k$, where $X_0 = X \times_C k$.*

Proof. As in the proof of (2.6) the exact sequence

$$0 \to J \otimes \mathcal{O}_X \to \mathcal{O}_{X'} \to \mathcal{O}_X \to 0$$

gives rise to an exact sequence of abelian groups

$$0 \to J \otimes \mathcal{O}_X \to \mathcal{O}_{X'}^* \to \mathcal{O}_X^* \to 0,$$

except that this time there is in general no splitting. Taking cohomology we obtain

$$0 \to H^0(J \otimes \mathcal{O}_X) \to H^0(\mathcal{O}_{X'}^*) \to H^0(\mathcal{O}_X^*) \to H^1(J \otimes \mathcal{O}_X)$$
$$\to H^1(\mathcal{O}_{X'}^*) \to H^1(\mathcal{O}_X^*) \to H^2(J \otimes \mathcal{O}_X) \to \cdots .$$

The invertible sheaf $\mathcal{L}$ on X gives an element in $H^1(\mathcal{O}_X^*)$. Its image δ in $H^2(J \otimes \mathcal{O}_X)$ is the obstruction, which is 0 if and only if $\mathcal{L}$ is the restriction of an element of $H^1(\mathcal{O}_{X'}^*)$, i.e., an invertible sheaf $\mathcal{L}'$ on X' with $\mathcal{L}' \otimes \mathcal{O}_X \cong \mathcal{L}$. Clearly $H^1(J \otimes \mathcal{O}_X)$ acts on the set of such $\mathcal{L}'$, but we cannot assert that it is a torsor unless the previous map $H^0(\mathcal{O}_{X'}^*) \to H^0(\mathcal{O}_X^*)$ is surjective.

Suppose now that $H^0(\mathcal{O}_{X_0}) = k$. Using induction on the length of C, one concludes that $H^0(\mathcal{O}_X) = C$ and $H^0(\mathcal{O}_{X'}) = C'$. Since $C'^* \to C^*$ is surjective, the conditions of (c) follow.

Remark 6.4.1. One can interpret $H^0(\mathcal{O}_X^*)$ as the group of automorphisms of the invertible sheaf $\mathcal{L}$. Thus the condition of (c) can be written $\operatorname{Aut} \mathcal{L}' \to \operatorname{Aut} \mathcal{L}$ is surjective. This type of condition on automorphisms appears frequently in deformation questions (cf. Chapter 3).

Remark 6.4.2. For an example of a line bundle on X that does not lift to X', see (Ex. 6.7).

References for this section. The obstructions to deforming a subscheme (6.2) and (6.3) appear in [45, exposé 221, 5.2]. The obstruction to deforming a line bundle (6.4) appears in [45, exposé 236, 2.10].

Exercises.

6.1. Let Y be a closed subscheme of X, where both Y and X are nonsingular and affine over k. Show that there are no obstructions to deforming Y as a subscheme of X. *Remark:* This will follow from (9.2), since Y is a local complete intersection in X, but see whether you can prove it directly using results from §4.

6.2. Plane curves. Let C be a curve in $\mathbb{P}_k^2$. Show that $H^1(\mathcal{N}_C) = 0$, confirming that the Hilbert scheme is nonsingular (Ex. 1.1).

6.3. Let C be a nonsingular curve of degree d and genus g in $\mathbb{P}_k^n$ that is *nonspecial*, i.e., $H^1(\mathcal{O}_C(1)) = 0$. Show that the Hilbert scheme is nonsingular at the corresponding point. This applies whenever $d > 2g - 2$. Note also that in $\mathbb{P}^3$, there exist nonspecial curves for any $d \geq g + 3$ [57, IV, 6.2].

6.4. Curves on a quadric surface. Let C be a nonsingular curve of bidegree (a, b) on a quadric surface Q in $\mathbb{P}_k^3$; cf. (Ex. 1.2).

(a) Show that $H^1(\mathcal{N}_{C/Q}) = 0$ for any a, b. Hence $H^1(\mathcal{N}_C) \cong H^1(\mathcal{N}_Q|_C) = H^1(\mathcal{O}_C(2))$.

(b) For $a = 1, 2$, or 3, show that $H^1(\mathcal{O}_C(2)) = 0$, so that $H^1(\mathcal{N}_C) = 0$ and the Hilbert scheme of curves in $\mathbb{P}_k^3$ is smooth at that point, if dimension $h^0(\mathcal{N}_C) = 4d$.

(c) However, if $a, b \geq 4$, show that $H^1(\mathcal{N}_C) \neq 0$. Thus the condition of (6.4) is not necessary for the Hilbert scheme to be smooth, as it is for these curves (Ex. 1.2).

(d) Now suppose $a = 1$ or 2 and $b \geq 4$. Show that the dimension of the family $\mathcal{C}$ of all curves C of bidegree (a, b) on quadric surfaces Q is $2b + 10$ for $a = 1$ and $3b + 11$ for $a = 2$, and that these numbers are strictly less than $4d$.

(e) Conclude that the family $\mathcal{C}$ of (d) above is a proper closed subset of an irreducible component $\mathcal{C}'$ of the Hilbert scheme that is smooth of dimension $4d$ at all points of $\mathcal{C}$.

6.5. For an example of the opposite kind to (Ex. 6.4c), consider $Y = \operatorname{Spec} B$, where $B = k[x, y, z]/\mathfrak{m}^2$, and $\mathfrak{m} = (x, y, z)$, as a subscheme of $\mathbb{P}_k^3$.

(a) Show that Y is a zero-dimensional scheme of length 4 that is in the closure of the irreducible component of the Hilbert scheme $H = \operatorname{Hilb}^4(\mathbb{P}^3)$ corresponding to sets of four distinct points. That component has dimension 12.

(b) Show that $H^0(\mathcal{N}_Y)$ has dimension 18, so that Y corresponds to a singular point of the Hilbert scheme (2.5). However, $H^1(\mathcal{N}_Y) = 0$, since Y is a scheme of dimension zero. We conclude from (6.4) that there must be local obstructions to deforming Y.

See [160, Ch. 4] for a detailed analysis of the abstract deformation space of this singularity.

6.6. Show that in (6.4) if instead of considering isomorphism classes of $\mathcal{L}'$ on X' whose restriction to X is isomorphic to $\mathcal{L}$, we consider the set of *deformations* of $\mathcal{L}$ over X', that is, pairs $\mathcal{L}', f$ where $f : \mathcal{L}' \to \mathcal{L}$ is a morphism inducing an isomorphism $\mathcal{L}' \otimes_{\mathcal{O}_{X'}} \mathcal{O}_X \xrightarrow{\sim} \mathcal{L}$, then the set of equivalence classes of such deformations (provided they exist) is always a torsor under the action of $H^1(J \otimes_C \mathcal{O}_X)$.

6.7. Let X be an integral projective scheme with an invertible sheaf $\mathcal{L}$ having a section s whose zero set is a divisor $Y \subseteq X$. Let X' be a deformation of X over the dual numbers.

(a) Generalize the method of (Ex. 2.5) to show that any deformation of Y to a closed subscheme Y' of X' gives rise to a deformation $\mathcal{L}'$ of $\mathcal{L}$ over X'. Conversely, however, a deformation of $\mathcal{L}$ to $\mathcal{L}'$ may not correspond to any deformation of Y.

(b) Extend the exact sequence of (Ex. 2.5) to obtain

$$\cdots \to H^1(\mathcal{L}) \to H^1(\mathcal{L}_Y) \xrightarrow{\delta} H^2(\mathcal{O}_X) \to \cdots,$$

where we can interpret $H^1(\mathcal{L}_Y)$ as containing the obstruction to deforming Y to a subscheme $Y' \subseteq X'$, and δ of that obstruction is the obstruction to extending $\mathcal{L}$ to an invertible sheaf $\mathcal{L}'$ on X'.

(c) For a particular example, let X be the quartic surface given by $f = x^4 + y^4 + xz^3 + yw^3$ in $\mathbb{P}_k^3$. Show that X is a nonsingular surface containing the line $Y : x = y = 0$, and let $\mathcal{L} = \mathcal{O}_X(Y)$ be the corresponding invertible sheaf. Consider the deformation X' of X over the dual numbers $k[t]/t^2$ given by $f' = f + tz^2w^2$. Show by a direct calculation that Y does not extend to X'. *Hint:* The inclusion $Y \hookrightarrow X$ corresponds to the ring homomorphism

$$k[x, y, z, w]/(f) \to k[z, w]$$

that sends x and y to zero. Show that there is no ring homomorphism

$$k[x, y, z, w, t]/(f', t^2) \to k[z, w, t]/(t^2)$$

restricting to the previous one when $t = 0$.

(d) Show that $H^1(X, \mathcal{L}) = 0$ and conclude that $\mathcal{L}$ does not extend to an invertible sheaf on X'.

6.8. Hilbert-flag schemes. Let Z be a fixed projective scheme over k (usually $Z = \mathbb{P}_k^n$). Let $Y \subseteq X$ be two closed subschemes of Z, with Hilbert polynomials P, Q, respectively. Then there is a projective scheme $H = \mathrm{Hilb}^{P,Q}$, called the *Hilbert-flag scheme* of the pair $Y \subseteq X$, parametrizing all such pairs of closed subschemes of Z having Hilbert polynomials P and Q. It has properties analogous to the usual Hilbert scheme (1.1a), namely, there are universal families $V \subseteq W$ in $Z \times H$, flat over H, such that the closed fibers over H are all pairs $Y \subseteq X$ as above. Furthermore, H is universal in the sense that given any families $Y' \subseteq X' \subseteq Z \times T$, flat over T, with the same Hilbert polynomials, there exists a unique morphism $T \to H$ such that Y' and X' are obtained by base extension from V and W. Since you have already accepted, dear reader, the existence of the Hilbert scheme, perhaps you will also accept without further question the existence of the Hilbert-flag scheme. For example, you can think of it as the relative Hilbert scheme of subschemes Y in the universal family of the Hilbert scheme of all X's (24.7). Or, you can refer to [152, §4.5] for more details.

(a) Assume that X is a local complete intersection, and that Y is a local complete intersection in X, so that the normal sheaves $\mathcal{N}_Y$, $\mathcal{N}_X$, and $\mathcal{N}_{Y/X}$ are locally free. By considering the normal sheaf sequence

$$0 \to \mathcal{N}_{Y/X} \to \mathcal{N}_Y \to \mathcal{N}_X \otimes \mathcal{O}_Y \to 0$$

and the restriction map $\mathcal{N}_X \to \mathcal{N}_X \otimes \mathcal{O}_Y$, show that the Zariski tangent space T_H to H at the point $Y \subseteq X$ is the fibered product of $H^0(\mathcal{N}_X)$ and $H^0(\mathcal{N}_Y)$ over $H^0(\mathcal{N}_X \otimes \mathcal{O}_Y)$.

(b) There are two successive obstructions to deforming the pair $Y \subseteq X$. First, the obstruction to deforming X, in $H^1(\mathcal{N}_X)$. Second, having chosen a deformation of X, an obstruction in $H^1(\mathcal{N}_{Y/X})$ to deforming Y in that deformation of X. This information is summarized in the diagram

$$
\begin{array}{ccc}
T_H & \to & H^0(\mathcal{N}_X) \\
\downarrow & & \downarrow \\
\end{array}
$$
$$0 \to H^0(\mathcal{N}_{Y/X}) \to H^0(\mathcal{N}_Y) \to H^0(\mathcal{N}_X \otimes \mathcal{O}_Y) \to H^1(\mathcal{N}_{Y/X}).$$

(c) There are forgetful morphisms from the Hilbert-flag scheme $H = \mathrm{Hilb}^{P,Q}$ to the Hilbert schemes Hilb^P parametrizing the Y's and to Hilb^Q parametrizing the X's.

6.9. Lines in cubic surfaces. Consider the Hilbert-flag scheme $H\{L, X\}$ of lines L in nonsingular cubic surfaces X in $\mathbb{P}_k^3$.

(a) In this case, $\mathcal{N}_{L,X} \cong \mathcal{O}_L(-1)$, so $H^0(\mathcal{N}_{L,X}) = H^1(\mathcal{N}_{L,X}) = 0$. Conclude that the natural map $T_H \to H^0(\mathcal{N}_X)$ is an isomorphism.

(b) Considering the forgetful morphism of $H\{L, X\}$ to $H\{L\}$, the Hilbert scheme of lines in $\mathbb{P}^3$, show that the latter is smooth of dimension 4, the fibers of the morphism are projective spaces of dimension 15, so that $H\{L, X\}$ is smooth of dimension 19. Since $H\{X\} = \mathbb{P}^{19}$ is also smooth of dimension 19, conclude that the morphism $H\{L, X\} \to H\{X\}$ is étale over a point corresponding to X. Thus a line extends uniquely to any deformation of X. (Knowing the structure of lines on the cubic surface [57, V, §4], we see that this is a finite étale morphism of degree 27 over the space of nonsingular cubic surfaces.)

6.10. Lines on quartic surfaces. We interpret the example of (Ex. 6.7) in terms of Hilbert-flag schemes. Let L be a line in a nonsingular quartic surface X in $\mathbb{P}^3$.

(a) Show that in this case there is a nontrivial obstruction in $H^1(\mathcal{N}_{L/X})$, so that while $H^0(\mathcal{N}_X)$ has dimension 34, the tangent space T to the Hilbert-flag scheme $H\{L, X\}$ has only dimension 33 at the corresponding point.

(b) Considering the projection to $H\{L\}$, show that $H\{L, X\}$ is smooth of dimension 33.

(c) Conclude that the morphism $H\{L, X\}$ to $H\{X\}$ is not surjective. On the contrary, its image is a subvariety of dimension 33 inside $H\{X\} \cong \mathbb{P}^{34}$. Thus we expect that for some deformations of X, the line will not extend, and (Ex. 6.7) above gave a particular example of this. For more about deforming curves on quartic surfaces, see (20.6).

7. Vector Bundles and Coherent Sheaves

Here we study Situation C, deformations of vector bundles, and more generally of coherent sheaves. Let X_0 be a scheme over k, and let $\mathcal{F}_0$ be a coherent sheaf on X_0. If X is a deformation of X_0 over a local Artin ring C, by a *deformation of $\mathcal{F}_0$ over X* we mean a coherent sheaf $\mathcal{F}$ on X, flat over C, together with a map $\mathcal{F} \to \mathcal{F}_0$ such that the induced map $\mathcal{F} \otimes_{\mathcal{O}_X} \mathcal{O}_{X_0} \to \mathcal{F}_0$ is an isomorphism. We have seen (2.7) that if C is the ring of dual numbers D, and $X = X_0 \times_k D$ is the trivial deformation of X_0, then such deformations $\mathcal{F}$ always exist, and they are classified by $\operatorname{Ext}^1_{X_0}(\mathcal{F}_0, \mathcal{F}_0)$.

Now we consider the more general situation of notation (6.1). Suppose we are given $X_0, \mathcal{F}_0, X, \mathcal{F}$ as above, and further suppose we are given an extension X' of X over C'. We ask for an *extension $\mathcal{F}'$ of $\mathcal{F}$ over C'*, that is, a coherent sheaf $\mathcal{F}'$ on X', flat over C', together with a map $\mathcal{F}' \to \mathcal{F}$ inducing an isomorphism $\mathcal{F}' \otimes_{C'} C \to \mathcal{F}$.

First we treat the case of a vector bundle, i.e., a locally free sheaf $\mathcal{F}_0$ on X_0, in which case $\mathcal{F}$ and $\mathcal{F}'$ will also be locally free (Ex. 7.1).

Theorem 7.1. *Let $X, \mathcal{F}$ be as above, and assume that $\mathcal{F}_0$ is locally free on X_0. Let $\mathcal{A}_0 = \mathcal{H}om(\mathcal{F}_0, \mathcal{F}_0)$ be the sheaf of endomorphisms of $\mathcal{F}_0$ (also sometimes written $\mathcal{E}nd\,\mathcal{F}_0$).*

(a) *If $\mathcal{F}'$ is an extension of $\mathcal{F}$ over X', then the group $\operatorname{Aut}(\mathcal{F}'/\mathcal{F})$ of automorphisms of $\mathcal{F}'$ inducing the identity automorphism of $\mathcal{F}$ is isomorphic to $H^0(X_0, \mathcal{A}_0 \otimes_k J)$.*

(b) *Given $\mathcal{F}$ on X, there is an obstruction in $H^2(X_0, \mathcal{A}_0 \otimes_k J)$ whose vanishing is a necessary and sufficient condition for the existence of an extension $\mathcal{F}'$ of $\mathcal{F}$ over X'.*

(c) *If an extension $\mathcal{F}'$ of $\mathcal{F}$ over X' exists, then the set of all such is a torsor under the action of $H^1(X_0, \mathcal{A}_0 \otimes_k J)$.*

Proof. (a) If $\mathcal{F}'$ is an extension of $\mathcal{F}$, because of flatness there is an exact sequence
$$0 \to J \otimes_k \mathcal{F}_0 \to \mathcal{F}' \to \mathcal{F} \to 0.$$
If $\sigma \in \mathrm{Aut}(\mathcal{F}'/\mathcal{F})$, then $\sigma - \mathrm{id}$ maps $\mathcal{F}'$ to $J \otimes_k \mathcal{F}_0$, and this map factors through the given maps $\pi : \mathcal{F}' \to \mathcal{F} \to \mathcal{F}_0$, thus giving a map of $\mathcal{F}_0$ to $J \otimes_k \mathcal{F}_0$. Conversely, if $\tau : \mathcal{F}_0 \to J \otimes_k \mathcal{F}_0$ is such a map, then $\mathrm{id} + \tau\pi$ is an automorphism of $\mathcal{F}'$ over $\mathcal{F}$. Thus $\mathrm{Aut}(\mathcal{F}'/\mathcal{F}) = H^0(X_0, \mathcal{A}_0 \otimes_k J)$. Note that this part of the proof does not use the hypothesis $\mathcal{F}_0$ locally free.

(b) Given $\mathcal{F}$ on X, choose a covering of X_0 by open sets U_i on which $\mathcal{F}$ is free. Let $\mathcal{F}'_i$ be free on X' over U_i, and $\mathcal{F}'_i \to \mathcal{F}|_{U_i}$ the natural map. Since the $\mathcal{F}'_i$ are free, we can choose isomorphisms $\gamma_{ij} : \mathcal{F}'_i|_{U_{ij}} \to \mathcal{F}'_j|_{U_{ij}}$ for each $U_{ij} = U_i \cap U_j$. On a triple intersection U_{ijk}, the composition $\delta_{ijk} = \gamma_{ik}^{-1}\gamma_{jk}\gamma_{ij}$ is an automorphism of $\mathcal{F}'_i|_{U_{ijk}}$, and so by (a) above gives an element of $H^0(U_{ijk}, \mathcal{A}_0 \otimes_k J)$. These form a Čech 2-cocycle for the covering $\mathcal{U} = \{U_i\}$, and so we get an element $\delta \in H^2(X_0, \mathcal{A}_0 \otimes_k J)$. If $\delta = 0$, then we can adjust the isomorphisms γ_{ij} so that they agree on U_{ijk}, and then we can glue the extensions $\mathcal{F}'_i$ to get a global extension $\mathcal{F}'$ of $\mathcal{F}$ over X'. Conversely, if $\mathcal{F}'$ exists, it is obvious that $\delta = 0$. So $\delta \in H^2(X_0, \mathcal{A}_0 \otimes_k J)$ is the obstruction to the existence of $\mathcal{F}'$.

(c) Let $\mathcal{F}'$ and $\mathcal{F}''$ be two extensions of $\mathcal{F}$ over X'. Since they are locally free, we can choose a covering $\mathcal{U} = \{U_i\}$ of X and isomorphisms $\gamma_i : \mathcal{F}'|_{U_i} \to \mathcal{F}''|_{U_i}$ for each i. On the intersection U_{ij} we find that $\delta_{ij} = \gamma_j^{-1}\gamma_i$ is an automorphism of $\mathcal{F}'|_{U_{ij}}$ and so determines an element of $H^0(U_{ij}, \mathcal{A}_0 \otimes_k J)$. These form a Čech 1-cocycle, and so we get an element $\delta \in H^1(X_0, \mathcal{A}_0 \otimes_k J)$. This element is zero if and only if the γ_i can be adjusted to agree on the overlaps and thus glue to give an isomorphism of $\mathcal{F}'$ and $\mathcal{F}''$ over $\mathcal{F}$. By fixing one $\mathcal{F}'$ then, we see that the set of extensions $\mathcal{F}'$, if nonempty, is a torsor under the action of $H^1(X_0, \mathcal{A}_0 \otimes_k J)$.

Next we consider the "embedded" version of this problem, which Grothendieck calls the Quot functor. Let $X_0, \mathcal{F}_0, X, X'$ be as before, but fix a locally free sheaf $\mathcal{E}_0$ on X_0 of which $\mathcal{F}_0$ is a quotient, fix a deformation $\mathcal{E}'$ of $\mathcal{E}_0$ over X', and let $\mathcal{E} = \mathcal{E}' \otimes_{\mathcal{O}_{X'}} \mathcal{O}_X$. A deformation of the quotient $\mathcal{E}_0 \to \mathcal{F}_0 \to 0$ over X is a deformation $\mathcal{F}$ of $\mathcal{F}_0$ on X, together with a surjection $\mathcal{E} \to \mathcal{F} \to 0$ compatible with the maps $\mathcal{E} \to \mathcal{E}_0$ and $\mathcal{F} \to \mathcal{F}_0$. For simplicity we will assume that the homological dimension of $\mathcal{F}_0$ (noted $\mathrm{hd}\, \mathcal{F}_0$) is at most 1, which in this case simply means that $\mathcal{Q}_0 = \ker(\mathcal{E}_0 \to \mathcal{F}_0)$ is locally free. This ensures that deformations exist locally, because then $\mathcal{Q}_0$ can be lifted locally to a locally free sheaf Q on X; and then lifting the map $\mathcal{Q}_0 \to \mathcal{E}_0$ any way locally to a map $Q \to \mathcal{E}$ will give a quotient $\mathcal{F}$, flat over C, locally on X, as required.

Theorem 7.2. *Given $X_0, \mathcal{E}_0 \to \mathcal{F}_0 \to 0$ in the situation as above, assuming $\mathcal{E}_0$ locally free, and $\mathrm{hd}\, \mathcal{F}_0 \le 1$, we have:*

(a) *There is an obstruction in $H^1(X_0, J \otimes_k \mathcal{H}om(Q_0, \mathcal{F}_0))$ whose vanishing is a necessary and sufficient condition for the existence of an extension $\mathcal{E}' \to \mathcal{F}' \to 0$ of $\mathcal{E} \to \mathcal{F} \to 0$ on X.*

(b) *If such extensions $\mathcal{E}' \to \mathcal{F}' \to 0$ exist, then the set of all such is a principal homogeneous space under the action of $H^0(X_0, J \otimes_k \mathcal{H}om(Q_0, \mathcal{F}_0))$.*

Proof. (a) Given $\mathcal{E} \to \mathcal{F} \to 0$, because of the hypothesis $\operatorname{hd} \mathcal{F}_0 \leq 1$, the kernel Q will be locally free. Therefore on a small open set U_i it can be lifted to a locally free subsheaf Q_i' of $\mathcal{E}'$, and we let $\mathcal{F}_i'$ be the quotient. Then on the open set U_i we have locally (suppressing subscripts U_i) a diagram

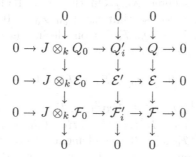

Now on U_{ij} we have two liftings Q_i' and Q_j' (restricted to U_{ij}). Take a local section x of Q. Lift it to sections $x' \in Q_i'$ and $x'' \in Q_j'$. The difference $x'' - x'$ is then a local section of $\mathcal{E}'$ that becomes zero in $\mathcal{E}$, hence lands in $J \otimes_k \mathcal{E}_0$. Let its image in $J \otimes_k \mathcal{F}_0$ be y. In this way we define an element $\gamma_{ij} \in H^0(U_{ij}, J \otimes_k \mathcal{H}om(Q_0, \mathcal{F}_0))$, since the map defined from Q factors through Q_0. The γ_{ij} define an element $\gamma \in H^1(X_0, J \otimes_k \mathcal{H}om(Q_0, \mathcal{F}_0))$. Now the usual argument shows that $\gamma = 0$ if and only if the Q_i' can be modified so as to patch together and thus define a global quotient $\mathcal{F}'$ of $\mathcal{E}'$.

(b) A similar argument shows that if one $\mathcal{F}'$ exists, then the set of all such is a torsor under the action of $H^0(X_0, J \otimes_k \mathcal{H}om(Q_0, \mathcal{F}_0))$. $\quad\blacksquare$

Remark 7.2.1. The hypothesis $\operatorname{hd} \mathcal{F}_0 \leq 1$ was used only to ensure the local existence of extensions. In the special case $\mathcal{E}_0 = \mathcal{O}_{X_0}$, the sheaf $\mathcal{F}_0$ is simply the structure sheaf of a closed subscheme $Y_0 \subseteq X_0$, and $\mathcal{H}om(Q_0, \mathcal{F}_0) = \mathcal{H}om(\mathcal{I}_{Y_0}, \mathcal{O}_{Y_0}) = \mathcal{N}_{Y_0/X_0}$, so we get the same result as in (6.3).

Now, armed with our discussion of the embedded case, we will tackle a more difficult case of the abstract deformation problem.

Theorem 7.3. *In the same situation as (7.1), instead of assuming $\mathcal{F}_0$ locally free, we will assume $\operatorname{hd} \mathcal{F}_0 \leq 1$ and X' projective. Then:*

(a) *If an extension $\mathcal{F}'$ of $\mathcal{F}$ over X' exists, then $\operatorname{Aut}(\mathcal{F}'/\mathcal{F}) = J \otimes_k \operatorname{Ext}^0_{X_0}(\mathcal{F}_0, \mathcal{F}_0)$.*

(b) *Given $\mathcal{F}$, there is an obstruction in $J \otimes_k \operatorname{Ext}^2_{X_0}(\mathcal{F}_0, \mathcal{F}_0)$ to the existence of $\mathcal{F}'$.*

(c) *If an $\mathcal{F}'$ exists, then the set of all such is a torsor under the action of* $J \otimes_k \mathrm{Ext}^1_{X_0}(\mathcal{F}_0, \mathcal{F}_0)$.

Proof. (a) The same as (7.1), since that step did not use the hypothesis $\mathcal{F}_0$ locally free, noting that $\mathrm{Ext}^0(\mathcal{F}_0, \mathcal{F}_0) = H^0(X_0, \mathcal{E}nd \, \mathcal{F}_0)$.

(b) Let $\mathcal{O}_{X'}(1)$ be an ample invertible sheaf on X', with restrictions $\mathcal{O}_X(1)$, $\mathcal{O}_{X_0}(1)$ to X and X_0. Given $\mathcal{F}$ on X, for any $a \gg 0$ we can find a surjection $\mathcal{E} = \mathcal{O}_X(-a)^q \to \mathcal{F} \to 0$ for some q, and this $\mathcal{E}$ lifts to $\mathcal{E}' = \mathcal{O}_{X'}(-a)^q$ on X'.

Note that $\mathrm{Ext}^i(\mathcal{E}_0, \mathcal{F}_0) \cong H^i(X_0, \mathcal{F}_0(a))^q$. So taking $a \gg 0$, we may assume that these groups are zero for $i > 0$. (Here we use Serre's vanishing theorem on the projective scheme X_0.) On the other hand, since $\mathrm{hd}\,\mathcal{F}_0 \leq 1$, we see that Q_0 is locally free, so that $\mathcal{E}xt^i(Q_0, \mathcal{F}_0) = 0$ for $i > 0$, and hence $\mathrm{Ext}^1(Q_0, \mathcal{F}_0) \cong H^1(X_0, \mathcal{H}om(Q_0, \mathcal{F}_0))$. Considering the exact sequence of Ext for homomorphisms of the sequence

$$0 \to Q_0 \to \mathcal{E}_0 \to \mathcal{F}_0 \to 0$$

into $\mathcal{F}_0$, we obtain $\mathrm{Ext}^1(Q_0, \mathcal{F}_0) \cong \mathrm{Ext}^2(\mathcal{F}_0, \mathcal{F}_0)$.

Thus the obstruction $\delta \in H^1(X_0, J \otimes_k \mathcal{H}om(Q_0, \mathcal{F}_0))$ of (7.2) gives us an element $\delta \in J \otimes_k \mathrm{Ext}^2(\mathcal{F}_0, \mathcal{F}_0)$. If this element is zero, then an extension $\mathcal{F}'$ of $\mathcal{F}$ in the embedded sense exists, and we just forget the embedding.

Conversely, if $\mathcal{F}'$ exists, we could have chosen a large enough that $\mathcal{F}'$ is a quotient of $\mathcal{E}'$, and this shows that $\delta = 0$.

(c) Given any two deformations $\mathcal{F}', \mathcal{F}''$, we can again choose a large enough that both of them appear as quotients of $\mathcal{E}'$. The embedded deformations are classified by $J \otimes_k \mathrm{Hom}(Q_0, \mathcal{F}_0)$, and the ambiguity of the quotient map $\mathcal{E}' \to \mathcal{F}'$ is resolved by $J \otimes_k \mathrm{Hom}(\mathcal{E}_0, \mathcal{F}_0)$. Then the long exact sequence of Ext's shows us that the (abstract) extensions $\mathcal{F}'$ are classified by $J \otimes_k \mathrm{Ext}^1(\mathcal{F}_0, \mathcal{F}_0)$.

Remark 7.3.1. If $\mathcal{F}_0$ is locally free, this gives the same result as (7.1), because in that case $\mathrm{Ext}^i(\mathcal{F}_0, \mathcal{F}_0) \cong H^i(\mathcal{H}om(\mathcal{F}_0, \mathcal{F}_0))$.

Reference. The Quot functor appears in parallel with the Hilbert scheme in [45, exposé 221].

Exercises.

7.1. Let C be a local Artin ring with residue field k. Let X be a scheme flat over C, and let $X_0 = X \times_C k$. If $\mathcal{F}$ is a coherent sheaf on X that is flat over C, and if $\mathcal{F}_0 = \mathcal{F} \otimes_{\mathcal{O}_X} \mathcal{O}_{X_0}$ is locally free on X_0, show that $\mathcal{F}$ is locally free on X.

7.2. A vector bundle $\mathcal{E}$ on a projective variety X is *simple* if the only endomorphisms are scalars, i.e., $H^0(\mathcal{E}nd \, \mathcal{E}) = k$. If X is a nonsingular curve of genus g, and $\mathcal{E}$ is a simple vector bundle of degree d and rank r, show that there are no obstructions to deforming $\mathcal{E}$, and that $h^1(\mathcal{E}nd \, \mathcal{E}) = r^2(g-1) + 1$.

7.3. For the null-correlation bundle of (Ex. 2.6c), show that $h^2(\mathcal{E}nd \, \mathcal{E}) = 0$, so there are no obstructions to deforming $\mathcal{E}$.

7.4. With a little more homological algebra, extend the result of (7.3) to arbitrary coherent sheaves. For simplicity, let's look at the affine case. Let $X' = \operatorname{Spec} B'$ be flat over C', and let $B = B' \otimes_{C'} C$ and $B_0 = B \otimes_C k$, so that there is an exact sequence

$$0 \to J \otimes_k B_0 \to B' \to B \to 0.$$

Let M be a B-module, flat over C. We want to study possible B'-modules M', flat over C', with maps $M' \to M$ inducing an isomorphism $M' \otimes_{C'} C \overset{\sim}{\to} M$. Let $L_\bullet \to M \to 0$ be a resolution of M by a complex of free B-modules.

(a) Show that to find M' it is sufficient to find a complex $L'_\bullet$ of free B'-modules such that $L'_\bullet \otimes_{B'} B \cong L_\bullet$ and there is an exact sequence of complexes

$$0 \to J \otimes_k L_\bullet \to L'_\bullet \to L_\bullet \to 0.$$

In that case $L'_\bullet$ will be exact except at the last place, and $h_0(L'_\bullet) = M'$ will be flat over B' and satisfy $M' \otimes_{B'} B \cong M$.

(b) Now study the problem of defining the maps in the complex $L'_\bullet$ and thus show that there is an obstruction in $J \otimes_k \operatorname{Ext}^2_{B_0}(M_0, M_0)$ for their existence, and if they exist, the corresponding extensions M' are classified by $J \otimes_k \operatorname{Ext}^1_{B_0}(M_0, M_0)$.

7.5. Deformations of dual modules. Let A_0 be a Gorenstein local k-algebra with residue field k, and let M_0 be a Cohen–Macaulay module over A_0 (that is, depth $M_0 = \dim A_0$). Let A be a deformation of A_0 over an Artin ring C, i.e., A flat over C together with a homomorphism $A \to A_0$ inducing an isomorphism $A \otimes_C k \overset{\sim}{\to} A_0$, and let M be a flat deformation of M_0 over A. In this exercise we will show that $M^\vee = \operatorname{Hom}_A(M, A)$ is a deformation of $M_0^\vee = \operatorname{Hom}_{A_0}(M_0, A_0)$. This result will be used later in studying deformations of linked schemes (Ex. 9.3, Ex. 9.4).

(a) The first step is to show that forming dual modules commutes with base extension to quotient rings of C. For this purpose, consider an exact sequence

$$0 \to k \to C' \to C'' \to 0,$$

where C' and C'' are quotients of C. Tensoring with $M^\vee$ and taking Hom of M into the sequence

$$0 \to A_0 \to A' \to A'' \to 0,$$

where A' and A'' are the restrictions of A to C' and C'', we obtain a diagram

$$
\begin{array}{ccccccc}
M^\vee \otimes_C k & \overset{\delta}{\to} & M^\vee \otimes_C C' & \to & M^\vee \otimes_C C'' & \to & 0 \\
\downarrow \alpha & & \downarrow \beta & & \downarrow \gamma & & \\
0 \to \operatorname{Hom}_A(M, A_0) & \to & \operatorname{Hom}_A(M, A') & \to & \operatorname{Hom}(M, A'') & \to & \operatorname{Ext}^1_A(M, A_0).
\end{array}
$$

(b) Use the flatness of M to show that $\operatorname{Ext}^1_A(M, A_0) = \operatorname{Ext}^1_{A_0}(M_0, A_0)$. Then use the hypothesis M_0 Cohen–Macaulay to show that this Ext group is zero.

(c) Now use descending induction on length C', starting with $C' = C$, for which β is an isomorphism, to show that γ, and hence β, are surjective for all C'. When $C'' = k$, we find that α is also surjective. Then make another descending induction, using surjectivity of α, to show that γ, and hence β, are isomorphisms for all C'. In particular α is an isomorphism, so δ is injective.

(d) Use ascending induction on length C' now to show, using (2.2), that $M^\vee \otimes C'$ is flat over C'. In particular, $M^\vee$ is flat over C and $M^\vee \otimes_C k \cong M_0^\vee$, so $M^\vee$ is a deformation of $M_0^\vee$.

(e) As an extra dividend, show that M is reflexive, i.e., the natural map $M \to M^{\vee\vee}$ is an isomorphism. Indeed, since M_0 is Cohen–Macaulay, it is reflexive over A_0 [63, 1.9] and its dual $M_0^\vee$ is also Cohen–Macaulay [63, 1.14]. Thus $M^\vee$ satisfies the same hypotheses as M, so $M^{\vee\vee}$ is also flat and $M^{\vee\vee} \otimes_C k \cong M_0^{\vee\vee}$. Now use flatness and reflexivity of M_0 to show that M is reflexive (cf. (Ex. 4.2)).

(f) Finally, show also that $\mathrm{Ext}^1_A(M, A) = 0$. *Hint:* Consider an exact sequence $0 \to N \to L \to M \to 0$ with L free, compare $\cdot^\vee \otimes k$ and $(\cdot \otimes k)^\vee$ applied to this sequence, and use (b) above.

8. Cohen–Macaulay in Codimension Two

We have seen that nonsingular varieties have no nontrivial local deformations (4.8). So we often consider deformations of a closed subscheme of a given nonsingular variety, so-called embedded deformations. One case we can handle well is the case of a subscheme of codimension two that is Cohen–Macaulay, meaning that all its local rings are Cohen–Macaulay rings. A local ring B is a *Cohen–Macaulay ring* if depth $B = \dim B$. In this case there is a structure theorem that allows us to track the deformations nicely. We will see that a codimension-two Cohen–Macaulay scheme is locally defined by the $r \times r$ minors of an $r \times (r+1)$ matrix of functions, and that deforming the subscheme corresponds to deforming the entries of the matrix.

In this section we work over an algebraically closed base field k, so that a scheme X is smooth if and only if it is nonsingular. This is also equivalent to saying that all its local rings are regular local rings; cf. §4.

We begin with the structure theorem for a codimension 2 Cohen–Macaulay quotient of a regular local ring. This result is well known, but we include its proof, since we will use some of the same ideas in subsequent proofs.

Theorem 8.1 (Hilbert, Burch). *Let A be a regular local ring of dimension n. Let $B = A/I$ be a Cohen–Macaulay quotient of codimension 2. Then there is an $r \times (r + 1)$ matrix φ of elements of A whose $r \times r$ minors $f_1, \ldots, f_{r+1}$ minimally generate the ideal I, and there is a resolution*

$$0 \to A^r \xrightarrow{\varphi} A^{r+1} \xrightarrow{f} A \to B \to 0$$

of B over A.

Proof. We make use of the theorem that if M is a finitely generated module over a regular local ring A, then depth $M + \mathrm{hd}\, M = n$, where $\mathrm{hd}\, M$ is the homological dimension of n [104, 19.1]. Thus $\mathrm{hd}\, B = 2$ as an A-module. If we take a minimal set of generators $a_1, \ldots, a_{r+1}$ for I, we get a resolution

$$0 \to A^r \xrightarrow{\varphi} A^{r+1} \xrightarrow{\alpha} A \to B \to 0, \tag{1}$$

where φ is an $r \times (r+1)$ matrix of elements φ_{ij} in A, and α is the map defined by $a_1, \ldots, a_{r+1}$. Let f_i be $(-1)^i$ times the determinant of the ith $r \times r$ minor of the matrix φ, and let $f : A^{r+1} \to A$ be defined by the f_i. Then we obtain a complex

$$A^r \xrightarrow{\varphi} A^{r+1} \xrightarrow{f} A, \tag{2}$$

because evaluating the product $f \circ \varphi$ amounts to taking determinants of $(r+1) \times (r+1)$ matrices with a repeated column; hence the product is zero.

We will show that the map f is the same as α, up to a unit in A. Looking at the generic point of $\operatorname{Spec} A$, i.e., tensoring with the quotient field K of A, since φ is injective, it has rank r, and at least one of its $r \times r$ minors is nonzero. Thus the map f is nonzero. Then looking at the ranks of the modules in the complex (2), we see that the homology in the middle must have rank 0. But from (1) we know that coker φ has no torsion, hence (2) is exact in the middle. Therefore the ideal $I = (a_1, \ldots, a_{r+1})$ and the ideal $(f_1, \ldots, f_{r+1})$ are isomorphic as A-modules.

Since B has codimension 2, if we look at a point of codimension 1 in $\operatorname{Spec} A$, the resolution (1) is split exact, so one of the f_i's is a unit. Hence the support of $A/(f_1, \ldots, f_{r+1})$ is also in codimension ≥ 2. An isomorphism between ideals of A is given by multiplication by some nonzero element of A. Since both I and $(f_1, \ldots, f_{r+1})$ define subsets of codimension 2, this element must be a unit in A. So, the ideal I is equal to $(f_1, \ldots, f_{r+1})$. Thus we have proved the theorem.

Suppose now that X is a smooth affine scheme over a field k, and that $Y \subseteq X$ is a closed subscheme of codimension 2 that is Cohen–Macaulay. We cannot expect that Y should be globally defined by minors of a matrix of functions on X, but we can accomplish this on small open affines.

Proposition 8.2. *Let X be a smooth scheme over a field k, and let $Y \subseteq X$ be a closed Cohen–Macaulay subscheme of codimension 2. Then for each point $y \in Y$ there is an open affine neighborhood U of y in X and there is a matrix φ of regular functions on U such that the maximal minors f_i of φ generate the ideal of $Y \cap U$ and there is a resolution*

$$0 \to \mathcal{O}_U^r \xrightarrow{\varphi} \mathcal{O}_U^{r+1} \xrightarrow{f} \mathcal{O}_U \to \mathcal{O}_{Y \cap U} \to 0.$$

Proof. We apply (8.1) to the local ring $\mathcal{O}_{y,X}$ and its quotient $\mathcal{O}_{y,Y}$. This gives a matrix φ of elements of $\mathcal{O}_{y,X}$. These elements are all defined on some open affine neighborhood U of y and so determine a complex

$$\mathcal{O}_U^r \xrightarrow{\varphi} \mathcal{O}_U^{r+1} \xrightarrow{f} \mathcal{O}_U.$$

This complex is exact at y, hence also in a neighborhood of y. The elements f_i generate the ideal of Y at y, hence also in a neighborhood. So by shrinking U to a smaller affine neighborhood, we obtain the result.

Our next task is to study deformations of codimension 2 Cohen–Macaulay subschemes. We will be looking now at rings over an Artin ring as base, and we put conditions on them by specifying that they should be flat over the Artin ring and have desired properties along the closed fiber. We will see that the same structure theorem persists in this situation, so that we can always extend deformations of codimension 2 Cohen–Macaulay subschemes in this local setting, by lifting the elements of the matrix φ.

Here is the situation. Using the notation of (6.1), we have $0 \to J \to C' \to C \to 0$, where C' and C are local Artin rings with residue field k, and $\mathfrak{m}J = 0$. Suppose we are given A', a finite type C'-algebra, flat over C'. Let $A = A' \otimes_{C'} C$ and let $A_0 = A' \otimes_{C'} k$. We assume that A_0 is smooth over k. Suppose also that we are given $B = A/I$ flat over C, and let $B_0 = B \otimes_C k$. We assume that B_0 admits a resolution as in (8.2), i.e.,

$$0 \to A_0^r \xrightarrow{\varphi_0} A_0^{r+1} \xrightarrow{f_0} A_0 \to B_0 \to 0,$$

where f_0 is defined by the $r \times r$ minors of φ_0.

Theorem 8.3 (Schaps [144]). *In the above situation we have:*

(a) *There is an $r \times (r+1)$ matrix φ of elements of A whose $r \times r$ minors f_i generate I and give a resolution*

$$0 \to A^r \xrightarrow{\varphi} A^{r+1} \xrightarrow{f} A \to B \to 0.$$

(b) *If φ' is any lifting of the matrix φ to elements of A' and the f_i' are its minors, then the sequence*

$$0 \to A'^r \xrightarrow{\varphi'} A'^{r+1} \xrightarrow{f'} A' \to B' \to 0$$

is exact and defines a quotient B', flat over C', and $B' \otimes_{C'} C = B$.

(c) *Any lifting of B, i.e., a quotient $B' = A'/I'$, with B' flat over C' and $B' \otimes_{C'} C = B$ arises by lifting the matrix φ, as in (b).*

Proof. We start with the proof of (b), assuming (a). Let φ' be any lifting of φ. Then we can consider the complex

$$L'_\bullet : A'^r \xrightarrow{\varphi'} A'^{r+1} \xrightarrow{f'} A'.$$

This is a complex for the same reason as given in the proof of (8.1) above—composition of φ' and f' amounts to evaluating determinants with a repeated column.

Since A' is flat over C', we can tensor with the exact sequence

$$0 \to J \to C' \to C \to 0$$

and obtain an exact sequence of complexes

$$0 \to L'_\bullet \otimes J \to L'_\bullet \to L'_\bullet \otimes C \to 0. \tag{3}$$

Then, since φ' is a lifting of φ, the complex $L'_\bullet \otimes C$ is just the complex that appears in (a), namely

$$L_\bullet : A^r \xrightarrow{\varphi} A^{r+1} \xrightarrow{f} A,$$

and $L'_\bullet \otimes_{C'} J = L_\bullet \otimes_C J$.

Because we are assuming (a) for the moment, $L_\bullet$ is exact, with cokernel B. Since B is flat over C, the complex $L_\bullet \otimes_C J$ is exact with cokernel $B \otimes_C J$. Now the long exact sequence of homology of the sequence of complexes (3) shows that $L'_\bullet$ is exact. We call its cokernel B':

$$0 \to A'^r \xrightarrow{\varphi'} A'^{r+1} \xrightarrow{f'} A' \to B' \to 0. \tag{4}$$

The homology sequence of (3) also shows that B' belongs to an exact sequence

$$0 \to B \otimes_C J \to B' \to B \to 0. \tag{5}$$

Since $L'_\bullet \otimes_{C'} C = L_\bullet$, we find that $B' \otimes_{C'} C = B$. Now by the local criterion of flatness (2.2), B' is flat over C', as required. This completes the proof of (b), assuming (a).

Next we prove (c), assuming (a) and (b). Let $B' = A'/I'$ be a lifting of B. Lift the elements $f_i \in I$ to $g_i \in I'$. By Nakayama's lemma, these will generate I', so we can write a resolution, with kernel M,

$$0 \to M \to A'^{r+1} \xrightarrow{g} A' \to B' \to 0.$$

Since B' is flat over C', so is M. And since B' lifts B and the g_i lift f_i, $M \otimes_{C'} C \cong A^r$. Hence M is free, equal to A'^r, so we get a resolution

$$0 \to A'^r \xrightarrow{\varphi'} A'^{r+1} \xrightarrow{g} A' \to B' \to 0$$

for a suitable matrix φ' lifting φ. But from (b) we also have

$$0 \to A'^r \xrightarrow{\varphi'} A'^{r+1} \xrightarrow{f'} A' \to B'' \to 0,$$

where B'' is another lifting of B. We must show $B' = B''$. First we need a lemma.

Lemma 8.4. *Let A be a C-algebra flat over C, with $A \otimes_C k$ normal. Let $Z \subseteq X = \mathrm{Spec}\, A$ be a subset of codimension ≥ 2. Then $H^0(X - Z, \mathcal{O}_X) = A$.*

Proof. By induction on length C, the case of length 1 being known, since then A is normal. The result follows inductively, using the sheaf sequence associated to the exact sequence of modules

$$0 \to A' \otimes_{C'} J \to A' \to A' \otimes_{C'} C \to 0.$$

To complete the proof of (c), we note from the sequences above that the ideals $I' = (g_1, \ldots, g_{r+1})$ and $I'' = (f_1', \ldots, f_{r+1}')$ are isomorphic as A'-modules. Let $X = \operatorname{Spec} A'$ and $Z = \operatorname{Supp} B$ (which is also the support of B' and B''). Restricting our isomorphism to $X - Z$ gives a section of $\mathcal{O}_X$, which by the lemma, is an element of A'. Since both B' and B'' have codimension 2, it must be a unit. So, up to change of basis, the $g_i = f_i'$ as required.

Finally, to prove (a), we use induction on the length of C' and use (c) at each step, starting from the case length 1, which we took as a hypothesis. Note that the induction step for (a) depends on (c) from the previous step, which in turn depends on (a) and (b) from the previous step.

Remark 8.4.1. Since we can always lift the deformations in this local affine case, we say that deformations of codimension 2 Cohen–Macaulay subschemes are *locally unobstructed*. However, in higher codimension, they may be obstructed (Ex. 6.5), (Ex. 29.6).

Remark 8.4.2. The proof of (8.3) given by Ellingsrud [27], following Peskine and Szpiro [132], is slightly different. By formulating the Hilbert–Burch theorem over a more general local ring, and using the more subtle theorem of Auslander that for a module M of finite type and finite homological dimension over a local ring A, depth $M + \operatorname{hd} M = \operatorname{depth} A$, they avoid some of the complicated induction in the proof of (8.3). I preferred to limit the homological algebra to the regular local ring, and then proceed by induction.

Remark 8.4.3. In case $X = \mathbb{A}_k^n$, Schaps shows, using the fact that a projective module on a polynomial ring is stably free, that one can get a global resolution, as in (8.2), over all of X. She then shows that all infinitesimal deformations of Y in X are given by lifting the matrix φ, as in (8.3).

Corollary 8.5. *In the situation of* (6.2), *deformations of a closed subscheme Y_0 of a scheme X_0, assume that X_0 is nonsingular and that Y_0 is Cohen–Macaulay of codimension* 2. *Then the obstructions to deforming Y to a closed subscheme $Y' \subseteq X'$ lie in $H^1(Y_0, \mathcal{N}_0 \otimes_k J)$.*

Proof. Indeed, (8.2) and (8.3) tell us that deformations of Y over C exist on small enough affine open subsets of X_0, so (6.2) applies.

The Global Projective Analogue

This time we consider $X = \mathbb{P}_k^n$, the projective space over k, and a closed subscheme $Y \subseteq X$ of codimension 2. To obtain results analogous to the affine case, we must put a global hypothesis on Y.

Definition. We say that a closed subscheme $Y \subseteq \mathbb{P}_k^n$ is *arithmetically Cohen–Macaulay* (ACM) if its homogeneous coordinate ring R/I_Y is a (graded) Cohen–Macaulay ring. Here $R = k[x_0, \ldots, x_n]$, and I_Y is the saturated homogeneous ideal of Y.

Proposition 8.6. *Let Y be a closed subscheme of $X = \mathbb{P}_k^n$. If $\dim Y = 0$, then Y is ACM. If $\dim Y \geq 1$, the following conditions are equivalent:*

(i) *Y is ACM.*
(ii) *$R \to H_*^0(\mathcal{O}_Y)$ is surjective, and $H_*^i(\mathcal{O}_Y) = 0$ for $0 < i < \dim Y$.*
(iii) *$H_*^i(\mathcal{I}_Y) = 0$ for $0 < i \leq \dim Y$.*

Proof. (Here we use the notation, for any coherent sheaf $\mathcal{F}$ on X, $H_*^i(\mathcal{F}) = \bigoplus_{l \in \mathbb{Z}} H^i(X, \mathcal{F}(l))$.) Let $\mathfrak{m} = (x_0, \ldots, x_n)$ be the irrelevant prime ideal of R. If $\dim Y = 0$, then $\dim R/I_Y = 1$, and since I_Y is the saturated ideal, it does not have $\mathfrak{m}$ as an associated prime. Hence R/I_Y has depth 1, and is a Cohen–Macaulay ring, so Y is ACM.

For $\dim Y \geq 1$, we use the exact sequence of local cohomology [24, A4.1]

$$0 \to H_{\mathfrak{m}}^0(R/I_Y) \to R/I_Y \to H_*^0(\mathcal{O}_Y) \to H_{\mathfrak{m}}^1(R/I_Y) \to 0$$

and the isomorphisms for $i > 0$

$$H_*^i(\mathcal{O}_Y) \cong H_{\mathfrak{m}}^{i+1}(R/I_Y),$$

together with the local cohomology criterion for depth [24, A4.3], to obtain the equivalence of (i) and (ii).

The equivalence of (ii) and (iii) follows from the cohomology of the exact sequence

$$0 \to \mathcal{I}_Y \to \mathcal{O}_X \to \mathcal{O}_Y \to 0.$$

Proposition 8.7. *Let Y be an ACM closed subscheme of $X = \mathbb{P}_k^n$ of codimension 2. Then there is an $r \times (r+1)$ matrix φ of homogeneous elements of R whose $r \times r$ minors f_i minimally generate I_Y, giving rise to a resolution*

$$0 \to \overset{r}{\underset{i=1}{\bigoplus}} R(-b_i) \overset{\varphi}{\to} \overset{r+1}{\underset{i=1}{\bigoplus}} R(-a_i) \overset{f}{\to} R \to R/I_Y \to 0.$$

Proof. Since R/I_Y is Cohen–Macaulay and a quotient of codimension 2 of R, it has homological dimension 2 over R. The proof then follows exactly as in the proof of the local case (8.1), using the graded analogues of depth and homological dimension.

Although the structure theorem (8.7) follows exactly as in the local case, when it comes to deformations, there is a new ingredient. We consider deformations of the subscheme Y, and some extra work is required to show that these give rise to deformations of the ring R/I_Y, so that we can apply the techniques we used in the local case. This extra work is contained in the following proposition, which applies only when $\dim Y \geq 1$.

Proposition 8.8. *Let $Y_0 \subseteq X_0 = \mathbb{P}^n_k$ be a closed subscheme, and assume that depth $R_0/I_0 \geq 2$, where $R_0 = k[x_0, \ldots, x_n]$ and I_0 is the homogeneous ideal of Y_0. Let C be a local Artin ring with residue field k, let $X = \mathbb{P}^n_C$, and let $Y \subseteq X$ be a closed subscheme, flat over C, with $Y \times_C k = Y_0$. Let $R = C[x_0, \ldots, x_n]$, and let $I \subseteq R$ be the ideal of Y. Then*

(a) $H^1_*(\mathcal{I}_Y) = 0$,
(b) $R/I \cong H^0_*(\mathcal{O}_Y)$,
(c) $R/I \otimes_C k = R_0/I_0$,
(d) R/I is flat over C.

Proof. (a) We use induction on the length of C. If $C = k$, the result follows from depth $R_0/I_0 \geq 2$ and the exact sequence in the proof of (8.6). For the induction step, let

$$0 \to J \to C' \to C \to 0$$

with $J \cong k$, and suppose $Y' \subseteq X' = \mathbb{P}^n_{C'}$, flat over C' as above. Then $\mathcal{I}_{Y'}$ is also flat, so we get an exact sequence

$$0 \to \mathcal{I}_{Y'} \otimes J \to \mathcal{I}_{Y'} \to \mathcal{I}_Y \to 0.$$

Now the exact sequence of H^1_* and the induction hypothesis show that $H^1_*(\mathcal{I}_{Y'}) = 0$.

(b) This follows from the exact sequence

$$0 \to \mathcal{I}_Y \to \mathcal{O}_X \to \mathcal{O}_Y \to 0,$$

which gives

$$0 \to H^0_*(\mathcal{I}_Y) \to R \to H^0_*(\mathcal{O}_Y) \to H^1_*(\mathcal{I}_Y) \to 0,$$

and (a) above, since $H^0_*(\mathcal{I}_Y) = I$.

For (c) and (d), we use the isomorphism of (b). Tensoring with k we obtain a diagram

$$\begin{array}{ccc} R/I \otimes_C k & \cong & H^0_*(\mathcal{O}_Y) \otimes_C k \\ \downarrow \alpha & & \downarrow \beta \\ R/I_0 & \cong & H^0_*(\mathcal{O}_{Y_0}) \end{array}$$

Note that α is surjective, since both terms are quotients of R. Hence β is surjective. It follows from cohomology and base change [57, III, 12.11] applied to the sheaves $\mathcal{O}_Y(\nu)$ for all ν, that β is an isomorphism, hence also α. This proves (c).

To show that R/I is flat over C, we proceed by induction on length C. For an exact sequence $0 \to J \to C' \to C \to 0$ and Y' over C', the flatness of Y' gives an exact sequence

$$0 \to J \otimes \mathcal{O}_{Y_0} \to \mathcal{O}_{Y'} \to \mathcal{O}_Y \to 0.$$

Then from (b) we obtain

$$0 \to J \otimes R_0/I_0 \to R'/I' \to R/I \to 0,$$

and the flatness of R'/I' follows from induction and the local criterion of flatness (2.2).

Now we can study deformations of codimension 2 ACM subschemes of $\mathbb{P}^n$.

Theorem 8.9 (Ellingsrud [27]). *Let Y_0 be an ACM closed subscheme of codimension 2 of $X_0 = \mathbb{P}^n_k$, and assume $\dim Y_0 \geq 1$. Using notation (6.1), suppose we are given a closed subscheme Y of $X = \mathbb{P}^n_C$, flat over C and with $Y \times_C k = Y_0$. Then:*

(a) *There is an $r \times (r+1)$ matrix φ of homogeneous elements of $R = C[x_0, \ldots, x_n]$ whose $r \times r$ minors f_i generate the ideal I of Y, and give a resolution*

$$0 \to \bigoplus R(-b_i) \xrightarrow{\varphi} \bigoplus R(-a_i) \xrightarrow{f} R \to R/I \to 0.$$

(b) *For any lifting φ' of φ to $R' = C'[x_0, \ldots, x_n]$, taking f' to be the $r \times r$ minors gives an exact sequence*

$$0 \to \bigoplus R'(-b_i) \xrightarrow{\varphi'} \bigoplus R'(-a_i) \xrightarrow{f'} R' \to R'/I' \to 0$$

and defines a closed subscheme Y' of $X' = \mathbb{P}^n_{C'}$, flat over C', with $Y' \times_{C'} C = Y$.

(c) *Any lifting Y' of Y to X', flat over C' with $Y' \times_{C'} C = Y$, arises by lifting φ as in (b).*

Proof. Since Y_0 is ACM of dimension ≥ 1, it follows that R_0/I_0 is Cohen–Macaulay of dimension ≥ 2. In particular, it has depth ≥ 2, so we can apply (8.8) and thus reduce to studying deformations of R_0/I_0. Then using (8.7) we can adapt the proof of (8.3) to the graded case to prove the result.

Example 8.9.1. The conclusions of (8.9) are false in the case of a scheme Y_0 of dimension 0 in $\mathbb{P}^n$. For example, let Y_0 be a set of three collinear points in $\mathbb{P}^2$. Then there is a linear form in the ideal of Y_0. But as you deform the points in the direction of a set of three noncollinear points of $\mathbb{P}^2$, the linear form does not lift. So the deformations of Y_0 cannot all be obtained by lifting the elements of the corresponding matrix φ.

Corollary 8.10. *The Hilbert scheme at a point corresponding to a codimension 2 ACM closed subscheme $Y \subseteq \mathbb{P}^n_k$ is smooth.*

Proof. If $n = 2$ and Y is a zero-scheme, then (8.5) tells us that deformations extend, since there is no H^1 on Y. If $\dim Y \geq 1$, then (8.9) tells us similarly that deformations always extend. The infinitesimal lifting property (4.6) implies that the Hilbert scheme is smooth, as in the proof of (6.3).

Remark 8.10.1. We have shown that a Cohen–Macaulay codimension 2 subscheme of $\mathbb{P}^n$ corresponds to a smooth point of the Hilbert scheme. However, we have not said anything about existence of such schemes, what possible Hilbert polynomials they can have, whether there exist nonsingular ones, the dimension of the Hilbert scheme, and whether it is irreducible. From a presentation such as (8.7) one can compute (in principle) the numerical invariants of the scheme and the cohomology of its normal bundle. We will not go into details here, since it would carry us too far from the main purpose of this book. I would like to mention, however, that for ACM curves in $\mathbb{P}^3$, one can say exactly what the possible degree and genus pairs are (see note to (Ex. 8.11)). The family of ACM curves with given d and g is not in general irreducible (Ex. 8.12), but becomes so if one fixes one additional piece of combinatorial data, called by various authors "numerical character" or "postulation character" or "h-vector." Then one can give necessary and sufficient conditions for the existence of smooth curves with that data, and one can compute the dimension of the corresponding irreducible component of the Hilbert scheme. We refer to [27], [49], [133], [99, V, §1] for statements and proofs.

Remark 8.10.2. The Hilbert scheme of codimension 3 ACM subschemes of $\mathbb{P}^n$ is not necessarily smooth (Ex. 6.5), (Ex. 8.9).

Theorem 8.11. *For every $n > 0$, the Hilbert scheme $\mathrm{Hilb}^n(\mathbb{P}^2_k)$, parametrizing zero-dimensional subschemes of length n of $\mathbb{P}^2$, is irreducible.*

Proof. By induction on n, the case $n = 1$ being trivial. There is one obvious component, containing the sets of n distinct points, that is irreducible of dimension $2n$. Thus it will be sufficient to show that any zero-dimensional subscheme Z of $\mathbb{P}^2$ is a limit of a flat family of subschemes whose general member consists of n distinct points. Furthermore, if Z has support at several distinct points, we can deform the component of Z at each of these points independently. Thus, by induction on the length of a "punctual" zero-scheme concentrated at a single point, it will be sufficient to show that any punctual zero-scheme can be "pulled apart" into at least two pieces. This process is local, so we can work in the affine plane $\mathbb{A}^2$, and thus we reduce to proving the following lemma.

Lemma 8.12. *Let $\mathfrak{a} \subseteq A = k[x,y]$ be an ideal of finite colength n such that $Z = \mathrm{Spec}(A/\mathfrak{a})$ has support at the origin $(0,0)$. Then there is an ideal $\mathfrak{a}_t \subseteq A[t]$, defining a family Z_t of subschemes of $\mathbb{A}^2_T$, flat over T, where $T = \mathrm{Spec}\, k[t]$, whose fiber Z_0 for $t = 0$ is Z, and whose fiber Z_t for $t \neq 0$ is supported at two distinct points, $(0,0)$ and $(0,t)$.*

Proof of Lemma. Choose $f \in \mathfrak{a}$ of minimal order s, that is, $f \in \mathfrak{m}^s - \mathfrak{m}^{s+1}$ with s minimal, where $\mathfrak{m} = (x,y)$. Then, by a linear change of coordinates, we may assume that the leading form of f is $f_0 = x^s + \cdots$. Furthermore, since $\mathfrak{a}$ is of finite colength supported at $(0,0)$, $\mathfrak{a}$ contains $\mathfrak{m}^N$ for some N, so after subtracting various higher-degree multiples of f from itself, we may also assume that f, as a polynomial in x, is of degree s, with leading term x^s.

Now let $\mathfrak{b} = \{g \in A \mid yg \in \mathfrak{a}\}$. Then it is clear that $f \in \mathfrak{b}$ and $\mathfrak{a} = (f) + y\mathfrak{b}$. Indeed, any element of $\mathfrak{a}$ can be written as a multiple of f plus something with no pure powers of x, hence divisible by y. Note that $\mathfrak{b}$ is also of finite colength, supported at $(0,0)$, since it contains $\mathfrak{m}^{N-1}$.

Let $\mathfrak{a}_t = (f) + (y-t)\mathfrak{b}$. Then $\mathfrak{a}_t = (f, y-t) \cap \mathfrak{b}$. Indeed, the inclusion $\subseteq$ is obvious. For the opposite inclusion, let $uf + v(y-t) = b$ be an element of the intersection with $u, v \in A$ and $b \in \mathfrak{b}$. Then $v(y-t) \in \mathfrak{b}$. But $y - t$ is invertible in the local ring at the origin, so $v \in \mathfrak{b}$ also.

We take $Z_t = \operatorname{Spec} k[x,y,t]/\mathfrak{a}_t$. Then $Z_0 = \mathfrak{a}$, and for $t \neq 0$, Z_t consists of $\operatorname{Spec} A/\mathfrak{b}$ at the origin and $\operatorname{Spec} k[x,y,t]/(f, y-t)$ at the point $(0,t)$.

To show that Z_t is flat over T, we must check that the length of Z_t is constant in the family [57, III, 9.9]. Let $n = $ colength $\mathfrak{a}$. Then from the exact sequence

$$0 \to A/\mathfrak{b} \xrightarrow{y} A/\mathfrak{a} \to A/\mathfrak{a} + (y) \to 0$$

and the fact that $\mathfrak{a} + (y) = (x^s, y)$ we see that length $A/\mathfrak{b} = n - s$. On the other hand, for a particular t, $(f, y-t)$ has colength equal to the number of solutions of $f(x,t) = 0$, which is s, since f as a polynomial in x has degree s. Thus Z_t has length $(n-s) + s = n$ and the family is flat.

Corollary 8.13. $\operatorname{Hilb}^n \mathbb{P}^2$ *is smooth and irreducible.*

Proof. Combine (8.10) and (8.11).

Remark 8.13.1. The irreducibility of $\operatorname{Hilb}^n \mathbb{P}^2$ also follows from smoothness (8.10) together with the more general result that for any polynomial P, the Hilbert scheme of closed subschemes of $\mathbb{P}^n$ with Hilbert polynomial P is connected [53].

Remark 8.13.2. For $N \geq 3$, the Hilbert scheme of zero-dimensional subschemes of $\mathbb{P}^N$ is not in general irreducible (Ex. 5.8), (Ex. 8.10).

References for this section. The statement that under certain conditions an ideal I is generated by the $r \times r$ minors of an $r \times (r+1)$ matrix, now known as the Hilbert–Burch theorem, appears in many forms in the literature. Hilbert, in his fundamental paper [67], where he first proves the finite generation of an ideal in a polynomial ring ("Hilbert basis theorem"), the existence of a finite free resolution for a homogeneous ideal in a polynomial ring ("Hilbert syzygy theorem"), defines the characteristic function ("Hilbert function") of an ideal, and proves that it is a polynomial ("Hilbert polynomial") for large integers, as an application of his methods, gives the structure of a homogeneous ideal I in the polynomial ring $k[x_1, x_2]$ by showing that if it is generated by $f_1, \ldots, f_n$ with no common factor, then the f_i are, up to a scalar, the $(n-1) \times (n-1)$ minors of an $(n-1) \times n$ matrix.

Burch [14] proved the same structure theorem for an ideal of homological dimension one in a local ring, referring to an earlier paper of his for the case of a local domain.

Buchsbaum [11, 3.4] proved the result for an ideal of homological dimension one in a local UFD, as a consequence of a more general result of his.

As far as I can tell from the published record, these three discoveries are all independent of each other, except that Burch and Buchsbaum were both working within a larger context of rapidly developing results in homological algebra involving many other mathematicians.

Peskine and Szpiro [132, 3.3] extended Buchbaum's result to the case of an arbitrary local ring. Ellingsrud [27] used the result of Peskine–Szpiro in his study of Cohen–Macaulay schemes of codimension 2 in $\mathbb{P}^n$. The deformation theory in this case is new with Ellingsrud.

Meanwhile, Burch's result is given as an exercise in Kaplansky [76, p. 148]. Schaps [144] gives a proof of this exercise for the case of a Cohen–Macaulay subscheme of codimension 2 in $\mathbb{A}^n$, and studies the deformation theory, new with her in this case. Artin [8] gives an account of the result, with deformation theory, saying it is due to Hilbert and Schaps.

By the time of Eisenbud's book [24, p. 502] the result for an ideal of homological dimension one in a local ring appears as the "Hilbert–Burch" theorem. Eisenbud's proof is a consequence of his more general theory of "what makes a complex exact."

For a refinement of (8.8) of this section, see Piene and Schlessinger [134].

The previously unpublished proof of (8.11) given here was discovered by Hartshorne during the academic year 1961–62, when Grothendieck gave lectures on the Hilbert scheme at Harvard. This idea of pulling apart thick schemes later became one of the key ingredients in his thesis [53], in which he proved that for any polynomial P, the Hilbert scheme of closed subschemes of $\mathbb{P}^n$ with Hilbert polynomial P is connected.

Another more conceptual proof of (8.11), generalized to zero-schemes in any smooth surface, appears in Fogarty [33].

Exercises.

8.1. ACM curves on quadric and cubic surfaces.

(a) A curve Y of bidegree (a, b) on a nonsingular quadric surface in $\mathbb{P}^3$ is ACM if and only if $|a - b| \leq 1$.

(b) A curve Y on a nonsingular cubic surface in $\mathbb{P}^3$ is ACM if and only if it is linearly equivalent to $C + mH$, where H is a hyperplane section, $m \geq 0$ is an integer, and C is either a line, a conic, a twisted cubic, or a hyperplane section.

8.2. Give an example of a nonsingular ACM curve Y in $\mathbb{P}^3$, therefore corresponding to a smooth point of the Hilbert scheme, but for which $H^1(\mathcal{N}_Y) \neq 0$.

8.3. Show that the union of a plane cubic curve and a line described in (Ex. 1.4) is an ACM curve in $\mathbb{P}^3$. Hence by (8.10) it corresponds to a nonsingular point of the Hilbert scheme, giving another proof of (Ex. 1.5).

8.4. Linkage. For ease in computing examples, we introduce the notion of linkage. For simplicity, we treat only curves in $\mathbb{P}^3$, though it can be defined in all dimensions and codimensions. When speaking of linkage, a *curve* will mean a locally Cohen–Macaulay one-dimensional closed subscheme of $\mathbb{P}^3$. In other words, it may be singular, reducible, or nonreduced, but it must not have embedded points or isolated points. Two curves C and C' are *linked* by a complete intersection of surfaces $X \cap Y$ if $C' = X \cap Y - C$ as divisors on X. This is easiest to understand if X is nonsingular, in which case curves on X are Cartier divisors, but it also makes sense on an arbitrary surface X, using the theory of generalized divisors [63, §4].

(a) If C is a curve contained in a complete intersection $X \cap Y$, then there exists a curve C' such that C and C' are linked, and C' is also linked to C.
(b) If C has degree d and genus g, and the surfaces X, Y have degrees s, t, then the degree d' and genus g' of C' are given by

$$\begin{cases} d' = st - d, \\ g' = \frac{1}{2}(d' - d)(s + t - 4) + g. \end{cases}$$

(c) We define the *Rao module* of a curve C to be the graded $R = k[x, y, z, w]$-module $M = \bigoplus_{n \in \mathbb{Z}} H^1(\mathcal{I}_C(n))$. Using exact sequences and duality on X, show that the Rao module M' of C' is isomorphic to $M^*(s + t - 4)$. Here $*$ denotes the dual vector space, so this means that $H^1(\mathcal{I}_{C'}(n))$ is dual to $H^1(\mathcal{I}_C(s + t - 4 - n))$ for each n.
(d) As a consequence of (c), any curve linked to an ACM curve is also ACM.
(e) The *speciality* function $H^1_*(\mathcal{O}_{C'}(n))$ is related to the *postulation* function $H^0_*(\mathcal{I}_C(n))$ by the formula

$$h^1(\mathcal{O}_{C'}(s + t - 4 - n)) = h^1(\mathcal{I}_C(n)) - \binom{n - s + 3}{3} - \binom{n - t + 3}{3}$$
$$+ \binom{n - s - t + 3}{3},$$

where the binomial coefficient $\binom{a}{b}$ is taken to be 0 whenever $a < b$.

8.5. Twisted cubic curves. We have seen that the Hilbert scheme of all subschemes with Hilbert polynomial $3z + 1$, i.e., degree 3 and arithmetic genus 0 in $\mathbb{P}^3$, is reducible (Ex. 1.6). However, here we show that the family of ACM curves with degree 3 and genus 0 is an open subset of one irreducible component of the Hilbert scheme, containing the (nonsingular) twisted cubic curves.

(a) Let C be an ACM curve of degree 3 and genus 0. Then $h^0(\mathcal{I}_C(2)) = 3$. Deduce from this that C is contained in an irreducible quadric surface X, and hence in a complete intersection $X \cap Y$ of two quadric surfaces.
(b) Show that the linked curve is a line, and then by doing linkages in reverse conclude that the family of all ACM $(3, 0)$ curves is irreducible of dimension 12.

8.6. Biliaison. If a curve C' is linearly equivalent to $C + mH$ on a surface X in $\mathbb{P}^3$, where H is the hyperplane class, we say that C' is obtained from C by an *elementary biliaison* of height m.

(a) If C has degree d and genus g, and X has degree s, then the degree and genus of C' are given by

$$\begin{cases} d' = d + ms, \\ g' = g + md + \frac{d}{2}ms(m + s - 4). \end{cases}$$

(b) The Rao module M' of C' is isomorphic to $M(m)$.

8.7. ACM $(6, 3)$ curves in $\mathbb{P}^3$. We consider ACM curves of degree 6 and genus 3 in $\mathbb{P}^3$.

(a) Consider a curve of type $(4; 1^6)$ in the notation of [57, V, §4] on a nonsingular cubic surface X. This is the transform of a plane quartic curve passing through the six points that are blown up to obtain the cubic surface. The general such curve is nonsingular, of degree 6 and genus 3, and is ACM in $\mathbb{P}^3$. This shows existence.

(b) Now let C be any ACM $(6, 3)$ curve in $\mathbb{P}^3$, not necessarily nonsingular, irreducible, or reduced. Show that it cannot lie in a plane or any surface of degree 2: in the plane there are no curves of that degree and genus, nor are there any on a quadric cone [57, V, Ex. 2.9]. There are $(6, 3)$ curves on a nonsingular quadric, but they are not ACM (Ex. 8.1).

(c) Show that $h^0(\mathcal{I}_C(3)) = 4$. Moreover, $h^i(\mathcal{I}_C(3 - i)) = 0$ for $i > 0$, so that C is 3-*regular* in the sense of Castelnuovo–Mumford [115] and it follows that $\mathcal{I}_C(3)$ is generated by global sections.

(d) Use the Bertini theorem applied to $\mathbb{P}^3$ with the curve C blown up to show that C is contained in a nonsingular cubic surface X. It follows (Ex. 8.1) that C is linearly equivalent to $\Gamma + H$, where Γ is a twisted cubic curve and H is the hyperplane section. Since the twisted cubic curves in $\mathbb{P}^3$ form an irreducible family (Ex. 8.5), conclude that the ACM $(6, 3)$ curves form an open nonsingular subset of an irreducible component of the Hilbert scheme.

8.8. A remarkable family. We have seen earlier (Ex. 6.4) that the family of curves of bidegree $(2, 4)$ on quadric surfaces is irreducible of dimension 23 and must be contained in the closure of an irreducible component of dimension 24 of the Hilbert scheme. These are $(6, 3)$ curves that are not ACM.

(a) If C is a bidegree $(2, 4)$ curve on the nonsingular quadric surface Q, show that its Rao module is k in degree 2, and 0 elsewhere (use (Ex. 8.6), noting that C is obtained by biliaison from two skew lines).

(b) Show conversely that if C is a $(6, 3)$ curve in $\mathbb{P}^3$ with $h^1(\mathcal{I}_C(2)) = 1$, then C is contained in a quadric surface, and $\mathcal{I}_C(4)$ is generated by global sections. Then one can link by an intersection of surfaces of degrees 2 and 4 to a curve of degree 2 and genus -1. This must be two skew lines or a double line with $g = -1$, and these form an irreducible family. Hence the $(6, 3)$ curves with $h^1(\mathcal{I}_C(2)) = 1$ form an irreducible family containing the curves described in (a).

(c) If C_t is a flat family of $(6, 3)$ curves with C_0 on a quadric surface and the general C_t not on a quadric surface, show by semi-continuity that the general C_t is ACM.

(d) Conclude that the bidegree $(2, 4)$ curves on Q are in the closure of the irreducible component of ACM $(6, 3)$ curves (Ex. 8.7). Hence for any bidegree $(2, 4)$ curve C_0 on Q there exists a flat family C_t whose general member is a nonsingular ACM $(6, 3)$ curve and whose special fiber is C_0.

Note. This seems a rather mysterious way of proving the existence of a family, because we cannot see how it happens, but only deduce its existence from the deformation theory of the Hilbert scheme.

(e) Show that the nonsingular bidegree $(2,4)$ curves on Q are hyperelliptic, and every hyperelliptic genus 3 curve appears as one of these, while the ACM $(6,3)$ curves are not hyperelliptic. Thus the family of (d) shows that every hyperelliptic genus 3 curve is contained in a flat family of curves whose general member is not hyperelliptic.

Note. It is clear that there are irreducible families containing all hyperelliptic curves of a given genus g (since they can be represented in the plane by $y^2 = f(x)$, where $f(x)$ has degree $2g+1$ or $2g+2$), and it is clear that there is an irreducible family containing all nonhyperelliptic genus 3 curves (since they have a canonical embedding as plane quartic curves), but it is by no means obvious that there is a single irreducible family containing both types of curves, as we have just shown with the Hilbert scheme of all smooth $(6,3)$ curves in $\mathbb{P}^3$. For another proof of this fact, see (Ex. 27.2). Of course the existence of such families is a consequence of the irreducibility of the variety of moduli of curves $\mathcal{M}_g$. But that is a big theorem [21], beyond the scope of this book, so the point is to find a more direct proof.

(f) Generalize the above argument to show that any hyperelliptic curve of genus 4 is a limit of a family of nonhyperelliptic curves of genus 4.

8.9. Let Y be the curve in $X = \mathbb{P}^4$ consisting of the four coordinate axes of $\mathbb{A}^4$ through the point $(0,0,0,0,1)$. Thus, if $\mathbb{P}^4$ has coordinates x, y, z, w, t, then Y is the union of the four lines defined by the ideals (x,y,z), (x,y,w), (x,z,w), (y,z,w).

(a) Show that Y is an ACM curve of degree 4 and arithmetic genus 0 and that its ideal is $I = (xy, xz, xw, yz, yw, zw)$.
(b) Show that $\mathcal{O}_Y$ has a resolution
$$\mathcal{O}_X(-3)^{12} \to \mathcal{O}_X(-2)^6 \to \mathcal{O}_X \to \mathcal{O}_Y \to 0,$$
and use this to deduce an exact sequence
$$0 \to H^0(\mathcal{N}_Y) \to H^0(\mathcal{O}_Y(2))^6 \to H^0(\mathcal{O}_Y(3))^{12}.$$

(c) By explicit computation with the basis of I and their relations, using the sequence of (b), show that $h^0(\mathcal{N}_Y) = 24$.
(d) Construct flat families of curves in $\mathbb{P}^4$ to show that Y is in the closure of the irreducible component of the Hilbert scheme whose general point corresponds to a nonsingular rational normal quartic curve in $\mathbb{P}^4$, and show that this component has dimension 21. Conclude that Y corresponds to a singular point of Hilb.
(e) Using the exact sequence
$$0 \to \mathcal{T}_Y \to \mathcal{T}_X|_Y \to \mathcal{N}_Y \to \mathcal{T}_Y^1 \to 0$$
show that $H^1(\mathcal{N}_Y)) = 0$.
(f) Conclude that Y has obstructed local deformations at the singular point.

(g) Show that $H^0(\mathcal{N}_Y) \to H^0(\mathcal{T}_Y^1) \to 0$ is surjective, so every abstract deformation of the scheme Y can be realized in $\mathbb{P}^4$.

8.10. (Iarrobino [71]). The Hilbert scheme of zero-dimensional subschemes of $\mathbb{P}^3$ is not irreducible. In the affine 3-space $\mathbb{A}^3$ with coordinate ring $R = k[x, y, z]$, let $\mathfrak{m} = (x, y, z)$, let V be a 24-dimensional subscheme of $\mathfrak{m}^7/\mathfrak{m}^8$, and let $I = V + \mathfrak{m}^8$. We consider the subscheme Z concentrated at the origin defined by $B = R/I$.

(a) Show that the length of B is 96.
(b) The dimension of the component of $\mathrm{Hilb}^{96}(\mathbb{P}^3)$ containing sets of 96 distinct points is 288.
(c) The dimension of the family of subschemes of type Z, as V varies, and the support point varies, is 291.
(d) Therefore, the schemes of type Z are not in the closure of the component containing sets of distinct points, so $\mathrm{Hilb}^{96}(\mathbb{P}^3)$ is not irreducible.

8.11. The h-vector [105, §1.4]. One can define the h-vector for ACM schemes of any dimension in any projective space, but for simplicity we will treat the case of curves in $\mathbb{P}^3$ and let the reader generalize. Let C be an ACM curve in $\mathbb{P}^3$, with homogeneous coordinate ring R/I_C. This is a Cohen–Macaulay ring of dimension 2. Dividing by two general linear forms, we obtain a zero-dimensional graded ring R_0. The h-vector of C is just the Hilbert function of this ring: $h(n) = \dim_k(R_0)_n$. Equivalently, one can define $h(n)$ as the second difference function of the Hilbert function of C, $h^0(\mathcal{O}_C(n))$.

(a) Prove the following properties of h:
   (1) $h(0) = 1$.
   (2) There is an integer $s \geq 1$, equal to the least degree of a surface containing C, such that
$$\begin{cases} h(n) = n + 1 & \text{for } 0 \leq n \leq s - 1, \\ h(n) \geq h(n+1) & \text{for } n \geq s - 1, \\ h(n) = 0 & \text{for } n \gg 0. \end{cases}$$
   (3) $h^0(\mathcal{I}_C(s)) = s + 1 - h(s)$.
(b) One can recover the degree and (arithmetic) genus of C from the h-vector as follows:
$$d = \sum_{n \geq 0} h(n),$$
$$g = \sum_{n \geq 1} (n - 1)h(n).$$

(c) The h-vector is constant in a flat family of ACM curves. *Hint:* Use semicontinuity of $h^0(\mathcal{I}_C(n))$ and $h^0(\mathcal{O}_C(n))$.
(d) For example, show that:
   (1) A plane curve of degree d has $h = 1, 1, \ldots, 1$ (d times).
   (2) The twisted cubic curve has $h = 1, 2$.
   (3) The $(6, 3)$ curves of (Ex. 8.7) have $h = 1, 2, 3$.

(4) A complete intersection of surfaces of degrees s, t, with $s \leq t$, has symmetric h-vector $1, 2, 3, \ldots, s, s, \ldots, s-1, \ldots, 3, 2, 1$, where there are $t - s + 1$ copies of s in the middle.

Note. Some facts, which we will not use [105, pp. 95–97], are that:

(1) ACM curves exist with any h-vector satisfying the properties listed in (a) above.
(2) There exists a nonsingular curve with given h-vector if and only if it is *of decreasing type*, which means that if $h(n) > h(n+1)$ for some n, then the same is true for all $n' \geq n$ until $h(n') = 0$.
(3) The set of ACM curves with given h-vector forms an open subset of an irreducible component of the Hilbert scheme.

8.12.   The Hilbert scheme of ACM curves with given degree and genus in $\mathbb{P}^3$ is not always irreducible. The first case in which this happens for nonsingular ACM curves is for $d = 18$ and $g = 39$.

(a) If C is an ACM $(18, 39)$ curve, then C is not contained in any surface of degree ≤ 3. *Hint:* Use properties of the h-vector.
(b) **Case 1.** C is contained in a quartic surface X. Then C can be linked by a complete intersection of surfaces of degrees 4 and 6 to an ACM $(6, 3)$ curve (Ex. 8.4). Since these form an irreducible family (Ex. 8.7), show that these $(18, 39)$ ACM curves also form an irreducible family C_1, and that the dimension of this family is $4d = 72$. These curves have h-vector $1, 2, 3, 4, 4, 4$.
(c) **Case 2.** C is not contained in a quartic surface. Then $h^0(\mathcal{I}_C(5)) = 4$, and C can be linked by two surfaces of degree 5 to an ACM $(7, 6)$ curve (Ex. 8.4). Such curves lie on a quadric surface, are of bidegree $(3, 4)$, and form an irreducible family of dimension 28. Show therefore that these $(18, 39)$ curves also form an irreducible family C_2 of dimension 72. They have h-vector $1, 2, 3, 4, 5, 2, 1$.
(d) Since the h-vectors are different, we thus have two nonintersecting open subsets of the Hilbert scheme consisting of ACM curves of degree 18 and genus 39. This situation is used by Sernesi to construct singular points lying on two distinct irreducible components of the Hilbert scheme (Ex. 13.2).

9. Complete Intersections and Gorenstein in Codimension Three

As in the previous section, Cohen–Macaulay in codimension two, there are some other situations in which the particular structure of the resolution of an ideal allows us to show that all deformations have the same structure, and that deformations can always be extended by lifting the corresponding resolutions. These are the cases of complete intersections and Gorenstein schemes in codimension 3.

Complete Intersections

Proposition 9.1. *Let A be a local Cohen–Macaulay ring, let $a_1, \ldots, a_r$ be elements of A, let $I = (a_1, \ldots, a_r)$, and let $B = A/I$. The following conditions are equivalent:*

(i) $a_1, \ldots, a_r$ is a regular sequence in A.

(ii) $\dim B = \dim A - r$.

(iii) The Koszul complex $K_\bullet(a_1, \ldots, a_r)$ is exact and so gives a resolution of B over A.

Proof. [104, 16.5].

Definition. In case the equivalent conditions of the proposition are satisfied, we say that I is a *complete intersection ideal* in A, or that B is a *complete intersection quotient* of A.

Thus already in our definition of complete intersection, we have the resolution

$$0 \to \overset{r}{\bigwedge} A^r \to \overset{r-1}{\bigwedge} A^r \to \cdots \to A^r \to A \to B \to 0.$$

The *Koszul complex* of $a_1, \ldots, a_r$ is all of this except the B at the right.

We will see that deformations of a complete intersection correspond to lifting the generators of the ideal, and that the Koszul resolution follows along.

Theorem 9.2. Using notation (6.1), suppose we are given A' flat over C' such that $A_0 = A' \otimes_{C'} k$ is a local Cohen–Macaulay ring. Suppose also that $B = A/I$, a quotient of $A = A' \times_{C'} C$, flat over C, such that $B_0 = B \otimes_C k$ is a complete intersection quotient of A_0 of codimension r. Then:

(a) I can be generated by r elements $a_1, \ldots, a_r$, and the Koszul complex $K_\bullet(A; a_1, \ldots, a_r)$ gives a resolution of B.

(b) If $a'_1, \ldots, a'_r$ are any liftings of the a_i to A', then the Koszul complex $K_\bullet(A'; a'_1, \ldots, a'_r)$ is exact and defines a quotient $B' = A'/I'$, flat over C', with $B' \times_{C'} C = B$.

(c) Any lifting of B to a quotient B' of A', flat over C', such that $B' \times_{C'} C = B$ arises by lifting the a_i, as in (b).

Proof. The proof follows the plan of proof of (8.3) except that it is simpler.

For (b), suppose we are given the situation of (a) and let $a'_1, \ldots, a'_r$ be liftings of the a_i. Then we get an exact sequence of Koszul complexes

$$0 \to K_\bullet(A'; a'_1, \ldots, a'_r) \otimes J \to K_\bullet(A'; a'_1, \ldots, a'_r) \to K_\bullet(A; a_1, \ldots, a_r) \to 0,$$

and the one on the left is equal to $K_\bullet(A; a_1, \ldots, a_r) \otimes_C J$. Since B is flat over C, this complex is exact with quotient $B \otimes J$. The exact sequence of homology of the sequence of complexes shows that the middle one is exact, and that its cokernel B' belongs to an exact sequence

$$0 \to B \otimes_C J \to B' \to B \to 0.$$

Now the local criterion of flatness (2.2) shows that B' is flat over C'.

For (c), we just use Nakayama's lemma to show that I' can be generated by r elements $a'_1, \ldots, a'_r$. Then (a) follows from (b) and (c) by induction on length C.

We leave to the reader to formulate an affine version of this theorem similar to (8.3), in case A is a finitely generated ring over a field k whose localizations are all Cohen–Macaulay rings. In this case a complete intersection ideal would be $I = (a_1, \ldots, a_r)$, such that for every prime ideal $\mathfrak{p} \in \operatorname{Supp} A/I$, the a_i generate a complete intersection ideal in the local ring $A_\mathfrak{p}$. We can form a Koszul complex globally, and show that it is a resolution of $B = A/I$ by looking at its localizations. Deformations will behave exactly as in (9.2).

Corollary 9.3. *If Y is a locally complete intersection subscheme of $\mathbb{P}^n$, then obstructions to deforming Y as a subscheme of $\mathbb{P}^n$ lie in $H^1(\mathcal{N}_Y)$. In particular, this applies to nonsingular subschemes.*

Proof. Combine (9.2) with (6.2) and (4.3). This proves (1.1c). $\quad\blacksquare$

For the global projective case, we say that a closed subscheme $Y \subseteq X = \mathbb{P}^n_k$ is a *(global) complete intersection* if its homogeneous ideal $I_Y \subseteq R = k[x_0, \ldots, x_n]$ can be generated by $r = \operatorname{codim}(Y, X)$ homogeneous elements. These elements will then form a regular sequence in R, and the associated Koszul complex will give a resolution of R/I_Y over R.

To deal with deformations, we must again assume $\dim Y \geq 1$ so as to be able to apply (8.8).

Theorem 9.4. *Let C', J, C be as before. Let $Y \subseteq X = \mathbb{P}^n_C$ be a closed subscheme, flat over C, such that $Y \times_C k = Y_0 \subseteq X_0 = \mathbb{P}^n_k$ is a complete intersection of codimension r, and assume $\dim Y_0 \geq 1$. Then as in (8.3) (we abbreviate the statement),*

(a) *R/I_Y has a resolution by the Koszul complex of graded R-modules.*
(b) *Any lifting of the generators of I_Y gives a deformation $Y' \subseteq X'$.*
(c) *Any deformation $Y' \subseteq X'$ of Y arises as in (b).*

We could abbreviate this further by saying that any deformation of a complete intersection of $\dim \geq 1$ is again a complete intersection, and that these deformations are unobstructed.

Corollary 9.5. *If $Y_0 \subseteq X_0 = \mathbb{P}^n_k$ is a complete intersection, the Hilbert scheme at the corresponding point is smooth.*

Proof. If $\dim Y_0 \geq 1$, the result follows from (9.4), as in the proof of (8.10). If $\dim Y_0 = 0$, then Y_0 is contained in an affine n-space $\mathbb{A}^n$, and we can use (9.3) together with the fact that a zero-scheme has no H^1. $\quad\blacksquare$

Example 9.5.1. The conclusion of (9.3) is false for Y_0 of dimension 0. The same example mentioned in (8.9.1) of three collinear points in $\mathbb{P}^2$ is a complete intersection, but its general deformation to three noncollinear points is not a complete intersection.

Gorenstein in Codimension 3

A local ring A of dimension n with residue field k is a *Gorenstein ring* if $\mathrm{Ext}^n(k, A) = k$ and $\mathrm{Ext}^i(k, A) = 0$ for all $i \neq n$. The same definition applies to a graded ring R with $R_0 = k$. A scheme X is (locally) *Gorenstein* if all of its local rings are Gorenstein rings. A closed subscheme Y of $\mathbb{P}_k^n$ is *arithmetically Gorenstein* if its homogeneous coordinate ring $k[x_0, \ldots, x_n]/I_Y$ is a Gorenstein ring. We refer to [104] for general results about Gorenstein rings.

The study of deformations of Gorenstein schemes in codimension 3 is similar to that of Cohen–Macaulay in codimension 2, but more difficult. First we have a structure theorem due to Buchsbaum and Eisenbud [12, 2.1]. If ψ is a skew-symmetric matrix of even rank, its determinant is the square of a polynomial in its entries, called the *Pfaffian* of ψ. If φ is a skew-symmetric matrix of odd rank n, the Pfaffian of the rank $n - 1$ matrix obtained by deleting the ith row and the ith column is called the ith *Pfaffian* of φ.

Theorem 9.6. *Let A be a regular local ring, and let $B = A/I$ be a quotient that is Gorenstein and of codimension 3. Then there is a skew-symmetric matrix φ of odd order n of elements of A whose Pfaffians f_i generate the ideal I and that gives rise to a resolution*

$$0 \to A \xrightarrow{f^\vee} A^n \xrightarrow{\varphi} A^n \xrightarrow{f} A \to A/I \to 0.$$

Using techniques analogous to those in the Cohen–Macaulay codimension 2 case, one can show that deformations of B always extend, and have resolutions of the same type. We leave the details to the reader.

Theorem 9.7. *Let A_0 be a smooth algebra of finite type over k, let φ be a skew-symmetric matrix of odd order n with Pfaffians f_i, and suppose that $B_0 = A_0/I_0$ has a resolution as in (9.6). Then deformations of B_0 have similar resolutions: the analogues of statements* (a), (b), (c) *of* (8.3) *hold.*

The same kind of resolution holds also in the graded case, and using (8.8) to pass from deformations of projective schemes to the associated graded rings, one can prove in the same way the following theorem.

Theorem 9.8. *Let $Y_0 \subseteq X_0 = \mathbb{P}_k^n$ be an arithmetically Gorenstein closed subscheme of codimension 3. Then there is a skew-symmetric matrix φ of homogeneous polynomials such that the homogeneous coordinate ring R/I_{Y_0} of Y_0 has a graded resolution of the form (9.6), analogous to (8.7). Furthermore, assuming that $\dim Y_0 \geq 1$, all deformations of Y_0 have similar resolutions: the analogous statements to* (a), (b), (c) *of* (8.9) *hold.*

Corollary 9.9 (Miró–Roig [106]). *Let $Y \subseteq X = \mathbb{P}_k^n$ be an arithmetically Gorenstein scheme of codimension 3, and assume $\dim Y \geq 1$. Then the Hilbert scheme is smooth at the point corresponding to Y, and all nearby points also represent arithmetically Gorenstein schemes.*

References for this section. The relation between complete intersections, regular sequences, and the Koszul complex is by now classical—I first learned about it from Serre's *Algèbre Locale Multiplicités* [156].

As for the Gorenstein in codimension 3 case, the first paper was by Watanabe [173]. Then came the structure theorem of Buchsbaum and Eisenbud [12], on which all later results are based. The proof of the structure theorem (9.6) is rather subtle. But once one has that result, the implications we have listed for deformation theory follow quite easily using the methods of the previous section.

As general references for linkage (in the exercises), see the book of Migliore [105] for the global case, and the papers of Huneke and Ulrich, starting with [69], for the local case. Deformations of linkages are studied in the global case by Kleppe [83] and in the local case by Buchweitz [13]. One knows from the theorems of Gaeta and Watanabe (see for example [105]) that Cohen–Macaulay schemes in codimension 2 and Gorenstein schemes in codimension 3 are licci. Hence the interest in finding nonlicci schemes (Ex. 9.4), (Ex. 29.7).

Exercises.

9.1. Let C be the curve in $\mathbb{A}^3$ defined parametrically by $(x, y, z) = (t^3, t^4, t^5)$. Show that C is not a local complete intersection at the origin [57, I, Ex. 1.11]. Nevertheless, it is Cohen–Macaulay in codimension 2, so there are no local obstructions to deformations.

9.2. Consider canonical curves C of genus 5 and degree 8 in $\mathbb{P}^4$ [57, IV, Ex. 5.5].

(a) The general such curve is a complete intersection of three quadric hypersurfaces, so the corresponding point on the Hilbert scheme is nonsingular.
(b) But if the curve is trigonal, the three quadrics containing it intersect in a ruled cubic surface, so the curve is no longer a complete intersection. However, it is still ACM, and is arithmetically Gorenstein, since $\omega_C = \mathcal{O}_C(1)$, so it still corresponds to a smooth point on the Hilbert scheme.

9.3. Linkage of deformations. Let Z_0 be a Gorenstein scheme over k (meaning equidimensional and all local rings are Gorenstein) and let X_0 be a Cohen–Macaulay closed subscheme of the same dimension. Let $X \subseteq Z$ be deformations of X_0 and Z_0 over an Artin ring C. Define a sheaf of ideals $\mathcal{I}_Y = \mathcal{H}om_{\mathcal{O}_Z}(\mathcal{O}_X, \mathcal{O}_Z)$, and let $Y \subseteq Z$ be the subscheme defined by $\mathcal{I}_Y$. In this case we say that Y is *linked* to X by Z.

(a) Show that the sheaf $\mathcal{I}_X$ is a flat deformation of the sheaf $\mathcal{I}_{X_0}$, and that $\mathcal{I}_{X_0}$ is also a Cohen–Macaulay sheaf on Z_0.
(b) Dualize the sequence
$$0 \to \mathcal{I}_X \to \mathcal{O}_Z \to \mathcal{O}_X \to 0$$
and use (Ex. 7.5f) to show that $\mathcal{O}_Y = \mathcal{I}_X^\vee$. Then from (Ex. 7.5) show that Y is flat over C and is a deformation of Y_0, defined by $\mathcal{I}_{Y_0} = \mathcal{H}om_{\mathcal{O}_{Z_0}}(\mathcal{O}_{X_0}, \mathcal{O}_{Z_0})$, which is also a Cohen–Macaulay scheme.
(c) Using (Ex. 7.5e) show that $\mathcal{I}_X = \mathcal{H}om(\mathcal{O}_Y, \mathcal{O}_Z)$, so X is also linked to Y by Z. In other words, the relation of linkage is symmetric.

9.4. Extension of linked deformations. Let P_0 be the spectrum of a regular local ring over k, let Z_0 be a complete intersection subscheme of P_0, and let $X_0 \subseteq Z_0$ be a Cohen–Macaulay scheme of the same dimension. Suppose we are given deformations $X \subseteq Z \subseteq P$ over an Artin ring C, and using notation (6.1), suppose we are given extensions $X' \subseteq P'$ of X and P over C'.

(a) Show that there exists an extension Z' of Z (as in (9.2)) that contains X', and hence the scheme Y' linked to X' by Z' is an extension of Y linked to X by Z.

(b) Let Y_0 be linked to X_0 by Z_0, and conclude that Y_0 has obstructed deformations if and only if X_0 has obstructed deformations.

(c) We say that a subscheme $X_0 \subseteq P_0$ is *licci* (in the liaison class of a complete intersection) if there is a finite sequence of linkages X_0 to $X_1, \ldots, X_n$ by complete intersection schemes $Z_0, \ldots, Z_{n-1}$ such that X_n is itself a complete intersection. Conclude that if X_0 is licci it has unobstructed deformations. (This result is used in (Ex. 29.7) to give an example of a codimension 3 Cohen–Macaulay scheme that is not licci.)

(d) Prove a global analogue of this result, which says that if X_0 and Y_0 are ACM subschemes of $\mathbb{P}_k^n$ linked by a complete intersection scheme Z_0, and if x_0, y_0 are the corresponding points of their respective Hilbert schemes, then x_0 is a smooth point if and only if y_0 is a smooth point.

10. Obstructions to Deformations of Schemes

In §6 and §7 we have discussed higher-order deformations of closed subschemes, line bundles, and vector bundles—Situations A, B, and C. Now we continue with Situation D, higher-order deformations of schemes, whose first-order deformation we discussed in §5.

Recall from §5 that if X_0 is a scheme over k, then a deformation of X_0 over an Artin ring C is a scheme X, flat over C, together with a closed immersion $X_0 \hookrightarrow X$ inducing an isomorphism $X_0 \xrightarrow{\sim} X \times_C k$.

Definition. If C' is another Artin ring, together with a surjective map $C' \to C$, and if X is a deformation of X_0 over k, an *extension* of X over C' is a deformation X' of X_0 over C', together with a closed immersion $X \hookrightarrow X'$ inducing an isomorphism $X \xrightarrow{\sim} X' \times_{C'} C$. Two such extensions X' and X'' are *equivalent* if there is an isomorphism of deformations $X' \xrightarrow{\sim} X''$ compatible with the closed immersions of X into each.

We consider the affine case first. Using the notation of (6.1), suppose we are given a k-algebra B_0, and a deformation B of B_0 over C. We ask for extensions of B over C'.

Theorem 10.1. *In the above situation:*

(a) *There is an element $\delta \in T^2(B_0/k, B_0 \otimes J)$, called the obstruction, with the property that $\delta = 0$ if and only if an extension B' of B exists.*

(b) *If extensions exist, then the set of equivalence classes of such extensions is a torsor under the action of $T^1(B_0/k, B_0 \otimes J)$.*

Proof. (a) To define the obstruction, choose a polynomial ring $R = C[x_1, \ldots, x_n]$ and a surjective mapping $R \to B$ with kernel I. Let $f_1, \ldots, f_r \in I$ be a set of generators, let F be the free module R^r, and let Q be the kernel of the natural map $F \to I$:

$$0 \to Q \to F \to I \to 0.$$

The idea is to lift the f_i to elements $f_i' \in R' = C'[x_1, \ldots, x_n]$, define $B' = A'/I'$, where $I' = (f_i', \ldots, f_n')$, and investigate whether we can make B' flat over C'. Let $F' = R'^r$, and let $Q' = \ker(F' \to I')$. Tensoring with $0 \to J \to C' \to C \to 0$ we get a diagram

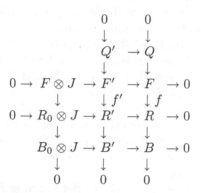

From the snake lemma there is a map $\delta_0 : Q \to B_0 \otimes J$, depending on the lifting f' of f. We know from the local criterion of flatness (2.2) that B' is flat over C' if and only if the map $B_0 \otimes J \to B'$ is injective, and this is equivalent to $\delta_0 = 0$. Any element in F_0 (in the notation of (3.1)) is of the form $f_j e_i - f_i e_j$ in F, and this lifts to $f_j' e_i' - f_i' e_j'$ in Q', so the map δ_0 factors through Q/F_0. Thus we get a homomorphism $\delta_1 \in \operatorname{Hom}(Q/F_0, B_0 \otimes J)$. And this, using the definition of $T^2(B/C, B_0 \otimes J)$ as $\operatorname{coker}(\operatorname{Hom}(F/IF, B_0 \otimes J) \to \operatorname{Hom}(Q/F_0, B_0 \otimes J))$, gives us the desired element $\delta \in T^2(B_0/k, B_0 \otimes J)$, which is the same as $T^2(B/C, B_0 \otimes J)$, by Base Change II (Ex. 3.8).

We must show that δ is independent of all the choices made. If we make a different choice of lifting f_i'' of the f_i, then the $f_i' - f_i''$ define a map from F' to $R_0 \otimes J$ and hence from F/IF to $B_0 \otimes J$, and these go to zero in T^2. If we choose a different polynomial ring $R^* \to B$, then as in the proof of (3.3) we reduce to the case $R^* = R[y_1, \ldots, y_s]$, and the y_i go to zero in B. The contribution to δ of the y_i is then zero. Thus δ depends only on the initial situation B/C and $C' \to C$.

If the extension B'/C' does exist, we can start by picking a polynomial ring R' over C' that maps surjectively to B', and use $R = R' \otimes_{C'} C$ mapping to B in the above construction. Then the generators of I' descend to generators of I, and the diagram above shows that $\delta_0 = 0$, so a fortiori $\delta = 0$.

Conversely, suppose $\delta = 0$. Then we must show that an extension B' exists. In fact, we will show something apparently stronger, namely that having made a choice $R \to B \to 0$ as above, we can lift B to a quotient of the corresponding R'. In fact, we will show that it is possible to lift the f_i to f'_i in such a way that the map $\delta_0 : Q \to B_0 \otimes J$ is zero.

Our hypothesis is only that $\delta \in T^2$ is zero. By definition of T^2, this means that the map $\delta_1 \in \mathrm{Hom}(Q/F_0, B_0 \otimes J)$ lifts to a map $\gamma : F/IF \to B_0 \otimes J$. This defines a map $F \to B_0 \otimes J$, and since F is free, it lifts to a map $F \to R_0 \otimes J$, defined by $g_1, \ldots, g_r \in R_0 \otimes J$. Now take $f''_i = f'_i - g_i$. The g_i cancel out the images of δ_1, and so we find the new $\delta_0 = 0$, so the new B' is flat over C'.

(b) Suppose one such extension B'_1 exists. Let $R_0 = k[x_1, \ldots, x_n]$ be a polynomial ring of which B_0 is a quotient, with kernel I_0. Then the map $R_0 \to B_0$ lifts to a map of $R = C[x_1, \ldots, x_n]$ to B and to a map of $R' = C'[x_1, \ldots, x_n]$ to B'_1, compatible with the map to B. These maps $R \to B$ and $R' \to B'_1$ are surjective, by Nakayama's lemma.

For any other extension B'_2 of B, the map $R \to B$ lifts to a map $R' \to B'_2$. Thus every abstract deformation is also an embedded deformation (in many ways perhaps), and we know that the embedded deformations are a torsor under the action of $\mathrm{Hom}(I/I^2, B \otimes J)$ by (6.2). So comparing with the fixed one B'_1, we get an element of this group corresponding to B'_2. Now if two of these, B'_2 and B'_3, happen to be equivalent as abstract extensions of B choose an isomorphism $B'_2 \cong B'_3$. Using this we obtain two maps $R' \rightrightarrows B'_2$, and hence a derivation of R' to $B \otimes J = B_0 \otimes J$, by (4.5), which can be regarded as an element of $\mathrm{Hom}(\Omega_{R'/C'}, B_0 \otimes J)$.

Now we use the exact sequence (3.10) determining T^1,

$$\mathrm{Hom}(\Omega_{R/C}, B \otimes J) \to \mathrm{Hom}(I/I^2, B \otimes J) \to T^1(B/C, B \otimes J) \to 0,$$

to see that the ambiguity of embedding is exactly resolved by the image of the derivations, and so the extensions B' of B, up to equivalence, form a torsor under $T^1(B/C, B_0 \otimes J)$, which is equal to $T^1(B_0/k, B_0 \otimes J)$ by Base Change II (Ex. 3.8).

Remark 10.1.1. For future reference, we note that given an extension B' of B over C', the group of automorphisms of B' lying over the identity of B is naturally isomorphic to the group $T^0(B_0/k, J \otimes B_0)$ of derivations of B_0 into $J \otimes B_0$. Indeed, we have only to apply (4.5) taking $R = B'$.

Example 10.1.2. It may happen that the obstructions vanish even when the group T^2 is nonzero: (Ex. 3.3) gives an example with $T^2 \neq 0$, but there are no obstructions, since it is Cohen–Macaulay in codimension 2 (§8).

Now we pass to the global case. With the notation of (6.1), we suppose we are given a scheme X_0 over k and a deformation X of X_0 over C. We look for extensions of X to a deformation X' over C'. We recall the notation $\mathcal{T}^i_{X_0} = T^i(X_0/k, \mathcal{O}_{X_0})$.

Theorem 10.2. *In the above situation:*

(a) *There are three successive obstructions to be overcome for the existence of an extension X' of X over C', lying in $H^0(X_0, T_{X_0}^2 \otimes J)$, $H^1(X_0, T_{X_0}^1 \otimes J)$, and $H^2(X_0, T_{X_0}^0 \otimes J)$.*

(b) *Let $\mathrm{Def}(X/C, C')$ be the set of all such extensions X', up to equivalence. Having fixed one such X_1', there is an exact sequence*

$$0 \to H^1(X_0, T_{X_0}^0 \otimes J) \to \mathrm{Def}(X/C, C') \to H^0(X_0, T_{X_0}^1 \otimes J) \to$$
$$\to H^2(X_0, T_{X_0}^0 \otimes J).$$

Proof. (a) Suppose we are given X. For each open affine subset $U_i \subseteq X$ there is an obstruction lying in $H^0(U_i, T_{U_i}^2 \otimes J)$ for the existence of a deformation U_i' over U_i, by (10.1). These patch together to give a global obstruction $\delta_1 \in H^0(X_0, T_{X_0}^2 \otimes J)$.

If this obstruction vanishes, then for each U_i there exists a deformation U_i' over U_i. For each $U_{ij} = U_i \cap U_j$ we then have two deformations $U_i'|_{U_{ij}}$ and $U_j'|_{U_{ij}}$. By (10.1) again, their difference gives an element in $H^0(U_{ij}, T^1 \otimes J)$. The difference of three of these is zero on U_{ijk}, so we get the second obstruction $\delta_2 \in H^1(X_0, T_{X_0}^1 \otimes J)$.

If this obstruction vanishes, then we can modify the deformations U_i' so that they become equivalent on the overlap U_{ij}. Choose isomorphisms $\varphi_{ij} : U_i'|_{U_{ij}} \overset{\sim}{\to} U_j'|_{U_{ij}}$ for each ij. On the triple intersection U_{ijk}, composing three of these gives an automorphism of $U_i'|_{U_{ijk}}$ by (10.1.1), which gives an element in $H^0(U_{ijk}, T^0 \otimes J)$. On the fourfold intersection, these agree, so we get an obstruction $\delta_3 \in H^2(X_0, T_{X_0}^0 \otimes J)$.

If this last obstruction also vanishes, we can modify the isomorphisms φ_{ij} so that they agree on the U_{ijk}, and then we can glue the extensions U_i' to get a global extension X'.

(b) Suppose now we are given one fixed extension X_1' of X over C'. If X_2' is another extension, then on each open affine U_i we have two, and their difference gives an element of $H^0(U_i, T^1 \otimes J)$ by (10.1). These glue together to give a global element of $H^0(X_0, T_{X_0}^1 \otimes J)$. We have already seen in part (a) above that conversely, given a global element of $H^0(X_0, T_{X_0}^1 \otimes J)$, it gives extensions over the open sets U_i that are isomorphic on the intersections U_{ij}, and in that case there is an obstruction in $H^2(X_0, T_{X_0}^0 \otimes J)$ to gluing these together to get a global extension.

Now suppose two extensions X_2' and X_3' give the same element in $H^0(X_0, T_{X_0}^1 \otimes J)$. This means that they are isomorphic on each open affine U_i. Choose isomorphisms $\varphi_i : X_2'|_{U_i} \overset{\sim}{\to} X_3'|_{U_i}$. On the intersection U_{ij}, we get $\psi_{ij} = \varphi_j^{-1} \circ \varphi_i$, which is an automorphism of $X_2'|_{U_{ij}}$ and so defines a section of $T^0 \otimes J$ over U_{ij} by (10.1.1). These agree on the triple overlap, so we get an element of $H^1(X_0, T_{X_0}^0 \otimes J)$. The vanishing of this element is equivalent to the possibility of modifying the isomorphisms φ_i so that they will agree on the overlap, which is equivalent to saying that X_2' and X_3' are globally isomorphic. Thus we get the exact sequence for $\mathrm{Def}(X/C, C')$ as claimed.

Remark 10.2.1. This theorem suggests the existence of a spectral sequence beginning with $H^p(X, T^q \otimes J)$ and ending with some groups of which the H^2 would contain the obstruction to lifting X, and the $H^1 = \mathrm{Def}(X/C, C')$ would classify the extensions if they exist. But I will not attempt to say where such a spectral sequence might come from. If you really want to know, you will have to look in another book! Try [73] or [92].

Remark 10.2.2. As in (10.1.1) we observe that given an extension X' of X over C', the group of automorphisms of X' lying over the identity automorphism of X is naturally isomorphic to the group $H^0(X_0, T^0_{X_0} \otimes J)$. Just apply (10.1.1) and glue.

Corollary 10.3. *If X_0 is nonsingular, then*

(a) *There is just one obstruction in $H^2(X_0, \mathcal{T}_{X_0} \otimes J)$ for the existence of an extension X' of X over C'.*

(b) *If such extensions exist, their equivalence classes form a torsor under $H^1(X_0, \mathcal{T}_{X_0} \otimes J)$.*

Proof. In this case the sheaves $\mathcal{T}^1_{X_0}$ and $\mathcal{T}^2_{X_0}$ are zero (4.9).

Using this local study of abstract deformations, we can complete the results of §6 in case there are local obstructions.

Proposition 10.4. *Let Y_0 be a closed subscheme of X_0 over k, and let X, Y, X', C, C', J be as in (6.2). Then there is an obstruction $\beta \in H^0(Y_0, \mathcal{T}^2_{Y_0/k} \otimes_k J)$ for the local existence of extensions of Y over C'. If this obstruction vanishes, then as in (6.2b) there is the obstruction $\alpha \in H^1(Y_0, \mathcal{N}_0 \otimes_k J)$ for the global existence of Y', and the conclusions of (6.2) hold.*

Proof. If we examine the proof of (6.2), we see that what was missing was the existence of an affine covering of Y_0 where local extensions exist. Since for affine schemes abstract and embedded obstructions are the same (Ex. 10.1), the obstruction for each open affine subset lies in the corresponding T^2 group. These glue together to form the $\mathcal{T}^2$ sheaf, so the obstruction β is a global section of this sheaf. Once β vanishes, the rest of the proof of (6.2) carries through.

Reference. The applications of the T^i functors to deformation theory are in the paper of Lichtenbaum and Schlessinger [96].

Exercises.

10.1. Let Y_0 be a closed subscheme of an affine nonsingular variety X_0 over k. Let $Y \subseteq X$ be a deformation of $Y_0 \subseteq X_0$ over C, and let X' be an extension of X to C', in the notation of (6.1). Show that an extension Y' of Y over C' as an abstract scheme exists if and only if an extension Y' exists as a closed subscheme of X'. In fact, any abstract extension Y' can be embedded in X'. *Hint:* One direction is obvious. For the other, use the infinitesimal lifting property of smooth morphisms (§4). Note that the hypothesis X_0 affine is essential.

10.2. If X_0 over k is a locally complete intersection scheme, or is locally isomorphic to a Cohen–Macaulay codimension 2 scheme or a Gorenstein codimension 3 scheme, then the first of the three obstructions in (10.2a) vanishes.

10.3. In (5.3.1) we called a scheme *rigid* if all of its deformations over the dual numbers were trivial. Show that every deformation of a rigid scheme over any Artin ring is trivial.

10.4. If X is a reduced locally complete intersection curve over k, then there are no obstructions to deformations of X. Furthermore, every local deformation of a singular point of X can be extended to a global deformation of X (as an abstract variety).

10.5. Let X be a nonsingular projective surface and let $Y \subseteq X$ be an exceptional curve of the first kind (cf. (6.2.2)). Let X_0 be the nonsingular surface obtained by blowing down Y to a point $P \in X_0$ [57, V, 5.7]. We will show that any deformation of X arises from deforming X_0 or moving the point $P \in X_0$, or both.

(a) First show that there is an exact sequence relating the tangent sheaves of X and X_0,
$$0 \to \mathcal{T}_X \to f^*\mathcal{T}_{X_0} \to \mathcal{O}_Y(1) \to 0.$$

(b) Show that $R^1 f_* \mathcal{T}_X = 0$, so there is an exact sequence
$$0 \to f_*\mathcal{T}_X \to \mathcal{T}_{X_0} \to k^2 \to 0,$$
from which we deduce a sequence of cohomology
$$0 \to H^0(X, \mathcal{T}_X) \to H^0(X_0, \mathcal{T}_{X_0}) \to k^2 \to H^1(X, \mathcal{T}_X) \to$$
$$\to H^1(X_0, \mathcal{T}_{X_0}) \to 0$$
and an isomorphism
$$H^2(X, \mathcal{T}_X) \cong H^2(X_0, \mathcal{T}_{X_0}).$$

(c) Interpret this as follows. First of all, the obstructions to deforming X are the same as the obstructions to deforming X_0. Next, think of k^2 as the deformations of P inside X_0. If the infinitesimal automorphisms of X_0 (that is, $H^0(X_0, \mathcal{T}_{X_0})$) map surjectively to deformations of P in X_0, then deformations of X are just given by deformations of X_0. Otherwise, moving P gives nontrivial deformations of X.

(d) Apply this to the nonsingular cubic surface in $\mathbb{P}^3$, obtained by blowing up six points of $\mathbb{P}^2$ [57, V, §4]. The projective space $\mathbb{P}^2$ itself is rigid ($H^1(\mathcal{T}_{\mathbb{P}^2}) = 0$) and it has a group of automorphisms of dimension 8. Thus any four points in general position can be moved to any other four points in general position. We find that $\mathbb{P}^2$ with four points blown up is still rigid, but has no infinitesimal automorphisms. Thus the deformations of the cubic surface depend on the position of the fifth and sixth points. We find that the cubic surface, as an abstract surface, is unobstructed, and has a four-dimensional space of deformations. (See (20.2.2) for another approach.)

10.6. Deforming a scheme with a line bundle. Suppose we are given a nonsingular scheme X_0 over k and a line bundle L_0 on X_0. We consider deformations of the pair (X_0, L_0), that is, for an Artin ring C, pairs (X, L), where X is a deformation of X_0 over C and L is a deformation of L_0 on X.

(a) With the usual notation (6.1), show that if (X', L') over C' is an extension of (X, L) over C, then the automorphisms of (X', L') over (X, L) are given by $H^0(X_0, \mathcal{P}_{L_0})$, where, for any nonsingular scheme X with a line bundle L, we define the *sheaf of principal parts* $\mathcal{P}_L$ as an extension

$$0 \to \mathcal{O}_X \to \mathcal{P}_L \to \mathcal{T}_X \to 0$$

defined by the cohomology class $c(L) \in H^1(X, \Omega_X) = \text{Ext}^1(\mathcal{T}_X, \mathcal{O}_X)$ and $c(L)$ is defined as the image of the class of L in $H^1(X, \mathcal{O}_X^*)$ under the map $d \log : \mathcal{O}_X^* \to \Omega_X^1$ (cf. [57, V, Ex. 1.8]).

(b) Show that the obstruction to extending a deformation (X, L) over C to one over C' is in $H^2(X_0, \mathcal{P}_{L_0} \otimes J)$ and that when extensions exist, they are classified by $H^1(X_0, \mathcal{P}_{L_0} \otimes J)$.

(c) In the cohomology sequence

$$\cdots \to H^1(\mathcal{O}_{X_0}) \to H^1(\mathcal{P}_{L_0}) \to H^1(\mathcal{T}_{X_0}) \xrightarrow{\delta} H^2(\mathcal{O}_{X_0}) \to \cdots$$

coming from the exact sequence of (a) above, we can interpret $H^1(\mathcal{P}_{L_0})$ as deformations of the pair (X_0, L_0), and its image in $H^1(\mathcal{T}_{X_0})$ as the deformation of the underlying scheme X_0. For a given deformation of X_0, we can interpret its image under δ as the obstruction to extending the line bundle L_0. Show that the map δ is given by cup product with the cohomology class $c(L_0) \in H^1(\Omega^1_{X_0})$: $H^1(\mathcal{T}_{X_0}) \times H^1(\Omega^1_{X_0}) \to H^2(\mathcal{O}_{X_0})$, using the pairing $\mathcal{T}_{X_0} \otimes \Omega^1_{X_0} \to \mathcal{O}_{X_0}$ of dual sheaves.

(d) If X_0 is embedded in $\mathbb{P}^n_k$, show that there is an exact sequence

$$0 \to \mathcal{P}_{L_0} \to \mathcal{O}_{X_0}(1)^{n+1} \to \mathcal{N}_{X_0/\mathbb{P}^n} \to 0,$$

where $L_0 = \mathcal{O}_{X_0}(1)$. Thus we obtain a map $H^0(\mathcal{N}_{X_0/\mathbb{P}^n}) \to H^1(\mathcal{P}_{L_0})$ expressing the deformation of the pair (X_0, L_0) induced by a deformation of X_0 in $\mathbb{P}^n$.

For more details on deformations of pairs (X, L), see [152, 3.3.3].

10.7. Abelian surfaces. Let X be an abelian surface. Then $h^i(\mathcal{O}_X) = 1, 2, 1$ for $i = 0, 1, 2$ respectively. The tangent bundle $\mathcal{T}_X$ is isomorphic to $t_e \otimes \mathcal{O}_X$, where t_e is the tangent space at the origin $e \in X$. Thus $H^2(\mathcal{T}_X) \neq 0$ and it appears as if there might be obstructions to deforming X. However, assuming char $k \neq 2$, the obstructions vanish [129, p. 237].

(a) If $\alpha : X_1 \to X_2$ is an isomorphism of nonsingular varieties, then α sends the obstruction to deforming X_1 to the obstruction to deforming X_2.

(b) Now let $i : X \to X$ be the automorphism of our abelian surface that sends x to $-x$ in the group law. If $\tau \in H^2(\mathcal{T}_X)$ is the obstruction to deforming X, then $i^* \tau = \tau$.

(c) Observe that i^* acts by -1 on the tangent space t_e at the origin. Hence also it acts by -1 on $H^1(\mathcal{O}_X)$, which is isomorphic to the tangent space of the dual abelian surface X^* at the origin.

(d) On the abelian surface X, $H^2(\mathcal{T}_X) \cong t_e \otimes H^2(\mathcal{O}_X)$. Furthermore, $H^2(\mathcal{O}_X) \cong \bigwedge^2 H^1(\mathcal{O}_X)$. Therefore i^* acts by $+1$ on $H^2(\mathcal{O}_X)$, and hence by -1 on $H^2(\mathcal{T}_X)$. So $i^* \tau = -\tau$. This shows that $\tau = 0$ (at least assuming char $k \neq 2$).

11. Obstruction Theory for a Local Ring

Let $(A, \mathfrak{m})$ be a local ring with residue field k. We want to investigate properties of A in terms of homomorphisms of A to local Artin rings and how these homomorphisms lift to larger Artin rings. This is useful for studying local properties of the scheme representing a functor, for example the Hilbert scheme, because it allows us to translate properties of the local ring on the representing scheme into properties of the functor applied to local Artin rings.

We have already seen one case of this kind of analysis (4.7), where the property of A being a regular local ring is characterized by always being able to lift maps into Artin rings—the infinitesimal lifting property of smoothness (or regularity). (For simplicity we assume k algebraically closed.)

In this section we take this analysis one step further, by considering cases in which the homomorphisms do not always lift. For this purpose we define the notion of an obstruction theory.

Definition. Let $(A, \mathfrak{m})$ be a local ring with residue field k. We will consider sequences $0 \to J \to C' \to C \to 0$ as in (6.1). An *obstruction theory* for A is a vector space V over k, together with, for every sequence $0 \to J \to C' \to C \to 0$ as above, and for every homomorphism $u : A \to C$,

$$A$$
$$\downarrow u$$
$$0 \to J \to C' \xrightarrow{u} C \to 0$$

an element $\varphi(u, C') \in V \otimes J$, satisfying two properties:

(a) $\varphi(u, C') = 0$ if and only if u lifts to a map $u' : A \to C'$,
(b) φ is functorial in the sense that if $K \subseteq J$ is a subspace, then the element $\varphi(u, C'/K)$ associated with u and the sequence $0 \to J/K \to C'/K \to C \to 0$ is just the image of $\varphi(u, C')$ under the natural map $V \otimes J \to V \otimes J/K$.

Example 11.0.1. We have already seen a typical example of an obstruction theory in studying the Hilbert scheme. Let Y_0 be a closed subscheme of $X_0 = \mathbb{P}_k^n$, and assume that Y_0 has no local obstructions to its deformations (e.g., Y_0 is nonsingular, or locally complete intersection, ...). Then by (6.2), obstructions to deforming Y_0 lie in $H^1(Y_0, \mathcal{N}_{Y_0/X_0} \otimes J)$. Let y be the corresponding point of the Hilbert scheme H, and let A be the local ring of y on H. Then a homomorphism of A to a local Artin ring C corresponds to a deformation of Y_0 over C as a closed subscheme of $\mathbb{P}^n$, according to the universal property of the Hilbert scheme (1.1a). To lift u to a homomorphism $u' : A \to C'$ corresponds to extending the deformation Y to a deformation Y' over C'. Thus if we take $V = H^1(Y_0, \mathcal{N}_{Y_0/X_0})$, we have an obstruction theory for the local ring A.

Example 11.0.2. Suppose that A is a quotient of a regular local ring P by any ideal I, and assume that $I \subseteq \mathfrak{m}_P^2$. Then we can construct what we might call the "canonical" obstruction theory for A as follows. Take V_A to be the dual vector space $(I/\mathfrak{m}_P I)^*$. Given a diagram as in the definition, we can always lift u to a homomorphism $f : P \to C'$, since P is regular (4.4):

$$
\begin{array}{ccccccccc}
0 & \to & I & \to & P & \to & A & \to & 0 \\
& & \downarrow \bar{f} & & \downarrow f & & \downarrow u & & \\
0 & \to & J & \to & C' & \to & C & \to & 0
\end{array}
$$

This induces a map $\bar{f} : I \to J$, which factors through $I/\mathfrak{m}I$, since J is a k-vector space. This gives us an element $\varphi \in \operatorname{Hom}(I/\mathfrak{m}I, J) \cong V_A \otimes J$.

We need to show that φ is independent of the choice of lifting f. So let $f' : P \to C'$ be another lifting. Then by (4.5), $f' - f$ is a derivation of P to J. Since $I \subseteq \mathfrak{m}^2$, it follows that $(f' - f)(I) \subseteq \mathfrak{m}J = 0$.

Now condition (a) is clear: if $\varphi(u, C') = 0$, then f factors through A, so that u lifts. Conversely, if u lifts, this gives a lifting $f' : P \to C'$ that vanishes on I, so $\varphi = 0$.

Condition (b) is obvious by construction.

Note that in this example, $\dim V_A = \dim(I/\mathfrak{m}I)$, which is the minimal number of generators of I.

Our main result is a converse to this example.

Theorem 11.1. *Let A be a local ring that can be written as a quotient of a regular local ring P by an ideal $I \subseteq \mathfrak{m}_P^2$, and let (V, φ) be an obstruction theory for A. Then there is a natural inclusion of V_A (11.0.2) into V. In particular, I can be generated by at most $\dim V$ elements.*

Proof. Note first that we cannot expect to get the exact number of generators for I, because if (V, φ) is an obstruction theory, any bigger vector space V' containing V will also be one.

We apply the obstruction theory V to a particular case. Take $0 \to J \to C' \to C \to 0$ to be the sequence, for any integer n,

$$
0 \to (I + \mathfrak{m}^n)/(\mathfrak{m}I + \mathfrak{m}^n) \to P/(\mathfrak{m}I + \mathfrak{m}^n) \to A/\mathfrak{m}^n \to 0,
$$

and take the natural quotient map $u : A \to A/\mathfrak{m}^n$. By a standard isomorphism theorem,

$$
(I + \mathfrak{m}^n)/(\mathfrak{m}I + \mathfrak{m}^n) \cong I/I \cap (\mathfrak{m}I + \mathfrak{m}^n) = I/(\mathfrak{m}I + (I \cap \mathfrak{m}^n)).
$$

By the Artin–Rees lemma, we have $I \cap \mathfrak{m}^n \subseteq \mathfrak{m}I$ for $n \gg 0$. So for n sufficiently large we can write

$$
\begin{array}{ccccccccc}
0 & \to & I & \to & P & \to & A & \to & 0 \\
& & \downarrow \bar{f} & & \downarrow f & & \downarrow u & & \\
0 & \to & I/\mathfrak{m}I & \to & P/(\mathfrak{m}I + \mathfrak{m}^n) & \to & A/\mathfrak{m}^n & \to & 0
\end{array}
$$

and our obstruction theory gives us an element $\varphi \in V \otimes I/\mathfrak{m}I \cong \operatorname{Hom}_k(V_A, V)$. I claim that φ is injective as a homomorphism of V_A to V. Indeed, let $w \in V_A$,

$w \neq 0$. Then w gives a surjective homomorphism $w : I/\mathfrak{m}I \to k$. If we let $K = \ker w$, and divide the exact sequence by K, as in the definition of the obstruction theory above, the composed map $\bar{f} : I \to k$ is still nonzero. So u does not lift; hence the image of φ in $V \otimes k$ is nonzero. Thus $\varphi : V_A \to V$ is injective as claimed. Since the minimal number of generators of I is equal to $\dim V_A$, the conclusion follows.

Corollary 11.2. *Let $(A, \mathfrak{m})$ be a local ring that can be written as a quotient of a regular local ring P of dimension $n = \dim \mathfrak{m}/\mathfrak{m}^2$. If A has an obstruction theory in a vector space V, then $\dim A \geq n - \dim V$. Furthermore, if equality holds, then A is a local complete intersection.*

Proof. Indeed, $\dim A \geq \dim P - \#$ generators of I, and equality makes A a local complete intersection ring by definition.

Applying this to the Hilbert scheme, we obtain the proof of (1.1d):

Theorem 11.3. *Let Y be a locally complete intersection subscheme of the projective space $X = \mathbb{P}^n_k$. Then the dimension of the Hilbert scheme H at the point $y \in H$ corresponding to Y is at least $h^0(Y, \mathcal{N}) - h^1(Y, \mathcal{N})$. Furthermore, in case of equality, H is a locally complete intersection scheme at that point.*

Proof. Let A be the local ring of y on H. Then $V = H^1(Y, \mathcal{N})$ gives an obstruction theory for A (11.0.1). On the other hand, A has embedding dimension equal to $h^0(Y, \mathcal{N})$ by (2.4). Hence the result follows from (11.2).

Remark 11.3.1. As an application of this result, we will prove in the next section the classical result, stated by M. Noether [125, I, §2] for nonsingular curves in $\mathbb{P}^3$, that every component of the Hilbert scheme of locally Cohen–Macaulay curves of degree d in $\mathbb{P}^3$ has dimension at least $4d$.

Remark 11.3.2. The corollary (11.2) generalizes the result (4.7) that a local ring is regular if it has the infinitesimal lifting property, because in that case we can take $V = 0$ as an obstruction theory.

References for this section. The results of this section are certainly consequences of the general deformation theories of Laudal [92], Illusie [73], and Rim [141], but for a more direct approach, I have given a simplified version of the proof due to Mori [109, Prop. 3]. A slightly different proof appears in the book of Kollár [88, p. 32].

Exercise.

11.1. Show that the canonical obstruction space V_A (11.0.2) is isomorphic to $T^1(A/k, k)$.

12. Dimensions of Families of Space Curves

A classical problem, studied by G. Halphen and M. Noether in the 1880s, and the subject of considerable activity one hundred years later, is the problem of classification of algebraic space curves. Here we mean closed one-dimensional subschemes of $\mathbb{P}^n$, while the classical case was irreducible nonsingular curves in $\mathbb{P}^3$. See [57, IV, §6], [59], [61], [51] for some surveys of the problem.

To begin with, let us focus our attention on nonsingular curves in $\mathbb{P}^3$. These form an open subset of the Hilbert scheme, so the problem is to find which pairs (d, g) can be the possible degree and genus of a curve, and then for each such (d, g) to find the irreducible components and the dimensions of the corresponding parameter space $H_{d,g}$. The problem of determining the possible (d, g) pairs has been solved by Gruson and Peskine [50] (see also [62]). Easy examples show that $H_{d,g}$ need not be irreducible [57, IV, 6.4.3]. So we consider the problem of dimension.

Max Noether's approach to the problem [125, pp. 18, 19] goes like this. The choice of an abstract curve of genus g depends on $3g - 3$ parameters. The choice of a divisor of degree d on the curve (up to linear equivalence) is another g parameters. If $d \geq g + 3$, a general such divisor D is nonspecial, so by the Riemann–Roch theorem, the dimension of the complete linear system $|D|$ is $d - g$. Here we must choose a 3-dimensional subsystem, and such a choice depends on $4(d - g - 3)$ parameters (the dimension of the Grassmann variety of $\mathbb{P}^3$'s in a $\mathbb{P}^{d-g}$). Now add 15 parameters for an arbitrary automorphism of $\mathbb{P}^3$. Putting these together, we find that the dimension of the family of general curves of genus g, embedded with a general linear system of degree $d \geq g + 3$, is $4d$.

Refining his argument (but still speaking always of curves that are general in the variety of moduli, and general linear systems on these), Noether claimed that:

(a) If $d \geq \frac{3}{4}(g + 4)$, the family has dimension $4d$.
(b) If $d < \frac{3}{4}(g + 4)$, the family has dimension $\geq 4d$.

These methods do not take into account curves whose moduli are special, and therefore may have linear systems of kinds that do not appear on general curves, and of course they do not apply to singular or reducible curves. Furthermore, Noether's method depends on knowing the dimension of the variety of moduli $(3g - 3)$, the dimension of the Jacobian variety (g), the dimension of Grassmann varieties, and also depends on having confidence in the method of "counting constants," which sometimes seems more like an art than a science.

In this section we will use an entirely different method, the infinitesimal study of the Hilbert scheme, to prove the following theorem.

Theorem 12.1. *Every irreducible component of the Hilbert scheme of locally Cohen–Macaulay curves of degree d and arithmetic genus g in $\mathbb{P}^3$ has dimension $\geq 4d$.*

Proof. These curves are locally Cohen–Macaulay and of codimension 2 in $\mathbb{P}^3$, so there are no local obstructions to embedded deformations, and hence the obstructions to global deformations lie in $H^1(C, \mathcal{N}_C)$ (8.5). Now using the dimension theorem for a local ring with an obstruction theory (11.0.1) and (11.2), we obtain

$$\dim_C \text{Hilb} \geq h^0(C, \mathcal{N}_C) - h^1(C, \mathcal{N}_C).$$

Thus we are reduced to the problem of evaluating the cohomology of the normal sheaf $\mathcal{N}_C$.

We do an easy case first. If C is irreducible and nonsingular, then $\mathcal{N}_C$ is locally free and is just the usual normal bundle to the curve. It belongs to an exact sequence

$$0 \to \mathcal{T}_C \to \mathcal{T}_{\mathbb{P}^3}|_C \to \mathcal{N}_C \to 0,$$

where $\mathcal{T}$ denotes the tangent bundle of C (resp. $\mathbb{P}^3$). On the other hand, the tangent bundle of $\mathbb{P}^3$ belongs to an exact sequence

$$0 \to \mathcal{O}_{\mathbb{P}^3} \to \mathcal{O}_{\mathbb{P}^3}(1)^4 \to \mathcal{T}_{\mathbb{P}^3} \to 0.$$

From these two sequences we obtain $\deg \mathcal{N}_C = 4d + 2g - 2$, and so by the Riemann–Roch theorem $\chi(\mathcal{N}_C) = h^0(\mathcal{N}_C) - h^1(\mathcal{N}_C) = 4d$. This proves the theorem for nonsingular C.

The general case is a bit more technical, because we do not have the same simple relationship between the normal sheaf and the tangent sheaves. But we have assumed that C is locally Cohen–Macaulay, so there is a resolution

$$0 \to \mathcal{E} \to \mathcal{F} \to \mathcal{I}_C \to 0, \tag{6}$$

where $\mathcal{E}, \mathcal{F}$ are locally free sheaves on $\mathbb{P}^3$. Taking $\mathcal{H}om(\cdot, \mathcal{O}_{\mathbb{P}})$, we find that

$$0 \to \mathcal{O}_{\mathbb{P}} \to \mathcal{F}^\vee \to \mathcal{E}^\vee \to \mathcal{E}xt^1(\mathcal{I}_C, \mathcal{O}_{\mathbb{P}}) \to 0. \tag{7}$$

Since C has codimension 2, this $\mathcal{E}xt^1(\mathcal{I}_C, \mathcal{O}_{\mathbb{P}})$ is isomorphic to $\mathcal{E}xt^2(\mathcal{O}_C, \mathcal{O}_{\mathbb{P}})$, which is just $\omega_C(4)$, where ω_C is the dualizing sheaf of C.

Tensoring the sequence (7) with $\mathcal{O}_C$, we therefore get

$$\mathcal{F}^\vee \otimes \mathcal{O}_C \to \mathcal{E}^\vee \otimes \mathcal{O}_C \to \omega_C(4) \to 0. \tag{8}$$

On the other hand, applying $\mathcal{H}om(\cdot, \mathcal{O}_C)$ to the sequence (6) we obtain

$$0 \to \mathcal{N}_C \to \mathcal{F}^\vee \otimes \mathcal{O}_C \to \mathcal{E}^\vee \otimes \mathcal{O}_C \to \mathcal{E}xt^1(\mathcal{I}_C, \mathcal{O}_C) \to 0. \tag{9}$$

Comparing (8) and (9), we see that $\mathcal{E}xt^1(\mathcal{I}_C, \mathcal{O}_C) \cong \omega_C(4)$.

Now we take Euler characteristics and obtain

$$\chi(\mathcal{N}_C) = \chi(\mathcal{F}^\vee \otimes \mathcal{O}_C) - \chi(\mathcal{E}^\vee \otimes \mathcal{O}_C) + \chi(\omega_C(4)).$$

Suppose $\mathcal{E}$ has rank r and $\mathcal{F}$ has rank $r + 1$. From the sequence (6) we see that they both have the same first Chern class $c_1(\mathcal{E}) = c_1(\mathcal{F}) = c$. Thus the degree of the locally free sheaves $\mathcal{F}^\vee \otimes \mathcal{O}_C$ and $\mathcal{E}^\vee \otimes \mathcal{O}_C$ on C is just $-cd$. Now applying the Riemann–Roch theorem (which works on any curve C for the restriction of locally free sheaves from $\mathbb{P}^3$), and using Serre duality to note that $\chi(\omega_C(4)) = -\chi(\mathcal{O}_C(-4))$, we get

$$\chi(\mathcal{N}_C) = cd + (r+1)(1-g) - [cd + r(1-g)] - [-4d + 1 - g] = 4d.$$

Remark 12.1.1. If we apply the same argument to (say) locally complete intersection curves of degree d and arithmetic genus g in $\mathbb{P}^4$ we obtain $\dim \mathrm{Hilb} \geq 5d + 1 - g$. (There is a similar formula for $\mathbb{P}^n$, $n > 4$.) This number can become negative for large values of g, making the result worthless. This led Joe Harris to ask:

(a) Can you find a better (sharp) lower bound for the dimension of the Hilbert scheme?
(b) Are there any "semirigid" curves in $\mathbb{P}^n$, i.e., curves whose only deformations come from automorphisms of $\mathbb{P}^n$, besides the rational normal curves of degree n?

References for this section. Since the work of Noether, the fact that the dimension of the families of curves of degree d in $\mathbb{P}^3$ is at least $4d$ seems to have passed into folklore. The first complete proof in the case of locally Cohen–Macaulay curves, as far as I know, is the one due to Ein [23, Lemma 5], as explained to me by Rao and reproduced here. Theorem 12.1 has been generalized by Rao and Oh to the case of one-dimensional closed subschemes of $\mathbb{P}^3$ that may have isolated or embedded points [128].

Exercises.

12.1. If C is a curve in $\mathbb{P}^3$ with $\dim_C \mathrm{Hilb} = 4d$, then the Hilbert scheme is a locally complete intersection at the point corresponding to C.

12.2. Curves passing through m general points in $\mathbb{P}^3$. Let H be an irreducible component of the Hilbert scheme of curves of degree d and genus g in $\mathbb{P}^3$, and let m be an integer. We ask whether a general curve in H can be made to pass through m general points in $\mathbb{P}^3$.

If H has dimension N, then the dimension of the family of pairs $Z \subseteq C$, where $C \in H$ and Z is a set of m points on C, is $N + m$. On the other hand, the dimension of the Hilbert scheme of sets of m points in $\mathbb{P}^3$ is $3m$. In order for a general Z to lie on a curve $C \in H$, we must have $N + m \geq 3m$, i.e., $N \geq 2m$.

This condition in general is not sufficient. For example, the nonsingular curves of degree 5 and genus 2 in $\mathbb{P}^3$ form an irreducible component H of dimension 20. They are all ACM curves, the general one being a curve of bidegree $(2, 3)$ on a nonsingular quadric surface. However, one of these curves C can never contain 10 general points, because C lies on a quadric surface, and a quadric surface can contain only 9 general points of $\mathbb{P}^3$.

Thus it is reasonable to ask, if $N \geq 2m$, and if the question is not impossible by reason of the degree of surfaces containing a general curve C, does the general curve $C \in H$ pass through m general points in $\mathbb{P}^3$? This problem is studied in the thesis of D. Perrin [131]. The answer is not known in general, but it seems likely to be true at least for ACM curves. Here we treat a special case.

(a) For any integer $r \geq 1$, we consider a nonsingular curve C with a resolution of the form
$$0 \to \mathcal{O}_{\mathbb{P}}(-r-1)^r \to \mathcal{O}_{\mathbb{P}}(-r)^{r+1} \to \mathcal{I}_C \to 0,$$
This is called a *linear resolution* because the entries in the matrix defining the first map are all linear forms. Show that C is an ACM curve of degree $d = \frac{1}{2}r(r+1)$ and genus $g = \frac{1}{6}(r-1)(r-2)(2r+3)$.

(b) By taking $\mathcal{H}om$ of the resolution into $\mathcal{O}_C$, show that one obtains an exact sequence
$$0 \to \mathcal{N}_C \to \mathcal{O}_C(r)^{r+1} \to \mathcal{O}_C(r+1)^r \to \omega_C(4) \to 0.$$
Use the fact that $h^0(\mathcal{I}_C(r-1)) = 0$ and C is ACM to show that $H^0(\mathcal{N}_C(-2)) = 0$.

(c) Consider the Hilbert-flag scheme $H\{Z,C\}$ (Ex. 6.8) parametrizing pairs $Z \subseteq C$, where C is one of the curves in (a), and we take Z to be a set $m = 2d$ distinct points on C, linearly equivalent to twice the hyperplane section. Then there is an exact sequence
$$0 \to \mathcal{N}_C(-2) \to \mathcal{N}_C \to \mathcal{N}_C|_Z \to 0.$$
Show that $\chi(\mathcal{N}_C(-2)) = 0$ and use b) above to conclude that $H^1(\mathcal{N}_C(-2)) = 0$, and so $H^0(\mathcal{N}_C) \to H^0(\mathcal{N}_C|_Z)$ is an isomorphism. Since $\mathcal{N}_C|_Z$ is a rank 2 vector bundle on a set of $2d$ points, its $h^0 = 4d$, $h^1 = 0$. We conclude from this that $h^0(\mathcal{N}_C) = 4d$, $h^1(\mathcal{N}_C) = 0$, and that $T \to H^0(\mathcal{N}_Z)$ is also an isomorphism, where T is the tangent space to $H\{Z,C\}$ (Ex. 6.8a).

(d) From (c) it follows that the Hilbert scheme of the curves C is nonsingular of dimension $4d$. Since $H\{Z,C\}$ is fibered over this by sets of $2d$ points, it is also smooth, of dimension $6d = 3m$. Now the "forgetful" map from $H\{Z,C\}$ to the Hilbert scheme $H\{Z\}$ of Z's is a morphism from one smooth scheme of dimension $3m$ to another smooth scheme of the same dimension, inducing an isomorphism on the Zariski tangent spaces at the points we have considered. It follows that it is étale and dominant, so a general set of $m = 2d$ points is contained in a curve C of the family of these ACM curves, which has dimension $N = 4d = 2m$.

13. A Nonreduced Component of the Hilbert Scheme

In the classification of algebraic space curves, the idea that curves form algebraic families, and that one could speak of the irreducible components and dimensions of these families, goes back well into the nineteenth century. The observation that the family of curves of given degree and genus in $\mathbb{P}^3$ need not be irreducible is due to Weyr (1873). Tables of families with their irreducible components and dimensions were computed (independently) by Halphen and Noether in 1882. Now these notions can be made rigorous by speaking of the Chow variety or the Hilbert scheme, and for the question of irreducible components and dimensions of families of *smooth* space curves, the answer is the same in both cases.

With Grothendieck's construction of the Hilbert scheme, a new element appears: the families of curves, which up to then were described only as algebraic varieties, now have a scheme structure. In 1962, only a few years after Grothendieck introduced the Hilbert scheme, Mumford [112] surprised everyone by showing that even for such nice objects as irreducible nonsingular curves in $\mathbb{P}^3$, there may be irreducible components of the Hilbert scheme that are generically nonreduced, that is to say that as schemes, they have nilpotent elements in their structure sheaves at all points of the scheme.

In this section we will give Mumford's example.

Theorem 13.1. *There is an irreducible component of the Hilbert scheme of smooth irreducible curves in $\mathbb{P}^3$ of degree 14 and genus 24 that is generically nonreduced.*

Proof. The argument falls into three parts:

(a) We construct a certain irreducible family U of smooth curves of degree 14 and genus 24, and show that the dimension of the family is 56.
(b) For any curve C in this family, we show that $H^0(C, \mathcal{N}_C)$ has dimension 57. This gives the Zariski tangent space to the Hilbert scheme at the point C.
(c) We show that the family U is not contained in any other irreducible family of curves with the same degree and genus, of dimension > 56.

Property (c) shows that the family U is actually an open subset of an irreducible component of the Hilbert scheme, of dimension 56. Hence the scheme U_{red} is integral, and therefore nonsingular on some open subset $V \subseteq U_{\mathrm{red}}$. If $C \in V$, then property (b) shows that U is not smooth at the point C; hence $U \neq U_{\mathrm{red}}$ at the point C. In other words, U is nonreduced along the open set V.

Step (a). The construction. Let X be a nonsingular cubic surface in $\mathbb{P}^3$, let H denote the hyperplane section of X, and let L be one of the 27 lines on X. We consider curves C in the linear system $|4H + 2L|$. If we take L to be the sixth exceptional curve E_6, then in the notation of [57, V, §4], the divisor class is $(12; 4, 4, 4, 4, 4, 2)$, and by [loc. cit. 4.12] this class is very ample, so the linear system contains irreducible nonsingular curves C. The formulas of [loc. cit.] show that the degree is $d = 14$ and the genus is $g = 24$. The family U we wish to consider consists of all nonsingular curves C in the above linear system, for all choices of X a smooth cubic surface and L a line on X.

The cubic surfaces move in an irreducible family of dimension 19, and as they move, the lines on them are permuted transitively, so that U is an irreducible family of curves. Since $\mathrm{Pic}\,X$ is discrete, the only algebraic families of curves on X are the linear systems. So to find the dimension of U we must add 19 to the dimension of the linear system $|C|$, which is $h^0(\mathcal{O}_X(C)) - 1$.

(Note that since $d > 9$, each of our curves C is contained in a unique cubic surface.) Consider the exact sequence

$$0 \to \mathcal{O}_X \to \mathcal{O}_X(C) \to \mathcal{O}_C(C) \to 0.$$

Since $h^0(\mathcal{O}_X) = 1$ and $h^1(\mathcal{O}_X) = 0$, we obtain $\dim |C| = h^0(\mathcal{O}_C(C))$. The linear system $\mathcal{O}_C(C)$ on C has degree C^2, which can be computed by [loc. cit.] as 60. This is greater than $2g - 2$, so the linear system is nonspecial on C, and by Riemann–Roch its dimension is $60 + 1 - 24 = 37$.

Adding, we obtain $\dim U = 19 + 37 = 56$.

Step (b). Computation of $h^0(C, \mathcal{N}_C)$. We use the exact sequence of normal bundles for the nonsingular curve C on the nonsingular surface X,

$$0 \to \mathcal{N}_{C/X} \to \mathcal{N}_C \to \mathcal{N}_X|_C \to 0.$$

Now $\mathcal{N}_{C/X} = \mathcal{O}_C(C)$, which as we saw above is nonspecial, with $h^0 = 37$. Since X is a cubic surface, $\mathcal{N}_X = \mathcal{O}_X(3)$, so $\mathcal{N}_X|_C = \mathcal{O}_C(3)$, and we obtain

$$h^0(\mathcal{N}_C) = 37 + h^0(\mathcal{O}_C(3)).$$

By Riemann-Roch,

$$h^0(\mathcal{O}_C(3)) = 3 \cdot 14 + 1 - 24 + h^1(\mathcal{O}_C(3)) = 19 + h^1(\mathcal{O}_C(3)).$$

By duality on C, $h^1(\mathcal{O}_C(3)) = h^0(\omega_C(-3))$. By the adjunction formula on X, $\omega_C = \mathcal{O}_C(C + K_X) = \mathcal{O}_C(C - H) = \mathcal{O}_C(3H + 2L)$. Thus $h^1(\mathcal{O}_C(3)) = h^0(\mathcal{O}_C(2L))$. Now we use the sequence

$$0 \to \mathcal{O}_X(2L - C) \to \mathcal{O}_X(2L) \to \mathcal{O}_C(2L) \to 0.$$

Note that $2L - C = -4H$, which has $h^0 = h^1 = 0$. Hence $h^0(\mathcal{O}_C(2L)) = h^0(\mathcal{O}_X(2L)) = 1$, since the divisor $2L$ is effective, but does not move in a linear system.

Thus $h^1(\mathcal{O}_C(3)) = 1$, $h^0(\mathcal{O}_C(3)) = 20$, and $h^0(\mathcal{N}_C) = 57$.

Step (c). To show that U is not contained in a larger family of dimension > 56, we proceed by contradiction. If $C' \in U'$ were a general curve in this supposed larger family U', then C' would be smooth, still of degree 14 and genus 24, but would not be contained in any cubic surface, because our family U contains all those curves that can be obtained by varying X and varying C on X. From the exact sequence

$$0 \to \mathcal{I}_{C'} \to \mathcal{O}_{\mathbb{P}^3} \to \mathcal{O}_{C'} \to 0,$$

twisting by 4, we obtain

$$0 \to H^0(\mathcal{I}_{C'}(4)) \to H^0(\mathcal{O}_{\mathbb{P}^3}(4)) \to H^0(\mathcal{O}_C(4)) \to \cdots.$$

The dimension of the middle term is 35; that of the term on the right, by Riemann–Roch, 33. Hence $h^0(\mathcal{I}_{C'}(4)) \geq 2$. Take two independent quartic surfaces F, F' containing C'. Since C' is not contained in a cubic (or lesser-degree) surface, F, F' are irreducible and distinct, so their intersection has dimension 1, and provides us with a linkage from C' to the residual curve $D = F \cap F' - C'$. Computation of degree and genus (Ex. 8.4) shows that $\deg D = 2$, and $p_a(D) = 0$. Note that D need not be irreducible or reduced. However, one knows that locally Cohen–Macaulay curves D with $d = 2$, $p_a = 0$ are just the plane conics (possibly reducible). In particular, D is ACM. This property is preserved by linkage, so C' is also ACM (Ex. 8.4d).

One can compute the dimension of the family of ACM curves C' as follows. The Hilbert scheme of plane conics D in $\mathbb{P}^3$ has dimension 8. The vector space of equations of quartics F containing D is $H^0(\mathcal{I}_D(4))$, which has dimension 26. The choice of a two-dimensional subspace (generated by F, F') is a Grassmann variety of dimension 48. Thus the dimension of the family of curves C' as above is $8 + 48 = 56$ and so this family cannot contain the family U.

This completes the proof of Mumford's example.

Example 13.1.1. A nonsingular 3-fold with obstructed deformations. In the same paper, Mumford observed that the above example, by blowing up the curve, produces a 3-fold with obstructed deformations. We outline the argument.

Let $C \subseteq \mathbb{P}^3$ be a nonsingular curve. Let $f : X \to \mathbb{P}^3$ be obtained by blowing up C. Let $E \subseteq X$ be the exceptional divisor. Then $f : E \to C$ is the projective space bundle $\mathbb{P}(\mathcal{I}/\mathcal{I}^2)$ over C, where $\mathcal{I} = \mathcal{I}_{C/\mathbb{P}^3}$.

We make use of the sequence of differentials

$$0 \to f^*\Omega^1_{\mathbb{P}^3} \to \Omega^1_X \to \Omega^1_{X/\mathbb{P}^3} \to 0,$$

the identification $\Omega^1_{X/\mathbb{P}^3} \cong \Omega^1_{E/C}$, and the Euler sequence

$$0 \to \Omega^1_{E/C} \to f^*(\mathcal{I}/\mathcal{I}^2)(-1) \to \mathcal{O}_E \to 0.$$

Dualizing we obtain sequences

$$0 \to \mathcal{T}_X \to f^*\mathcal{T}_{\mathbb{P}^3} \to \mathcal{T}_{E/C}(-1) \to 0$$

and

$$0 \to \mathcal{O}_E(-1) \to f^*\mathcal{N}_C \to \mathcal{T}_{E/C}(-1) \to 0.$$

Now the cohomology of $f^*\mathcal{T}_{\mathbb{P}^3}$ is the same as $\mathcal{T}_{\mathbb{P}^3}$; the cohomology of $\mathcal{O}_E(-1)$ is zero, and $f^*\mathcal{N}_C$ has the same cohomology as $\mathcal{N}_C$. Making the substitutions we obtain an exact sequence

$$0 \to H^0(\mathcal{T}_X) \to H^0(\mathcal{T}_{\mathbb{P}^3}) \to H^0(\mathcal{N}_C) \to H^1(\mathcal{T}_X) \to 0$$

and an isomorphism

$$H^1(\mathcal{N}_C) \cong H^2(\mathcal{T}_X).$$

The interpretation is that every infinitesimal deformation of X comes from a deformation of C, and that those deformations of C coming from automorphisms of $\mathbb{P}^3$ have no effect on X. The obstructions to deforming X are equal to the obstructions to deforming C in $\mathbb{P}^3$.

If we now take C to the curve of Mumford's example, it has obstructed deformations, and so the 3-fold X also has obstructed deformations (10.3).

Remark 13.1.2. (13.1) gives an example of an obstructed embedded deformation that is unobstructed as an abstract deformation.

References for this section. Mumford's example appeared in his second "pathologies" paper [112]. We have simplified his argument somewhat by using liaison theory instead of some delicate arguments on families of quartic surfaces. The same paper also gives (13.1.1).

Kleppe devoted his thesis [81] to an analysis and expansion of Mumford's example. He generalized the Hilbert scheme to the Hilbert-flag scheme (Ex. 6.8), parametrizing (for example) pairs of a curve in a surface in $\mathbb{P}^3$, and made an infinitesimal study of these schemes using Laudal's deformation theory [92]. Generalizing Mumford's example, he found for every $d \geq 14$ a suitable g, and a family of nonsingular curves of degree d and genus g on nonsingular cubic surfaces that form a nonreduced irreducible component of the Hilbert scheme [81, 3.2.10, p. 192]. Further study of this situation appears in his papers [82] and [84].

Gruson and Peskine [50] give an example of a family of nonsingular curves of degree 13 and genus 18 lying on ruled cubic surfaces that has dimension 52 and $h^0(\mathcal{N}_C) = 54$ for C in the family. They state that this is a nonreduced component of the Hilbert scheme, but say only "on peut alors montrer" to justify the hard part, which is to show that it is not contained in any larger family. I have not yet seen a complete proof of this statement.

Martin–Deschamps and Perrin [100] show that when one considers curves that need not be irreducible or reduced, then the Hilbert scheme is "almost always" nonreduced. More precisely, they show the following. Excluding the plane curves, one knows that the Hilbert scheme of locally Cohen–Macaulay curves in $\mathbb{P}^3$ of degree d and arithmetic genus g is nonempty if and only if $d \geq 2$ and $g \leq \frac{1}{2}(d-2)(d-3)$. They show that for $d \geq 6$ and $g \leq \frac{1}{2}(d-3)(d-4)+1$, the corresponding Hilbert scheme $H_{d,g}$ has at least one nonreduced irreducible component. For $d < 6$ there are also exact statements. The nonreduced components correspond to the "extremal" curves having the largest possible Rao module. These are almost always reducible, nonreduced curves. The example of smallest degree that they find is for $d = 3$, $g = -2$. In this case the Hilbert scheme $H_{3,-2}$ has two irreducible components. One component of dimension 12 has as its general curve the disjoint union of three lines. The other, of dimension 13, has as its general curve a double structure of arithmetic genus -3 on a line, plus a reduced line meeting the first with multiplicity 2. This component is nonreduced. By performing liaisons starting with nonreduced components of singular curves, Martin–Deschamps and Perrin obtain further examples of nonreduced irreducible components of the Hilbert scheme of smooth curves [100, 5.4]. Their first example is for $(d, g) = (46, 213)$.

The example of Sernesi (Ex. 13.2) first appeared in [150], where he identified one of the families of ACM curves specializing to his curves, and concluded that Hilb is singular there. It appears in [99, X, 5.8] as an intersection of two components of Hilb, and again in [66, 5.18] with further explanations.

Recently, Ravi Vakil [164] surprised everyone again by showing that any singularity you like will appear as a singularity of some Hilbert scheme. However, his construction, being quite general, does not show that every singularity can necessarily occur on a Hilbert scheme of nonsingular curves in $\mathbb{P}^3$.

Exercises.

13.1. Show that the hypothetical ACM curves C' of degree 14 and genus 24 not contained in cubic surfaces, mentioned in step (c) of (13.1), actually exist. The general such curve is irreducible and nonsingular. They form a nonsingular open subset of an irreducible component of the Hilbert scheme of dimension 56, and have h-vector $1, 2, 3, 4, 3, 1$ (Ex. 8.11).

13.2. An example of Sernesi. In this and the following exercise we give another example of nonsingular curves in $\mathbb{P}^3$ corresponding to singular points on the Hilbert scheme. In this case the family in question is contained in the intersection of two irreducible components of the Hilbert scheme.

(a) Let X be a nonsingular quartic surface containing two disjoint lines L_1, L_2. Let C be a curve in the linear system $4H + L_1 + L_2$ on X. Show that the linear system $2H + L_1 + L_2$ has no base points, so $4H + L_1 + L_2$ is very ample. Hence there exist irreducible nonsingular curves C in this linear system. Since C is obtained by biliaison (Ex. 8.6) from the two skew lines L_1, L_2, show that it has $(d, g) = (18, 39)$ and that $h^1(\mathcal{I}_C(4)) = 1$. In fact, the Rao module is just k in degree 4.
(b) Show that $h^1(\mathcal{O}_C(4)) = 1$, and then from the exact sequence of normal bundles for C on X, show that $H^1(\mathcal{N}_C) \neq 0$ and $h^0(\mathcal{N}_C) > 72$.
(c) Now let C be any integral $(18, 39)$ curve with $h^0(\mathcal{I}_C(3)) = 0$ and $h^1(\mathcal{I}_C(4)) = 1$.

Case 1. If $h^0(\mathcal{I}_C(4)) = 0$, then C can be linked by 5×5 (meaning an intersection of two surfaces of degree 5) to a $(7, 6)$ (meaning degree 7 and genus 6) curve D with $h^1(\mathcal{I}_D(2)) = 1$ and $h^1(\mathcal{O}_D(2)) = 0$ (Ex. 8.4). Conclude from this that $h^0(\mathcal{I}_D(2)) = 2$, and then show that this is impossible: there is no $(7, 6)$ curve contained in two distinct surfaces of degree 2.

Case 2. If $h^0(\mathcal{I}_C(4)) \neq 0$, then C, being integral, is contained in an irreducible quartic surface X. Show that $h^1(\mathcal{O}_C(4)) = 1$. Consider the linear system $C - 4H$. From the exact sequence [63, 2.10]

$$0 \to \mathcal{O}_X \to \mathcal{L}(C - 4H) \to \omega_C(-4) \to 0$$

and duality on C, conclude that $h^0(\mathcal{L}(C - 4H)) \neq 0$, so there is an effective divisor $D \sim C - 4H$. In other words, C is obtained by biliaison from D. Since D is a $(2, -1)$ curve, and these form an irreducible family, show that the family $\mathcal{C}_0$ of all integral $(18, 39)$ curves with $h^0(\mathcal{I}_C(3)) = 0$ and $h^1(\mathcal{I}_C(4)) = 1$ is irreducible of dimension 71. This family includes the nonsingular curves described in (a).

(d) Since the dimension of the family $\mathcal{C}_0$ is less than $4d = 72$, conclude (12.1) that this family must lie in the closure of some larger irreducible component of the Hilbert scheme.

13.3. Sernesi's example, continued.

(a) If C_t is a flat family of $(18, 39)$ curves whose special curve C_0 lies in the family $\mathcal{C}_0$ described in the previous exercise, but the general C_t does not lie in that family, show that the general C_t must be an ACM curve, hence in one of the two families described in (Ex. 8.12). We will show in fact that the special family of (Ex. 13.2) lies in the closure of both of the families of (Ex. 8.12).

(b) Let C_0 be a general curve in the family $\mathcal{C}_0$ of (Ex. 13.2). We can link it (Ex. 8.4) by 4×6 to a $(6, 3)$ curve D_0 with $h^1(\mathcal{I}_D(2)) = 1$ and $h^1(\mathcal{O}_D(2)) = 0$. Then D_0 is one of the curves described in (Ex. 8.8). Since we can link back in the other direction, we may assume that D_0 is general in that family and nonsingular. From (Ex. 8.8) we know there is a flat family D_t specializing to D_0, where the general D_t is ACM. Since $h^0(\mathcal{I}_{D_t}(4))$ and $h^0(\mathcal{I}_{D_t}(6))$ are constant in the family, we can link by a family of complete intersections 4×6 to obtain a family C_t specializing to C_0, where the general C_t is ACM and lies on a quartic surface. Such ACM curves C_t belong to the family $\mathcal{C}_1$ of Case 1 of (Ex. 8.12).

(c) On the other hand, for a general C_0 as above, put it on a nonsingular sextic surface X_0'. Show by an argument similar to the one used in (Ex. 13.2c) above that the linear system $C_0 - 2H$ on X_0' is effective, and contains a $(6, 3)$ curve D_0 as before. Since we can recover $C_0 = 2H + D_0$, we may assume that D_0 is general in its family. Consider again a family D_t specializing to D_0 where the general D_t is ACM. Put these curves in a family of sextic surfaces X_t', and let $C_t = 2H + D_t$ be obtained by biliaison on X_t'. Then the general C_t is ACM. This time, since $h^0(\mathcal{I}_{D_t}(2)) = 0$, we obtain $h^0(\mathcal{I}_{C_t}(4)) = 0$. So C_t belongs to the family of ACM curves $\mathcal{C}_2$ of Case 2 of (Ex. 8.12).

(d) Thus we have shown that the family $\mathcal{C}_0$ of (Ex. 13.2) is contained in the intersection of the closures of two families $\mathcal{C}_1$ and $\mathcal{C}_2$ of ACM curves, which are open subsets of irreducible components of the Hilbert scheme, each of dimension 72. In particular, the curves in the family $\mathcal{C}_0$ all correspond to singular points of the Hilbert scheme.

(e) Show that the Hilbert scheme just described, though singular, is reduced along $\mathcal{C}_0$. *Hint:* Use (Ex. 12.1).

13.4. Sernesi's example, continued some more.

(a) Use the fact that the Hilbert scheme of $(6, 3)$ curves is nonsingular at the points corresponding to curves D_0 on quadric surfaces (Ex. 6.4) and the liaisons of (Ex. 13.3b) to show that the closure of the component $\mathcal{C}_1$ mentioned there is still nonsingular at the special points C_0, at least when C_0 is nonsingular and contained in a nonsingular quartic surface X_0. This closure corresponds to $(18, 39)$ curves with the added condition $h^0(\mathcal{I}_C(4)) \neq 0$.

(b) Make a similar argument to show that the closure of the family $\mathcal{C}_2$ in (Ex. 13.3c) is nonsingular. These correspond to $(18, 39)$ curves with the extra condition $h^1(\mathcal{O}_C(4)) \neq 0$.

(c) Thus the Hilbert scheme at a point corresponding to a general curve of the family $\mathcal{C}_0$ is reduced, and has two irreducible components, both nonsingular of dimension 72, meeting in a subset of codimension 1. One can still ask, how do these two components meet? I don't know the answer to that.

3

Formal Moduli

In the previous two chapters we have studied infinitesimal deformations of scheme and sheaves in preparation for the study of global moduli questions. The ideal or gold standard for a moduli space is the model represented by the Hilbert scheme (1.1): there is a scheme whose points parametrize the objects in question, there is a universal family over this scheme, and any other family is obtained by a unique base extension from the universal family. In general, moduli problems do not have such a nice answer, and even if they do, the existence may be quite subtle.

In this chapter we deal with a situation intermediate between infinitesimal and global, namely the study of formal moduli, a situation that is usually more tractable than the difficult global questions. The object is to gather together all possible infinitesimal deformations into one structure, a "formal moduli space," so that any given infinitesimal deformation is obtained, hopefully uniquely, from this formal space. The precise formulation of this concept will be explained in Section 15.

We start with the case of plane curve singularities (Section 14), where we can compute by hand, to illustrate the kind of result we may hope to achieve in general. Then we formulate the general problem in terms of functors of Artin rings (Section 15), and give Schlessinger's criterion for pro-representability (Section 16), which is the principal technical result of this chapter. In the following sections, Sections 17, 18, 19, we apply this theory to each of our Situations A, B, C, D, and give examples and applications. In Section 20 we compare embedded and abstract deformations of projective varieties and prove a theorem of Noether that a general surface in $\mathbb{P}^3_{\mathbb{C}}$ of degree ≥ 4 contains only complete intersection curves. In Section 21 we discuss the problem of *algebraization*, that is, passage from a formal moduli space to a more global object. As an application (Section 22), we study the question of lifting varieties from characteristic p to characteristic 0.

R. Hartshorne, *Deformation Theory*, Graduate Texts in Mathematics 257, DOI 10.1007/978-1-4419-1596-2_4, © Robin Hartshorne 2010

14. Plane Curve Singularities

For plane curve singularities, one can ask questions similar to the ones we have been asking for global objects. Can one describe the set of possible singularities up to isomorphism, and can one find moduli spaces parametrizing them?

First we have to decide what we mean by isomorphism. We do not mean equal as subschemes of the plane (that question is answered by the Hilbert scheme) because it is the type of the singularity, not its embedding, that we are after. Nor do we mean isomorphism of a neighborhood of the point on the curve in the Zariski topology, because that already determines the birational equivalence class of the global curve. We want a purely local notion, and for the moment analytic isomorphism seems to be a reasonable choice. This means we ask for isomorphism of the completion of the local rings of the points on the curves (Ex. 14.1).

Right away we see that we cannot expect to have a moduli space of all curve singularities, because of *jump phenomena*, i.e., families where all fibers except one are isomorphic to each other. The family $xy - t = 0$ is nonsingular for all $t \neq 0$ but gives a node for $t = 0$. The family $y^2 - tx^2 - x^3 = 0$ gives a node for all $t \neq 0$ but a cusp for $t = 0$. Since all smooth points are analytically isomorphic, and all nodes are analytically isomorphic, there cannot be a coarse moduli space. This is because there would have to be a morphism from the t-line to the coarse moduli space M, sending all points $t \neq 0$ to one point of M, while $t = 0$ goes to a different point of M, and this is impossible.

One can improve the situation by considering only "equisingular" families, meaning families with roughly the same type of singularity. Then, for example, the analytic isomorphism types of ordinary fourfold points can be distinguished by a cross-ratio, leading to a moduli space similar to the j-invariant of an elliptic curve (Ex. 14.2). Equisingular deformations have been studied by Wahl [172]. See also [42].

For the moment, instead of looking for a moduli space of singularities, we will focus our attention on a single singularity, and attempt to describe all possible local deformations of this singularity. Our goal is to find a deformation over a suitable local parameter space that is "complete" in the sense that any other local deformation can be obtained (up to isomorphism) by base extension from this one, and that is "minimal" in the sense that it is the smallest possible. The completeness is expressed by saying it is a *versal* deformation space for the singularity, and if it is minimal, we call it *miniversal*. (The precise definition of these notions will come in §15 when we interpret the problem in terms of functors of Artin rings.) One could also ask for a family that is *universal* in the sense that any other family is obtained by a *unique* base extension from this one, but as we shall see, this rarely exists (14.0.4).

It turns out that our goal of finding such a versal or miniversal deformation of a given singularity can be accomplished only for strictly local deformations. This means over parameter spaces that are artinian or complete local rings.

(There is also a complex-analytic version using convergent power series over $\mathbb{C}$ [42], which we do not discuss in this book, and then there are algebraization theorems in certain circumstances allowing one to extend the formal family to an algebraic family; see §21).

So in this section, before introducing the general theory of formal deformations, we will construct explicitly some deformation spaces of plane curve singularities and prove directly their versal property. This will serve as an introduction to the general theory and will illustrate some of the issues we must deal with in studying local deformations.

Example 14.0.1. We start with a node, represented by the equation $xy = 0$ in the plane $\mathbb{A}^2 = \operatorname{Spec} k[x, y]$. We consider the family X given by $xy - t = 0$ in $\mathbb{A}^3 = \operatorname{Spec} k[x, y, t]$, together with its map to the parameter space $T = \operatorname{Spec} k[t]$. For $t \neq 0$ the fiber is a nonsingular hyperbola. For $t = 0$ we recover the original nodal singularity.

We will show that this family X/T has a versal property, at least in a formal sense. Let X'/S be any flat deformation of the node over the spectrum S of a complete local ring. We will assume for simplicity that $S = \operatorname{Spec} k[[s]]$, since the case of more variables or the quotient of a complete regular local ring can be handled similarly. We will also assume for this example that X' is defined by a single equation $g(x, y, s) = 0$ in the ring $k[[s]][x, y]$ with $g(x, y, 0) = xy$, though this property might not hold for an arbitrary X'/S.

We would like to show that there exists a morphism $S \to T$, i.e., a homomorphism $k[t] \overset{\varphi}{\to} k[[s]]$ given by a power series $\varphi(t) = T(s)$ with $T(0) = 0$, such that the base extension of the family X becomes isomorphic to the family X'. To do this, it will be sufficient to find functions $X(x, y, s)$ and $Y(x, y, s)$ reducing to x and y for $s = 0$, and a unit $U(x, y, s)$ reducing to 1 for $s = 0$, such that

$$U(XY - T) = g(x, y, s). \qquad (*)$$

We will construct T, X, Y, U as power series in s degree by degree. The constant terms (for $s = 0$) have already been prescribed: $T(0) = 0$, $X(0) = x$, $Y(0) = y$, $U(0) = 1$, and so the equation $(*)$ is satisfied for $s = 0$.

Let us write

$$T = \sum_{i \geq 1} a_i s^i,$$

$$X = x + \sum_{i \geq 1} b_i s^i,$$

$$Y = y + \sum_{i \geq 1} c_i s^i,$$

$$U = 1 + \sum_{i \geq 1} u_i s^i,$$

$$g = xy + \sum_{i \geq 1} g_i s^i,$$

where $a_i \in k$ and $b_i, c_i, u_i, g_i \in k[x, y]$. Substituting and looking at the degree-1 part of the equation (∗) (the coefficient of s), we obtain

$$xc_1 + yb_1 - a_1 + xyu_1 = g_1.$$

Now g_1 is a given polynomial in x, y. Any polynomial can be expressed as a constant term $-a_1$ plus polynomial multiples of x, y, and xy. Thus we can find a_1, b_1, c_1, u_1 to make this equation hold, and then the equation (∗) is valid for the coefficients of s.

Note that in making these choices, a_1 is uniquely determined, but there is considerable flexibility in choosing polynomials b_1, c_1, u_1.

We proceed inductively. Suppose that a_i, b_i, c_i, u_i have been chosen for all $i < n$ so that (∗) is satisfied for all coefficients of s^i with $i < n$. We write out the coefficient of s^n and obtain

$$h(x, y) + xc_n + yb_n - a_n + xyu_n = g_n,$$

where $h(x, y)$ is a polynomial consisting of all the cross products involving a_i, b_i, etc., with $i < n$, which are already determined. Then as before we can find a_n, b_n, c_n, u_n to satisfy this equation, and it follows that T, X, Y, U will satisfy (∗) up through the coefficient of s^n.

Proceeding in this manner, we find functions T, X, Y, U that are power series in s, with coefficients that are polynomials in x, y, i.e., elements of the ring $k[x, y][[s]]$, that make the equation (∗) hold.

This is not quite what we were hoping for, since the ring $k[x, y][[s]]$ is bigger than $k[[s]][x, y]$. So we have not found an isomorphism of X' with $X \times_T S$, but only an isomorphism of their formal completions [57, II, §9] along the closed fiber at $s = 0$. Thus the versality property is true only in this formal sense. In other words, we have shown that the family X/T has the following property: given a flat deformation X'/S as above of the node over a complete local ring S, there exists a morphism $S \to T$ such that the base extension $X \times_T S$ and X' have isomorphic formal completions along the fiber over the closed point of S.

Remark 14.0.2. While we wrote this example using polynomials in x and y, everything works just as well using power series in x and y, and in that case we obtain an analogous versal deformation property.

Remark 14.0.3. We saw in the calculation above that the linear coefficient a_1 of T was uniquely determined. Thus the morphism $S \to T$ induces a unique map on Zariski tangent spaces, and this implies that T is as small as possible, i.e., it is a miniversal deformation space for the node.

Remark 14.0.4. On the other hand, the higher coefficients of T are not uniquely determined. For example, let $u \in k[[s]]$ be a unit with constant term 1, and let u^{-1} be its inverse. Since

$$u^{-1}((xu)y - su) = xy - s,$$

we get an isomorphism of X with itself by taking $U = u^{-1}$, $X = xu$, $Y = y$, $T = su$. If we set $u = 1 - s$, $u^{-1} = 1 + s + s^2 + \cdots$, this gives $T = s - s^2$, so the coefficient a_2 is -1, while for the trivial isomorphism, $a_2 = 0$. This shows that the morphism $S \to T$ is not unique, even at the power series level, and so the deformation is not universal.

Remark 14.0.5. Note also that in the above proof we could take $u_n = 0$ for all $n \geq 1$. So why include U? We include U here, because in the generalization (14.1) below it becomes necessary when the equation $f(x, y) = 0$ is not homogeneous.

We will now generalize the above argument to an arbitrary isolated plane curve singularity, given by an equation $f(x, y) = 0$. This may be either a polynomial or a power series. We assume that it has an isolated singularity at the origin, so that the ideal $J = (f, f_x, f_y)$, where f_x, f_y are the partial derivatives, will be primary for the maximal ideal $\mathfrak{m} = (x, y)$.

To guess the versal deformation space of this singularity, we take a hint from the calculation of the T^1-functor, which parametrizes deformations over the dual numbers (5.2). Let $R = k[x, y]$ and $B = R/(f)$. Recall (Ex. 3.2) that $T^1(B/k, B)$ is just $B/(f_x, f_y) = R/J$.

Take polynomials $g_1, \ldots, g_r \in R$ whose images in R/J form a vector space basis. Then we take r new variables $t_1, \ldots, t_r$ and define a deformation X over $T = \operatorname{Spec} k[t_1, \ldots, t_r]$ by

$$F(x, y, t) = f(x, y) + \sum_{i=1}^{r} t_i g_i(x, y) = 0.$$

Theorem 14.1. *Given an isolated plane curve singularity $f(x, y) = 0$, the deformation X/T defined above is miniversal in the following sense:*

(a) *For any other deformation X'/S, with S the spectrum of a complete local ring, there is a morphism $\varphi : S \to T$ such that X' and $X \times_T S$ become isomorphic after completing along the closed fiber over zero, and*

(b) *although φ may not be unique, the induced map on Zariski tangent spaces of S and T is uniquely determined.*

Proof. The proof is a generalization of the one given in (14.0.1) above, once we make clear the role of the partial derivatives and of the basis g_i of R/J.

First of all, let $S = \operatorname{Spec} C$, where C is a complete local ring. Since any infinitesimal deformation of a complete intersection is a complete intersection (9.2), the fiber X'_n over any artinian quotient C_n of C is a complete intersection in $\mathbb{A}^2_{C_n}$, defined by a single equation. Taking the inverse limit over n, we find a single function $G(x, y, s)$ in $C[x, y]^\wedge$, the $\mathfrak{m}_C$-adic completion, that defines the completion of X' along the closed fiber. (This is a little weaker than the assumption we made in (14.0.1), but sufficient for our proof.) Second, writing C as a quotient of a formal power series $k[[s_1, \ldots, s_m]]$, we can lift the equation G to the power series ring. It will thus be sufficient to prove the theorem for the case $C = k[[s_1, \ldots, s_m]]$.

To establish the isomorphism required in (a) we will find power series T_i, $i = 1, \ldots, r$, with $T_i(0) = 0$ in C and X, Y, U in $k[x, y][[s_1, \ldots, s_m]]$ restricting to $x, y, 1$ respectively for $s_i = 0$, such that

$$UF(X, Y, T) = G(x, y, s), \qquad (*)$$

where T stands for $T_1, \ldots, T_r$. We will construct T, X, Y, U step by step as before.

Suppose inductively that we have constructed partial power series $T^{(\nu)}$, $X^{(\nu)}, Y^{(\nu)}, U^{(\nu)}$ so that the equation $(*)$ holds modulo $s^{\nu+1}$. (Here we will abbreviate $s_1, \ldots, s_n$ to simply s, leaving the reader to supply missing indices as needed—so for example $s^{\nu+1}$ means the ideal $(s_1, \ldots, s_n)^{\nu+1}$.) This has just been done for $\nu = 0$.

We define a new function

$$H^{(\nu)} = U^{(\nu)} F(X^{(\nu)}, Y^{(\nu)}, T^{(\nu)}) - G(x, y, s).$$

By construction this function lies in the ideal $(s^{\nu+1})$. Thus $H^{(\nu)} \bmod(s^{\nu+2})$ is homogeneous in s of degree $\nu + 1$, so we can write

$$H^{(\nu)} \equiv f(x, y)\Delta U + f_x \Delta X + f_y \Delta Y + \sum_{i=1}^{r} g_i \Delta T_i \quad (\bmod s^{\nu+2}),$$

where the ΔT_i are polynomials in s, and $\Delta U, \Delta X, \Delta Y$ are polynomials in x, y, s, all of these being homogeneous of degree $\nu + 1$ in s. This is possible, because the coefficient of each monomial in s in H^ν is a polynomial in x, y, which can be expressed as a combination of linear multiples of the g_i and polynomial multiples of f, f_x, f_y, since the g_i form a basis for R/J.

Now we define

$$T_i^{(\nu+1)} = T_i^{(\nu)} - \Delta T_i,$$

$$X^{(\nu+1)} = X^{(\nu)} - \Delta X,$$

$$Y^{(\nu+1)} = Y^{(\nu)} - \Delta Y,$$

$$U^{(\nu+1)} = U^{(\nu)} - \Delta U,$$

and I claim that these new functions will satisfy the equation $(*)$ mod $s^{\nu+2}$.

This is a consequence of the following lemma.

Lemma 14.2. Let $F(x_1, \ldots, x_n)$ be a polynomial or power series. Let $h_1, \ldots, h_n$ be new variables. Then

$$F(x_1 + h_1, \ldots, x_n + h_n) \equiv F(x_1, \ldots, x_n) + \sum_{i=1}^{n} h_i \frac{\partial F}{\partial x_i}(x_1, \ldots, x_n) \quad (\bmod(h)^2).$$

The proof of the lemma is elementary, and we leave to the reader the simple verification of the claim made above, applying the lemma to the function $UF(X, Y, T_1, \ldots, T_r)$.

Thus we have constructed power series $T \in S$ and X, Y, U in $k[x, y][[s_1, \ldots, s_m]]$ making the required isomorphism. It is clear from the proof, as in (14.0.1) before, that the linear parts of the functions $T_1, \ldots, T_r$ are uniquely determined, and so the map $S \to T$ is unique on the Zariski tangent spaces.

Remark 14.2.1. Exactly the same proof works for an isolated hypersurface singularity in any dimension. So if $f(x_1, \ldots, x_n) = 0$ in $\mathbb{A}^n$ has an isolated singularity at the origin, then the ideal $J = (f, f_{x_1}, \ldots, f_{x_n})$ will be primary for the maximal ideal $\mathfrak{m} = (x_1, \ldots, x_n)$. We take polynomials g_i to form a k-basis of R/J, and then the versal deformation space is defined by $F(x_1, \ldots, x_n, t_1, \ldots, t_r) = f - \Sigma t_i g_i = 0$.

Example 14.2.2. Let us study the cusp defined by $f(x, y) = y^2 - x^3$. The partial derivatives are $2y$ and $3x^2$, so (assuming char $k \neq 2, 3$) we can take $1, x$ as a basis for R/J, and the versal deformation is defined by $F(x, y, t, u) = y^2 - x^3 + t + ux = 0$. Here the parameter space is two-dimensional, given by t, u. For general values of t, u, the nearby curve will be nonsingular, but for special nonzero values of t, u it may be singular. Indeed, if we set F, F_x, and F_y equal to zero, we find a singular point at $t = -2x^3$, $u = 3x^2$. Hence there are singularities in the fiber over points on the discriminant locus $27t^2 - 4u^3 = 0$. It is easy to check that this singularity is a node when $t, u \neq 0$. So the general deformation is nonsingular, but some nearby deformations have nodes.

References for this section. My notes of lectures by Mike Schlessinger ca. 1972 (unpublished). See also Zariski [176], who discusses in detail the problem of constructing a moduli space of analytic isomorphism classes of plane curve singularities that have the same topological type. For a detailed study of deformations of plane curve singularities in the complex analytic case, see [42].

Exercises.

14.1. Analytic isomorphism. Two plane curves X, X', defined by polynomials $f(x, y) = 0$ and $f'(x, y) = 0$, having isolated singularities at the origin, are *analytically isomorphic* (at the origin) if the complete local rings $k[[x, y]]/(f)$ and $k[[x, y]]/(f')$ are isomorphic [57, I, Ex. 5.14]. Let $T^1 = T^1_{X/k}$, and let $\tau = \text{length } T^1$.
(a) If X, X' are analytically isomorphic, then $\tau = \tau'$.
(b) The converse is true for small values of τ. Show (assuming char. $k = 0$ if necessary) that:
   (1) If $\tau = 1$, then X is analytically isomorphic to the singularity $xy = 0$. These are *nodes*.
   (2) If $\tau = 2$, then X is analytically isomorphic to $y^2 - x^3 = 0$. These are *cusps*.
   (3) If $\tau = 3$, then X is analytically isomorphic to $y^2 - x^4 = 0$. These are *tacnodes*.
   (4) If $\tau = 4$, then X is analytically isomorphic to one of the following: (i) $y^2 - x^5 = 0$, a *higher-order cusp*, or (ii) $xy(y - x) = 0$, an *ordinary triple point*.

14.2. Ordinary fourfold points.

(a) A multiple point of a plane curve of *multiplicity* four (i.e., $f \in (x,y)^4 \setminus (x,y)^5$) with four distinct *tangent directions* (i.e., four distinct factors of $f \bmod (x,y)^5$), is analytically isomorphic to a curve defined by $xy(y-x)(y-\lambda x) = 0$ for some $\lambda \neq 0, 1$. Show that two of these are analytically isomorphic if and only if they have the same *j-invariant* (defined as in the case of an elliptic curve [57, IV, §4]).

(b) The ordinary fourfold point has $\tau = 9$. Can you show that for each $\tau = 5, 6, 7, 8$, there are only finitely many analytic isomorphism classes of singularities, and describe them?

14.3. Ordinary fivefold points.   Here we discuss plane curve singularities of multiplicity 5 with five distinct tangent vectors. Show that any such is analytically isomorphic to one of the two following types:

(i) the "straight" ones: $f = xy(y-x)(y-\lambda x)(y-\mu x)$, where $0, 1, \lambda, \mu$ are all distinct.
(ii) the "curly" ones: $f + g$, with f as above, and g a sufficiently general form of degree 6. In this case the analytic isomorphism class depends only on λ, μ, and not on g, as long as g is not in the ideal generated by f_x and f_y.

In either type (i) or (ii) two such are isomorphic if λ, μ are changed according to a certain finite group, corresponding to reordering $(0, 1, \lambda, \mu, \infty)$ and renormalizing by a fractional linear transformation. If g goes to zero, type (ii) specializes to type (i). So if there is a "moduli space," it consists of two copies of a 2-dimensional variety, with each point of the second copy specializing to the corresponding point of the first copy. Thus it cannot be represented by a scheme, or even an algebraic space. Note that $\tau = 16$ for type (i) and $\tau = 15$ for type (ii), and yet the deformations of any of these would be equisingular by anyone's definition.

14.4. τ is semicontinuous.

(a) Let X/S be a flat family of affine locally complete intersection schemes, with S a nonsingular curve. Show that for any point $s \in S$, $T^1_{X_s/k} \cong T^1_{X/S} \otimes_S k(s)$.

(b) Now suppose the fibers of X/S are plane curves with isolated singularities at the origin. Show that the function $\tau(X_s) = \text{length}(T^1_{X_s/k})$ is upper semicontinuous on S.

(c) Conclude for example that if X_0 has a tacnode for $0 \in S$, then there is a nonempty Zariski open subset of $S \setminus \{0\}$ where all the X_s have analytically isomorphic singularities, and they are either all smooth, all nodes, all cusps, or all tacnodes.

15. Functors of Artin Rings

In this section we will formalize the idea of studying local deformations of a fixed object. We introduce the notion of pro-representable functors and of versal deformation spaces, and in the next section we will prove the theorem of Schlessinger giving a criterion for the existence of a versal deformation space. This will give us a systematic way of dealing with questions of local deformations. Although it may seem rather technical at first, the formal local study is important because it gives necessary conditions for existence of global moduli, and is often easier to deal with than the global questions. Also it gives useful local information when there is no global moduli space at all.

The typical situation is to start with a fixed object X_0, which could be a projective scheme or an affine scheme with a singular point, or any other structure, and we wish to understand all possible deformations of X_0 over local Artin rings. We can consider the functor that to each local Artin ring associates the set of deformations (up to equivalence) of X_0 over that ring, and to each homomorphism of Artin rings associates the deformation defined by base extension. In this way we get a (covariant) functor from Artin rings to sets.

Now we describe the general situation that we will consider, which includes as a special case the deformations of a fixed object as above.

Let k be a fixed algebraically closed ground field, and let $\mathcal{C}$ be the category of local artinian k-algebras with residue field k. We consider a covariant functor F from $\mathcal{C}$ to (Sets). These hypotheses can be weakened (15.2.6), but we shall stick to this case for simplicity.

One example of such a functor is obtained as follows. Let R be a complete local k-algebra, and for each $A \in \mathcal{C}$, let $h_R(A)$ be the set of k-algebra homomorphisms $\mathrm{Hom}(R, A)$. For any morphism $A \to B$ in $\mathcal{C}$ we get a map of sets $h_R(A) \to h_R(B)$, so h_R is a covariant functor from $\mathcal{C}$ to (Sets).

Definition. A covariant functor $F : \mathcal{C} \to$ (sets) that is isomorphic to a functor of the form h_R for some complete local k-algebra R is called *pro-representable*.

To explain the nature of an isomorphism between h_R and F, let us consider more generally any homomorphism of functors $\varphi : h_R \to F$ for a complete local k-algebra R with maximal ideal $\mathfrak{m}$. In particular, for each n this will give a map $\varphi_n : \mathrm{Hom}(R, R/\mathfrak{m}^n) \to F(R/\mathfrak{m}^n)$, and the image of the quotient map of R to $R/\mathfrak{m}^n$ gives an element $\xi_n \in F(R/\mathfrak{m}^n)$. These elements ξ_n are compatible, in the sense that the natural map $R/\mathfrak{m}^{n+1} \to R/\mathfrak{m}^n$ induces a map of sets $F(R/\mathfrak{m}^{n+1}) \to F(R/\mathfrak{m}^n)$ that sends ξ_{n+1} to ξ_n. Thus the collection $\{\xi_n\}$ defines an element $\xi \in \varprojlim F(R/\mathfrak{m}^n)$. We will call such a collection $\xi = \{\xi_n\}$ a *formal family* of F over the ring R.

Here it is useful to introduce the category $\hat{\mathcal{C}}$ of complete local k-algebras with residue field k. The category $\hat{\mathcal{C}}$ contains the category $\mathcal{C}$, and we can extend any functor F on $\mathcal{C}$ to a functor $\hat{F}$ from $\hat{\mathcal{C}}$ to sets by defining $\hat{F}(R) = \varprojlim F(R/\mathfrak{m}^n)$ for any $R \in \hat{\mathcal{C}}$. In this notation, $\hat{F}(R)$ is the set of formal families of F over R.

Conversely, a formal family $\xi = \{\xi_n\}$ of $\hat{F}(R)$ defines a homomorphism of functors $\varphi : h_R \to F$ as follows. For any $A \in \mathcal{C}$ and any homomorphism $f : R \to A$, since A is artinian, it factors through $R/\mathfrak{m}^n$ for some n, say $f = g\pi$, where $\pi : R \to R/\mathfrak{m}^n$ and $g : R/\mathfrak{m}^n \to A$. Then let $\varphi(f)$ be the image of ξ_n under the map $F(g) : F(R/\mathfrak{m}^n) \to F(A)$. It is easy to check that these constructions are well-defined and inverse to each other, so we have the following:

Proposition 15.1. *If F is a functor from $\mathcal{C}$ to* (Sets) *and R is a complete local k-algebra with residue field k, then there is a natural bijection between the set $\hat{F}(R)$ of formal families $\{\xi_n \mid \xi_n \in F(R/\mathfrak{m}^n)\}$ and the set of homomorphisms of functors h_R to F.*

Thus, if F is pro-representable, there is an isomorphism $\xi : h_R \to F$ for some R, and we can think of ξ as an element of $\hat{F}(R)$. We say that the pair (R, ξ) *pro-represents* the functor F. One can verify easily that if F is pro-representable, the pair (R, ξ) is unique up to unique isomorphism (Ex. 15.1).

In many cases of interest, the functors we consider will not be pro-representable, so we define the weaker notions of having a versal family or a miniversal family.

Definition. Let $F : \mathcal{C} \to$ (Sets) be a functor. A pair (R, ξ) with $R \in \hat{\mathcal{C}}$ and $\xi \in \hat{F}(R)$ is a *versal family* for F if the associated map $h_R \to F$ is strongly surjective. Here we say that a morphism of functors $G \to F$ is *strongly surjective* if for every $A \in \mathcal{C}$, the map $G(A) \to F(A)$ is surjective, and furthermore, for every surjection $B \to A$ in $\mathcal{C}$, the map $G(B) \to G(A) \times_{F(A)} F(B)$ is also surjective. In our case this means that given a map $R \to A$ inducing an element $\eta \in F(A)$, and given $\theta \in F(B)$ mapping to η, one can lift the map $R \to A$ to a map $R \to B$ inducing θ. (Note: Some authors call this property of a morphism of functors $G \to F$ *smooth*, by analogy with the infinitesimal lifting property of smooth morphisms of schemes (Ex. 15.4).)

If in addition the map $h_R(D) \to F(D)$ is bijective, where $D = k[t]/t^2$ is the ring of dual numbers, we say that (R, ξ) is a *miniversal family*, or that the functor has a *pro-representable hull* (R, ξ). We say that (R, ξ) is a *universal family* if it pro-represents the functor F. The following proposition explains the significance of this terminology.

Proposition 15.2. *Let (R, ξ) be a formal family of the functor F.*

(a) *If (R, ξ) is a versal family, then for any other formal family (S, η), there is a ring homomorphism $f : R \to S$ such that the induced map $\hat{F}(R) \to \hat{F}(S)$ sends ξ to η.*

(b) *If (R, ξ) is miniversal, then for any (S, η) the map $f : R \to S$ of* (a) *induces a unique homomorphism $R/\mathfrak{m}_R^2 \to S/\mathfrak{m}_S^2$.*

(c) *If (R, ξ) is a universal family, then for any (S, η) as in* (a)*, the corresponding map $f : R \to S$ is unique.*

Proof. (a) Let (R, ξ) be a versal family, and let (S, η) be any formal family. Then by definition we have a strongly surjective morphism of functors $\varphi : h_R \to F$, determined by ξ. For each n, we have an element $\eta_n \in F(S/\mathfrak{m}_S^n)$. Since φ is surjective, we can lift it to an element $\theta_n \in h_R(S/\mathfrak{m}_S^n)$, i.e., a homomorphism $\theta_n : R \to S/\mathfrak{m}_S^n$ such that the induced map $F(R/\mathfrak{m}_R^n) \to F(S/\mathfrak{m}_S^n)$ sends ξ_n to η_n. Furthermore, because of the strong surjectivity of φ, starting from θ_1, we can lift successively and obtain a compatible family $\{\theta_n\}$, that is, so that each θ_n is obtained by θ_{n+1} followed by the natural map $S/\mathfrak{m}_S^{n+1} \to S/\mathfrak{m}_S^n$. Thus the maps θ_n determine a homomorphism $R \to \varprojlim S/\mathfrak{m}_S^n$, which is equal to S, since S is a complete local ring. This map f : $R \to S$ then induces the map $\hat{F}(R) \to \hat{F}(S)$ that sends ξ to η by construction.

(b) If (R, ξ) is miniversal, to show that the induced map $\bar{f} : R/\mathfrak{m}_R^2 \to S/\mathfrak{m}_S^2$ is unique, it is sufficient to show that for every map $S/\mathfrak{m}_S^2 \to D = k[t]/t^2$, the induced map $R/\mathfrak{m}_R^2 \to D$ is unique. But since $h_R(D) \to F(D)$ is bijective, by definition, this follows immediately.

(c) If (R, ξ) is a universal family, then $h_R \to F$ is an isomorphism, so η determines a unique map $h_S \to h_R$, and hence a unique $f : R \to S$.

Example 15.2.1. Suppose that $\mathcal{F}$ is a globally defined contravariant functor from (Sch $/k$) to (Sets). For example, think of the functor Hilb, which to each scheme S/k associates the set of closed subschemes of $\mathbb{P}_S^n$, flat over S. Given a particular element $X_0 \in \mathcal{F}(k)$, we can define a local functor $F : \mathcal{C} \to$ (Sets) by taking, for each $A \in \mathcal{C}$, the subset $F(A) = \mathcal{F}(\text{Spec } A)$ consisting of those elements $X \in \mathcal{F}(\text{Spec } A)$ that reduce to $X_0 \in \mathcal{F}(k)$.

If the global functor $\mathcal{F}$ is representable, then the local functor F will be pro-representable (23.3). Thus pro-representability of the local functor is a necessary condition for representability of the global functor.

Example 15.2.2. The converse of (15.2.1) is false: the local functor may be pro-representable when the global functor is not representable. Take for example deformations of $\mathbb{P}^1$. It is easy to see that this functor is not representable (25.2.1). But since all local deformations over Artin rings are trivial (5.3.1), (Ex. 10.3), the local functor is pro-represented by the ring k.

Example 15.2.3. It follows from (14.1) that the functor of local deformations of a plane curve singularity has a miniversal deformation space (Ex. 15.2). On the other hand, the functor is not pro-representable in general (14.0.4).

Example 15.2.4. For an example of a functor with no versal family, we note that if (R, ξ) is a versal family for the functor F, then the map $\text{Hom}(R, D) \to F(D)$ is surjective, so $F(D)$ is a quotient of a finite-dimensional vector space. If F is the functor of deformations of a k-algebra B, then $F(D)$ is given by $T^1_{B/k}$ (5.2). If $T^1_{B/k}$ is not finite-dimensional, F cannot have a versal deformation space. For example, let $B = k[x, y, z]/(xy)$. Then $T^1_{B/k} = k[z]$. The trouble is that B does not have isolated singularities.

Example 15.2.5. For an example of a functor with a versal family but no miniversal family, see (18.4.1) or (18.4.2).

Remark 15.2.6. There is considerable variation in the literature concerning the exact hypotheses and terminology in setting up this theory. One need not assume k algebraically closed, for example, and then there is a choice whether to stick with local k-algebras having residue field k or to allow finite field extensions. Also, one need not restrict to k-algebras. Sometimes it is convenient (e.g., for mixed characteristic cases; cf. §22) to take Artin algebras over a fixed ring such as the Witt vectors. Some people use "versal" to mean what we called "miniversal." Some call the latter "semi-universal." Some do not say universal but say only "weakly universal" for what we called universal, thinking more generally of the stack instead of the functor.

Reference for this section. Pro-representable functors were introduced by Grothendieck [45, exposé 195].

Exercises.

15.1. Verify that if the pair (R, ξ) pro-represents the functor F, then (R, ξ) is unique up to unique isomorphism. Show also that a miniversal family is unique up to isomorphism, but the isomorphism may not be unique.

15.2. Use (14.1) to verify that the ring $R = k[[t_1, \ldots, t_r]]$ and the formal family obtained by completing the family X/T given there form a miniversal family for the functor of deformations of the plane curve singularity over Artin rings.

15.3. Give an example to show that not every covariant functor $G : \hat{\mathcal{C}} \to$ (Sets) is of the form $\hat{F}$ for some functor $F : \mathcal{C} \to$ (Sets).

15.4. Let $f : X \to Y$ be a flat morphism of schemes of finite type over k, let $x \in X$ be a closed point, and let $y = f(x)$. For any Artin ring $A \in \mathcal{C}$, let $G(A)$ be the set of morphisms of Spec A to X sending the closed point to x, and let $F(A)$ be morphisms of Spec A to Y landing at y. Then G and F are covariant functors from $\mathcal{C}$ to (Sets), and composing with f gives a morphism of functors $f_* : G \to F$. Show that f is smooth at the point x if and only if the morphism of functors $G \to F$ is strongly surjective.

15.5. We define an *obstruction theory* for a functor $F : \mathcal{C} \to$ (Sets) to be a vector space V/k, together with, for every exact sequence $0 \to J \to C' \to C \to 0$ as in (6.1), and for every $u \in F(C)$, an element $\varphi(u, C') \in V \otimes J$ such that

(1) $\varphi(u, C') = 0$ if and only if u is the image of some $u' \in F(C')$, and
(2) formation of $\varphi(u, C')$ is functorial when we divide J by a subspace K, as in the definition of an obstruction theory for a local ring (§11).

(a) If F has a versal family (R, ξ), show that V acts as an obstruction theory for the local ring R also.
(b) In particular, if (R, ξ) is a miniversal family, then dim $R \geq \dim t_F - \dim V$, where $t_F = F(D)$ is the tangent space to F.
(c) Show that the following functors have obstruction theories:
  (1) Deformations of a locally complete intersection subscheme of $\mathbb{P}^n$.
  (2) Deformations of a line bundle on a projective variety.
  (3) Deformations of a locally free sheaf on a projective variety.
  (4) Deformations of an affine scheme over k.
  (5) Deformations of a nonsingular projective scheme over k.

15.6. We say that the functor $F : \mathcal{C} \to$ (Sets) is *unobstructed* if for every surjective map of Artin rings $A' \to A$, the induced map $F(A') \to F(A)$ is surjective.

(a) If F has an obstruction theory in a vector space V, and if $V = 0$, then F is unobstructed.

(b) Let F be a functor with a versal family (R, ξ). Show that F is unobstructed if and only if R is a regular local ring.

15.7. Let $F \to G$ be a strongly surjective morphism of functors, let $h_S \to F$ be a versal family for F, and let $h_R \to G$ be a miniversal family for G.

(a) Show that there is a strongly surjective morphism of functors $h_S \to h_R$ compatible with the map $F \to G$.

(b) Show that S is isomorphic to a ring $R[[t_1, \ldots, t_r]]$ of formal power series over R. *Hint:* Imitate the proof of (4.6).

15.8. Let $F \to G$ be a morphism of functors of Artin rings. Assume that F and G each have a miniversal family, that F is unobstructed, and that the map on tangent spaces $t_F \to t_G$ is surjective. Then show that the morphism $F \to G$ is strongly surjective and that G is unobstructed.

16. Schlessinger's Criterion

In this section we will prove Schlessinger's theorem [145], which gives criteria for a functor of Artin rings to have a versal family, a miniversal family, or to be pro-representable.

We keep the notation of the previous section: k is a fixed algebraically closed field, $\mathcal{C}$ is the category of local artinian k-algebras, and F is a covariant functor from $\mathcal{C}$ to (Sets). Note that the category $\mathcal{C}$ has fibered direct products. If $A' \to A$ and $A'' \to A$ are morphisms in $\mathcal{C}$, we take $A' \times_A A''$ to be the set-theoretic fibered product $\{(a', a'') \mid a' \text{ and } a'' \text{ have the same image in } A\}$. The ring operations extend naturally, giving another object of $\mathcal{C}$, and this object is also the categorical fibered direct product in $\mathcal{C}$.

It is convenient to introduce the notation t_F for $F(D)$, where D is the ring of dual numbers. We call this the *tangent space* of F. Similarly t_R denotes the tangent space of the functor h_R, which is just $\text{Hom}_k(R, D)$, and is equal to the dual vector space of $\mathfrak{m}_R/\mathfrak{m}_R^2$.

A *small extension* in $\mathcal{C}$ is a surjective map $A' \to A$ whose kernel I is a one-dimensional k-vector space.

We begin with some necessary conditions.

Proposition 16.1. *If F has a versal family, then:*

(a) $F(k)$ *has just one element.*

(b) *For any morphisms $A' \to A$ and $A'' \to A$ in $\mathcal{C}$, the natural map*

$$F(A' \times_A A'') \to F(A') \times_{F(A)} F(A'')$$

is surjective.

If furthermore F has a miniversal family, then:

(c) *For any $A \in \mathcal{C}$, considering the maps $A \to k$ and $D \to k$, the map of* (b) *above,*

$$F(A \times_k D) \to F(A) \times_{F(k)} F(D),$$

is bijective.

(d) *$F(D) = t_F$ has a natural structure of a finite-dimensional k-vector space.*

(e) *For any small extension $p : A' \to A$ and any element $\eta \in F(A)$, there is a transitive group action of the vector space t_F on the set $p^{-1}(\eta)$ (provided it is nonempty).*

Finally, if F is pro-representable, then:

(f) *The maps of* (b) *are all bijective and the action of* (e) *is bijective whenever $p^{-1}(\eta)$ is nonempty.*

Proof. (a) Since $\mathrm{Hom}(R, k) \to F(k)$ is surjective, and $\mathrm{Hom}(R, k)$ has just one element, so does $F(k)$.

(b) Given elements $\eta' \in F(A')$ and $\eta'' \in F(A'')$ mapping to the same element $\eta \in F(A)$, by the strong surjective property of a versal family, there are compatible homomorphisms of R to A', A, and A'' inducing these elements. Then there is a unique map of R to the product $A' \times_A A''$ inducing the given maps of R to A' and A''. This in turn defines an element of $F(A' \times_A A'')$ that restricts to η' and η'' as required. Note that although the map of R to $A' \times_A A''$ is uniquely determined by the maps of R to A' and A'', these latter may not be uniquely determined by η' and η'', and so the resulting element in $F(A' \times_A A'')$ may not be uniquely determined.

(c) Suppose we are given $\eta \in F(A)$ and $\xi \in F(D)$. We know from (b) that there are elements of $F(A \times_k D)$ lying over the pair (η, ξ). Suppose θ_1 and θ_2 are two such. Choose a homomorphism $u : R \to A$ inducing η. Since $A \times_k D = A[t]/t^2 \to A$ is surjective, we can lift u to v_1 and $v_2 : R \to A[t]/t^2$, inducing θ_1 and θ_2. Since θ_1 and θ_2 both lie over ξ, the projections of v_1 and v_2 to D both induce ξ. By the hypothesis of miniversality, $t_R \to t_F$ is bijective, so these restrictions are equal. Since v_1 and v_2 also induce the same map $u : R \to A$, we obtain $v_1 = v_2$ and hence $\theta_1 = \theta_2$.

(d) By miniversality, $t_R \to t_F$ is bijective, so we can just carry over the vector space structure on t_R to t_F. But this structure can also be recovered intrinsically, using only the functorial properties of F and condition (c) above (Ex. 16.1).

(e) Let $A' \to A$ be a small extension with kernel $I \cong k$. Note that $A' \times_A A' \cong A' \times_k k[I]$ by sending $(x, y) \mapsto (x, x_0 + y - x)$, where $x_0 \in k$ is the residue of $x \bmod \mathfrak{m}$. Consider the surjective map

$$F(A' \times_A A') \to F(A') \times_{F(A)} F(A')$$

of (b). Using the isomorphism above and condition (c), we can reinterpret the left-hand side as $F(A' \times_k k[I]) \cong F(A') \times t_F$, and we get a surjective map

$$F(A') \times t_F \to F(A') \times_{F(A)} F(A')$$

that is an isomorphism on the first factor. If we take $\eta \in F(A)$ and fix $\eta' \in p^{-1}(\eta)$ then we get a surjective map

$$\{\eta'\} \times t_F \to \{\eta'\} \times p^{-1}(\eta),$$

and this gives a transitive group action of t_F on $p^{-1}(\eta)$.

(f) If F is pro-representable the proof of (b) shows that these maps are all bijective. It follows that the action of (e) is also bijective.

Theorem 16.2 (Schlessinger's criterion). *The functor $F : \mathcal{C} \to$ (Sets) has a miniversal family if and only if:*

(H_0) $F(k)$ *has just one element.*
(H_1) $F(A' \times_A A'') \to F(A') \times_{F(A)} F(A'')$ *is surjective for every small extension $A'' \to A$.*
(H_2) *The map of H_1 is bijective when $A'' = D$ and $A = k$.*
(H_3) t_F *is a finite-dimensional k-vector space.*

Furthermore, F is pro-representable if and only if in addition:

(H_4) *For every small extension $p : A'' \to A$ and every $\eta \in F(A)$ for which $p^{-1}(\eta)$ is nonempty, the group action of t_F on $p^{-1}(\eta)$ is bijective.*

Proof. The necessity of conditions (H_i) has been seen in (16.1).

So now let F be a functor satisfying conditions H_0, H_1, H_2, H_3. First we will construct a ring R and a morphism $h_R \to F$. Then we will show that it has the versal family property.

We will define R and the map $h_R \to F$ as an inverse limit of rings R_q and elements $\xi_q \in F(R_q)$ that we construct inductively for $q \geq 0$. We take $R_0 = k$. Note that the vector space structure on t_F is already determined by conditions H_0, H_1, H_2, (Ex. 16.1), so that it makes sense to say in H_3 that it is finite-dimensional. Let $t_1, \ldots, t_r$ be a basis of the dual vector space t_F^*, let S be the formal power series ring $k[[t_1, \ldots, t_r]]$, with maximal ideal $\mathfrak{m}$, and take $R_1 = S/\mathfrak{m}^2$. Then $t_{R_1} \cong t_F$ by construction. Furthermore, by iterating the condition H_2, we obtain $F(R_1) = F(k[t_1] \times \cdots \times k[t_r]) \cong t_F \otimes_k t_F^*$. The natural element here gives $\xi_1 \in F(R_1)$, inducing the isomorphism $t_{R_1} \cong t_F$.

Now suppose we have constructed a compatible sequence (R_i, ξ_i) for $i = 1, \ldots, q$, with $\xi_i \in F(R_i)$, where $R_i = S/J_i$, and $\mathfrak{m}^{i+1} \subseteq J_i \subseteq J_{i-1}$, and for each i the natural map $R_i \to R_{i-1}$ sends ξ_i to ξ_{i-1}. Then, to construct R_{q+1}, we look at ideals J in S, with $\mathfrak{m}J_q \subseteq J \subseteq J_q$, and take J_{q+1} to be the minimal such ideal J with the property that $\xi_q \in F(R_q)$ lifts to an element $\xi' \in F(S/J)$. To show that there is a minimal such J, it will be sufficient to show that if J and K are two such, then their intersection $J \cap K$ is another one. By enlarging J or K we may assume without loss of generality that $J + K = J_q$. In that case $S/(J \cap K) = (S/J) \times_{(S/J_q)} (S/K)$. Now the existence of liftings of ξ_q over S/J and S/K implies by condition H_1 the existence of a lifting over $S/J \cap K$. Note that by iteration, H_1 implies surjectivity of the given map for any surjective $A'' \to A$, since any surjective map can be factored into a finite number of small extensions. Then we take $R_{q+1} = S/J_{q+1}$, and ξ_{q+1} any lifting of ξ_q, which exists by construction.

Thus we obtain a surjective system of rings R_q and compatible elements $\xi_q \in F(R_q)$. Let $J = \cap J_q$ and take R to be S/J. Then R is a complete local ring and $R_q = R/\bar{J}_q$, where $\bar{J}_q = J_q/J$. Since $J_q \supseteq \mathfrak{m}^{q+1}$ for each q, and conversely, for each s, some $\bar{J}_q \subseteq \mathfrak{m}_R^s$ (Ex. 16.2), it follows that the ideals $\bar{J}_q$ form a base for the $\mathfrak{m}_R$-adic topology of R. Therefore $R = \varprojlim R_q$, and we define $\xi = \varprojlim \xi_q \in \hat{F}(R)$. This gives the desired map $h_R \to F$ (15.1).

I claim that (R, ξ) is a miniversal family for F. Since $t_R \cong t_F$ by construction, we have only to show for any surjective map $A' \to A$ and any $\eta' \in F(A')$ restricting to $\eta \in F(A)$, and any map $R \to A$ inducing η, that there exists a lifting to a map $R \to A'$ inducing η'. Since any surjective map factors into a sequence of small extensions, it suffices to treat the case of a small extension $A' \to A$.

Let $u : R \to A$ induce η. It will be sufficient to show that u lifts to some map $v : R \to A'$. For then v will induce an element $\eta'' \in F(A')$ lying over η. Because of (16.1c), whose proof used only (16.1e), which is our H_2, there is an element of t_F sending η'' to η' by the group action. This same $t_F = t_R$ acts on the set of $v : R \to A'$ restricting to u, so then we can adjust v to a homomorphism $v' : R \to A'$ inducing η'.

Thus it remains to show that for a small extension $A' \to A$, the given map $u : R \to A$ lifts to a map $v : R \to A'$. Since A is an Artin ring, u factors through R_q for some q. On the other hand, R is a quotient of the power series ring S, and the map u will lift to a map of S into A'. Thus we get a commutative diagram

$$
\begin{array}{ccc}
S \xrightarrow{w} R_q \times_A A' & \to & A' \\
\downarrow & \downarrow p' & \downarrow p \\
R \to \quad R_q & \to & A
\end{array}
$$

Note that $p' : R_q \times_A A' \to R_q$ is also a small extension. If this map has a section $s : R_q \to R_q \times_A A'$, then using s and the second projection we get a map $R_q \to A'$ lifting u, and we are done.

If p' does not have a section, then I claim that the map w is surjective. Indeed, if w is not surjective, then $\operatorname{Im} w$ is a subring mapping surjectively to R_q. The kernel of $\operatorname{Im} w \to R_q$ must be strictly contained in $I = \ker p'$, which is a one-dimensional vector space, so this kernel is zero, the map $\operatorname{Im} w \to R_q$ is an isomorphism, and this gives a section. Contradiction!

Knowing thus that w is surjective, let $J = \ker w$. Then $J \subseteq J_q$, since S maps to R_q via w. On the other hand, $J \supseteq \mathfrak{m} J_q$, since p' is a small extension. But also we have $\xi_q \in F(R_q)$ and there is an $\eta' \in F(A')$ lying over $\eta \in F(A)$, so by H_1, there is a $\xi' \in F(R_q \times_A A')$ lying over both of these. Since $R_q \times_A A' = S/J$, this ideal J satisfies the condition imposed in the construction. Therefore

$J \supseteq J_{q+1}$ and w factors through R_{q+1}. This gives the required lifting of R to A', and completes the proof that (R, ξ) is a miniversal family for F.

Finally, suppose in addition that F satisfies H_4. To show that $h_R(A) \to F(A)$ is bijective for all A, it will be sufficient to show inductively, starting with $A = k$, that for any small extension $p : A' \to A$ we have that $\mathrm{Hom}(R, A') \to \mathrm{Hom}(R, A) \times_{F(A)} F(A')$ is bijective. So fix $u \in \mathrm{Hom}(R, A)$ and the corresponding $\eta \in F(A)$. If there is no map of R to A' lying over u, then there is also no $\eta' \in F(A')$ lying over η, and there is nothing to prove. On the other hand, if $p^{-1}(\eta)$ is nonempty, then the action of t_F on $p^{-1}(\eta)$ is bijective by H_4, and the action of t_R on the set of homomorphisms $R \to A'$ lying over u is bijective, since h_R is pro-representable, and $t_R \cong t_F$ by miniversality, so our map is bijective as required.

Remark 16.2.1. If (H_4) holds, then the map of (H_1) is bijective for all small extensions $A'' \to A$. In particular, (H_4) implies (H_2). Indeed, if $A'' \to A$ is a small extension, then $A' \times_A A'' \to A'$ is also a small extension. So the set of elements of $F(A' \times_A A'')$ going to a fixed element α' of $F(A')$ is in one-to-one correspondence with the set of elements of $F(A'')$ going to the image of α' in $F(A)$. Hence $F(A' \times_A A'') = F(A') \times_{F(A)} F(A'')$.

Next we include some technical results on fibered products and flatness that will be used in studying the pro-representability and existence of versal families for various functors. These results can be skipped at a first reading and referred to as needed.

We are dealing here with fibered products of sets. If A' and A'' are sets, with maps $A' \to A$, $A'' \to A$ to a set A, then the fibered product is

$$A' \times_A A'' = \{(a', a'') \mid a' \text{ and } a'' \text{ have the same image } a \in A\}.$$

If A, A', A'' have structures of abelian groups, or rings, or rings with identity and the maps respect these structures, then $A' \times_A A''$ has a structure of the same kind. This product is categorical; namely, given any set C together with maps $C \to A'$ and $C \to A''$ that compose to give the same map to A, there exists a unique map of C to $A' \times_A A''$ factoring the given maps.

Note that if we consider the schemes $\mathrm{Spec}\, A$, $\mathrm{Spec}\, A'$, $\mathrm{Spec}\, A''$, this is not related to the fibered product in the category of schemes. The arrows go in the opposite direction.

If M, M', M'' are modules over the rings A, A', A'' and we are given compatible maps of modules $M' \to M$, $M'' \to M$, then $M' \times_M M''$ is a module over $A' \times_A A''$.

If $\mathcal{F}, \mathcal{F}', \mathcal{F}''$ are sheaves of abelian groups on a fixed topological space X_0, and we are given maps $\mathcal{F}' \to \mathcal{F}$ and $\mathcal{F}'' \to \mathcal{F}$, the assignment of $\mathcal{F}'(U) \times_{\mathcal{F}(U)} \mathcal{F}''(U)$ to each open set U is a sheaf of abelian groups on X_0, which we will denote simply by $\mathcal{F}' \times_{\mathcal{F}} \mathcal{F}''$.

If $\mathcal{O}_X, \mathcal{O}_{X'}, \mathcal{O}_{X''}$ are sheaves of rings on X_0, together with maps $\mathcal{O}_{X'} \to \mathcal{O}_X$ and $\mathcal{O}_{X''} \to \mathcal{O}_X$, and $\mathcal{F}, \mathcal{F}', \mathcal{F}''$ are sheaves of modules with maps over the respective sheaves of rings, then $\mathcal{F}' \times_{\mathcal{F}} \mathcal{F}''$ is a sheaf of modules over the sheaf of rings $\mathcal{O}_{X'} \times_{\mathcal{O}_X} \mathcal{O}_{X''}$.

If $\mathcal{O}_X, \mathcal{O}_{X'}, \mathcal{O}_{X''}$ define scheme structures on the topological space X_0, then so does $\mathcal{O}_{X'} \times_{\mathcal{O}_X} \mathcal{O}_{X''}$. This will be a fibered sum in the category of scheme structures on X_0. One just has to check that localization is compatible with fibered product of rings and modules.

Lemma 16.3. *Let A, A', A'' be abelian groups, with maps $A' \to A$, $A'' \to A$. In the diagram*

$$
\begin{array}{ccccc}
0 \to \ker u' \to & A' \times_A A'' & \xrightarrow{u'} & A' \\
\downarrow & \downarrow & & \downarrow \\
0 \to \ker u \to & A'' & \xrightarrow{u} & A
\end{array}
$$

(a) *the natural map $\ker u' \to \ker u$ is bijective;*
(b) *if u is surjective, so is u'.*

Proof. Immediate diagram chasing.

Remark 16.3.1. The same applies to rings, modules, and sheaves on a fixed topological space.

Proposition 16.4. *Let A, A', A'' be rings with maps as before, and let $A^* = A' \times_A A''$. Let M, M', M'' be modules over A, A', A'' respectively, with compatible maps $M' \to M$ and $M'' \to M$, and assume that $M' \otimes_{A'} A \to M$ and $M'' \otimes_{A''} A \to M$ are isomorphisms. Let $M^* = M' \times_M M''$.*

(a) *Assume that $A'' \to A$ is surjective. Then $M^* \otimes_{A^*} A' \to M'$ is an isomorphism (and therefore also $M^* \otimes_{A^*} A \to M$ is an isomorphism).*
(b) *Now assume furthermore that $J = \ker(A'' \to A)$ is an ideal of square zero, and that M', M'' are flat over A', A'' respectively. Then M^* is flat over A^*, and also $M^* \otimes_{A^*} A'' \to M''$ is an isomorphism.*

Proof. (a) Since $A'' \to A$ is surjective and $M'' \otimes_{A''} A = M$, it follows that $M'' \to M$ is surjective. Then by Lemma 16.3, $M^* \to M'$ is surjective, and hence $M^* \otimes_{A^*} A' \to M'$ is surjective. To show injectivity, we consider an element $\Sigma \langle m_i', m_i'' \rangle \otimes b_i$ in the kernel of this map and show by usual properties of the tensor product that it is zero (Ex. 16.3).

(b) Since M'' is flat over A'' and $M'' \otimes_{A''} A \cong M$, and $J^2 = 0$, we have an exact sequence

$$0 \to J \otimes_A M \to M'' \to M \to 0$$

by the local criterion of flatness (2.2). From (16.3) it then follows that

$$0 \to J \otimes_A M \to M^* \to M' \to 0$$

is also exact. Now M' is flat over A' by hypothesis, and $M^* \otimes_{A^*} A' \cong M'$ by part (a) above, and since $M' \otimes_{A'} A \cong M$, we have also $J \otimes_A M = J \otimes_{A^*} M^* =$

$J \otimes_{A'} M'$. Now again by (2.2) it follows that M^* is flat over A^*. (Note that the kernel of $A^* \to A'$ is again J with $J^2 = 0$.)

For the last statement, we tensor the sequence $0 \to J \to A'' \to A \to 0$ with M^*, to obtain

$$0 \to J \otimes M^* \to M^* \otimes_{A^*} A'' \to M^* \otimes_{A^*} A \to 0.$$

On the right we have just M, because of part (a) and the hypothesis $M' \otimes_{A'} A \cong M$, and on the left we have $J \otimes M$, so comparing with the sequence for M'' above we obtain $M^* \otimes_{A^*} A'' \to M''$, also an isomorphism.

Example 16.4.1 (Schlessinger). Without the hypothesis $A'' \to A$ surjective, the proposition may fail. For example, take $A = k[t]/(t^3)$, take $A' = A'' = k[x]/(x^2)$, and for homomorphisms send x to t^2. Then $A^* = A' = A''$. Now take $M = A$, $M' = M'' = A' = A''$ (note that these are all flat), and for morphisms take $M' \to M$ the natural injection, but for $M'' \to M$ the natural injection followed by multiplication by the unit $1 + t$. Then $M' \otimes_{A'} A \cong M$ and $M'' \otimes_{A''} A \cong M$, but M^* is just $k \cdot x$, which is not flat over A^*, nor does its tensor product with A' or A'' give M' or M''.

References for this section. For the proof, I have mainly followed Schlessinger's original paper [145]. The proof is also given in the appendix of Sernesi's notes [151], and in abbreviated form in Artin's Tata lectures [8]. See also [152, §2.3]. The results on fibered products are also from [145], though he proves (16.4) only for free modules.

Exercises.

16.1. Let $F : \mathcal{C} \to$ (Sets) be a functor having properties (a) and (c) of (16.1). Show that $t_F = F(D)$ has a natural structure of a k-vector space as follows. For each $\lambda \in k$, the ring homomorphism $\lambda : D \to D$ sending t to λt induces a map $\lambda_* : t_F \to t_F$. Taking two copies of D, say $D_1 = k[t_1]/t_1^2$ and $D_2 = k[t_2]/t_2^2$, consider the ring homomorphism $D_1 \times_k D_2 \to D$ that sends $t_i \to t$, $i = 1, 2$. This induces a map $F(D_1 \times_k D_2) \to F(D)$. Using (a) and (c) this gives a map $\mu : t_F \times t_F \to t_F$. Then taking λ_* and μ as scalar multiplication and addition makes t_F into a k-vector space.

16.2.

(a) Let $(A, \mathfrak{m})$ be a complete local ring. Let $\{\mathfrak{a}_n\}$ be a descending sequence of ideals, with $\cap \mathfrak{a}_n = (0)$. Show that for each s, there is an $n(s)$ such that for $n \geq n(s)$, $\mathfrak{a}_n \subseteq \mathfrak{m}^s$. (This is called Chevalley's theorem in [175, Vol. II, VIII, Thm. 13, p. 270].)

(b) To show that this result fails if A is not complete, let A be a local ring of a node on an integral curve. Use the embedding of A in its normalization $\tilde{A}$ to make a counterexample.

16.3. Complete the proof of (16.4a).

16.4. We say that a functor F has a *tangent theory* if for every small extension $p : A' \to A$ and every $\eta \in F(A)$, there is a transitive action of t_F on $p^{-1}(\eta)$.

(a) Suppose that F has a tangent theory and an obstruction theory (Ex. 15.5), and that both of these are functorial for morphisms of small extensions. Then show that F satisfies (H_1) and (H_2) of (16.2).
(b) If in addition F satisfies (H_0) and (H_3), it has a miniversal family.

16.5. Let X_0 be a nonsingular projective variety over k, and for any $A \in \mathcal{C}$, let $F(A)$ be the set of isomorphism classes of deformations of X_0 over A. For any small extension $p : A' \to A$ and any $X \in F(A)$, let $\mathrm{Ex}(X, A')$ be the set of equivalence classes of extensions of X over A', as in §10. Show that there is a surjective map $\mathrm{Ex}(X, A') \to p^{-1}(X)$. Conclude from (10.3) that F has a tangent theory and an obstruction theory, and therefore has a miniversal family. (We prove a stronger result in (18.1), without the hypothesis X_0 nonsingular.)

17. Hilb and Pic are Pro-representable

There is a general theorem of Grothendieck (1.1), [45, exp. 221] that the Hilbert functor parametrizing closed subschemes of a given projective scheme over k is representable. From this it follows (15.2.1) that the local functor is pro-representable. However, the proof of existence of the Hilbert scheme is long and involved (and not given in this book), so it is of some interest to give an independent proof of pro-representability of the local Hilb functor.

Let X_0 be a given closed subscheme of $\mathbb{P}_k^n$. For each local artinian k-algebra A we let $F(A)$ be the set of deformations of X_0 over A, that is, the set of closed subschemes $X \subseteq \mathbb{P}_A^n$, flat over A, such that $X \times_A k \cong X_0$. (Here by abuse of notation, $X \times_A k$ means $X \times_{\mathrm{Spec}\, A} \mathrm{Spec}\, k$, the fibered product in the category of schemes, not the fibered product of sets!) Then F is a functor from the category $\mathcal{C}$ of local artinian k-algebras to (Sets), which we call the local Hilb functor of deformations of X_0.

Theorem 17.1. *For a given closed subscheme $X_0 \subseteq \mathbb{P}_k^n$, the local Hilb functor F is pro-representable.*

Proof. We apply Schlessinger's criterion (16.2). Condition (H_0) says that $F(k)$ should have just one element, which it does, namely X_0 itself.

Condition (H_1) says that for every small extension $A'' \to A$, and any map $A' \to A$, the map

$$F(A' \times_A A'') \to F(A') \times_{F(A)} F(A'')$$

should be surjective. So suppose we are given closed subschemes $X' \subseteq \mathbb{P}_{A'}^n$ and $X'' \subseteq \mathbb{P}_{A''}^n$, flat over A' and A'', respectively, and both restricting to $X \subseteq \mathbb{P}_A^n$. We let X^* be the scheme structure on the topological space X_0 defined by the fibered product of sheaves of rings $\mathcal{O}_{X'} \times_{\mathcal{O}_X} \mathcal{O}_{X''}$ (§16). Letting $A^* = A' \times_A A''$, we have surjective maps of sheaves $\mathcal{O}_{\mathbb{P}_{A^*}^n}$ to $\mathcal{O}_{X'}$ and to $\mathcal{O}_{X''}$, giving the same composed map to $\mathcal{O}_X$, hence a surjective map to $\mathcal{O}_{X^*}$. Therefore X^* is a closed subscheme of $\mathbb{P}^n_{A^*}$. It is flat over A^* and restricts to $\mathcal{O}_{X'}$ and $\mathcal{O}_{X''}$ over A' and A'' by (16.4). Thus X^* is an element of $F(A^*)$ mapping to X' and X'', and (H_1) is satisfied.

Condition (H_2) is a consequence of (H_4) (16.2.1).

For (H_3) we note that $t_F = F(k[t])$ is the set of deformations of X_0 over $k[t]$, which by (2.4) is $H^0(X_0, \mathcal{N}_{X_0/\mathbb{P}^n})$. Since X_0 is projective, this is a finite-dimensional vector space.

For (H_4), let $\eta \in F(A)$ be given by a deformation $X \subseteq \mathbb{P}^n_A$ of X_0. Then $p^{-1}(\eta)$ consists of subschemes $X' \subseteq \mathbb{P}^n_{A'}$, flat over A', with $X' \times_{A'} A \cong X$. If such exist, they form a torsor under the action of t_F by (6.2).

Thus all the conditions are satisfied and F is pro-representable.

There is also a theorem of Grothendieck [45, exp. 232] that the Picard functor is representable, from which it follows that the local functor is pro-representable, but here we give an independent proof.

Let X_0 be a given scheme over k, and $\mathcal{L}_0$ a given invertible sheaf on X_0. The local Picard functor F assigns to each local artinian k-algebra A the set of isomorphism classes of invertible sheaves $\mathcal{L}$ on $X = X_0 \times_k A$ for which $\mathcal{L} \otimes \mathcal{O}_{X_0} \cong \mathcal{L}_0$.

Theorem 17.2. *Assume X_0 is projective over k and that $H^0(X_0, \mathcal{O}_{X_0}) = k$. Then the local Picard functor for a given invertible sheaf $\mathcal{L}_0$ on X_0 is pro-representable.*

Proof. We apply Schlessinger's criterion. $F(k)$ consists of the one element $\mathcal{L}_0$, so (H_0) is satisfied. For (H_1), let invertible sheaves $\mathcal{L}'$ on X' and $\mathcal{L}''$ on X'' be given such that $\mathcal{L}' \otimes \mathcal{O}_X \cong \mathcal{L}'' \otimes \mathcal{O}_X \cong \mathcal{L}$ on X. Choose maps $\mathcal{L}' \to \mathcal{L}$ and $\mathcal{L}'' \to \mathcal{L}$ inducing these isomorphisms. Then we take $\mathcal{L}^* = \mathcal{L}' \times_{\mathcal{L}} \mathcal{L}''$ to be the fibered product of sheaves. It is an invertible sheaf on $X^* = X_0 \times_k A^*$, where $A^* = A' \times_A A''$, and by (16.4) it restricts to $\mathcal{L}'$ on X' and $\mathcal{L}''$ on X''. Thus (H_1) holds.

(H_2) is a consequence of (H_4) (16.2.1).

By (2.6), the tangent space t_F is $H^1(X_0, \mathcal{O}_{X_0})$, which is finite-dimensional, since X_0 is projective, so (H_3) holds. (H_4) is a direct consequence of (6.4), since we have assumed $H^0(\mathcal{O}_{X_0}) = k$. Thus F is pro-representable.

References for this section. The Hilbert scheme was first constructed by Grothendieck [45]. Other proofs of its existence can be found in Mumford [115] in a special case, in the lecture notes of Sernesi [151], and in the book of Kollár [88]. See also [152, Ch. 4]. Representability of the Picard functor was also proved by Grothendieck [45].

Exercise.

17.1. Verify similarly that the local Hilbert-flag functor of deformations of a pair of subschemes $Y_0 \subseteq X_0 \subseteq \mathbb{P}^n_k$ (Ex. 6.8) is pro-representable.

18. Miniversal and Universal Deformations of Schemes

In this section we will discuss the question of pro-representability or existence of a miniversal family of deformations of a scheme.

If we start with a global moduli problem, such as the moduli of curves of genus g, the global functor considers flat families X/S for a scheme S, whose geometric fibers are projective nonsingular curves of genus g, up to isomorphism of families. The formal local version of this functor around a given curve X_0/k would assign to each Artin ring A with residue field k the set $F_1(A)$ of isomorphism classes of flat families X/A such that $X \otimes_A k$ is isomorphic to X_0. We call this the *crude* local functor (cf. 23.3.1).

Since the functor F_1 is not well behaved, we consider instead the functor of local deformations of X_0, as in §10. Recall that a deformation of X_0 over A is a pair (X, i), where X is a scheme flat over A, and $i : X_0 \to X$ is a closed immersion such that the induced map $i \otimes k : X_0 \to X \otimes_A k$ is an isomorphism. We consider the functor $F(A)$ that to each A assigns the set of deformations (X, i) of X_0 over A, up to equivalence, where an equivalence of (X_1, i_1) and (X_2, i_2) means an isomorphism $\varphi : X_1 \to X_2$ compatible with the maps i_1, i_2 from X_0.

The effect of using the functor F instead of F_1 is to leave possible automorphisms of X_0 out of the picture and thus simplify the discussion. We will consider the relation between these two functors later (18.4).

Theorem 18.1. *Let X_0 be a scheme over k. Then the functor F (defined above) of deformations of X_0 over local Artin rings has a miniversal family under either of the two following hypotheses:*

(a) X_0 *is affine with isolated singularities.*
(b) X_0 *is projective.*

Proof. We verify the conditions of Schlessinger's criterion (16.2).

(H_0) $F(k)$ consists of the single object (X_0, id). If σ is an automorphism of X_0, the object (X_0, σ) is isomorphic to (X_0, id) by the map $\sigma : X_0 \to X_0$.

(H_1)   Suppose we are given a small extension $A'' \to A$ and any map $A' \to A$, and suppose we are given objects $X' \in F(A')$, $X'' \in F(A'')$ restricting to $X \in F(A)$. Then $X' \otimes_{A'} A \cong X$, the isomorphism being compatible with the maps from X_0, so we can choose a closed immersion $X \hookrightarrow X'$ inducing this isomorphism. Similarly choose $X \hookrightarrow X''$. Then we define X^* by the fibered product of sheaves of rings $\mathcal{O}_{X^*} = \mathcal{O}_{X'} \times_{\mathcal{O}_X} \mathcal{O}_{X''}$, and X^* will be an object of $F(A^*)$ reducing to X' and X'', where $A^* = A' \times_A A''$ (16.4).

(H_2)   Suppose $A = k$ in the situation of (H_1) (which effectively means $A'' \cong k[t]$) and let X^* be constructed as in (H_1). If W is any other object of $F(A^*)$ restricting to X' and X'' respectively, then we can choose immersions $X' \hookrightarrow W$ and $X'' \hookrightarrow W$ inducing these isomorphisms.

Since these maps are all compatible with the immersions from X_0, they agree with the chosen maps $X \hookrightarrow X'$ and $X \hookrightarrow X''$, since in this case $X = X_0$. Now by the universal property of fibered product of rings, there is a map $X^* \to W$ compatible with the above maps. Since X^* and W are both flat over A^*, and the map becomes an isomorphism when restricted to X_0, we find that X^* is isomorphic to W (Ex. 4.2), and hence they are equal as elements of $F(A^*)$.

(H_3) Here is the only place in the proof that we need the hypothesis (a) or (b).

(a) Let $X_0 = \operatorname{Spec} B$. Then $t_F = T^1_{B/k}$ by (5.2). This module is supported at the finite number of singular points of X_0 (Ex. 4.3), so has finite length, i.e., t_F is a finite-dimensional vector space.

(b) For arbitrary X_0, the tangent space t_F corresponds to deformations over the dual numbers D. Because of the exact sequence (Ex. 5.7)

$$0 \to H^1(X_0, \mathcal{T}_{X_0}) \to \operatorname{Def}(X_0/k, D) \to H^0(X_0, T^1_{X_0}) \to \cdots ,$$

we see that if X_0 is projective, the two outside groups are finite-dimensional vector spaces, and so $\operatorname{Def}(X_0/k, D)$ is also.

Thus conditions (H_0)–(H_3) are satisfied, and F has a miniversal family.

Examples 18.1.1. We have seen (15.2.3) that the plane curve singularity $xy = 0$ has a miniversal deformation space, but that the functor is not pro-representable. Thus we cannot expect to have a pro-representable functor without further hypotheses.

18.1.2. A rational ruled surface is an example of a nonsingular projective variety X_0 for which the functor is not pro-representable (Ex. 19.2).

Next, we consider conditions under which the functor F is actually pro-representable.

Theorem 18.2. Let X_0/k be given and assume the hypotheses of (18.1) satisfied. Then the functor F of deformations of X_0 is pro-representable if and only if for each small extension $A' \to A$, and for each deformation X' over A' restricting to a deformation X over A, the natural map $\operatorname{Aut}(X'/X_0) \to \operatorname{Aut}(X/X_0)$ of automorphisms of X' (and X) restricting to the identity automorphism of X_0 is surjective.

Proof. Suppose that $\operatorname{Aut}(X'/X_0) \to \operatorname{Aut}(X/X_0)$ is surjective for every X' lying over X. We need to verify condition (H_4) of (16.2). Let X_1', X_2' be elements of $F(A')$ inducing the same element X of $F(A)$. If $X \hookrightarrow X_1'$ and $X \hookrightarrow X_2'$ are maps inducing the isomorphisms $X_1' \otimes_{A'} A \cong X$ and $X_2' \otimes_{A'} A \cong X$, and if X_1' and X_2' are isomorphic as deformations of X_0, then I claim that the inclusions $X \hookrightarrow X_1'$ and $X \hookrightarrow X_2'$ are isomorphic as extensions of X over A'.

Indeed, let $u' : X_1' \to X_2'$ be an isomorphism over X_0. Then $u = u' \otimes_{A'} A$ is an automorphism of X over X_0. By hypothesis this lifts to an automorphism σ of X_1'. Then $v = u \circ \sigma^{-1} : X_1' \to X_2'$ is an isomorphism inducing the identity on X, so X_1' and X_2' are equivalent as extensions of X over A'.

Now by (Ex. 5.7) and (10.2), $\mathrm{Def}(X/A, A')$ is a principal homogeneous space under the action of t_F, so condition (H_4) of Schlessinger's criterion is satisfied, and F is pro-representable.

Conversely, suppose that F is pro-representable, let $X \in F(A')$ restrict to $X \in F(A)$, and choose a map $u : X \hookrightarrow X'$ inducing the isomorphism $X \overset{\sim}{\to} X' \otimes_{A'} A$. Let $\sigma \in \mathrm{Aut}(X/X_0)$. Then $u' = u \circ \sigma : X \hookrightarrow X'$ gives another element of $\mathrm{Def}(X/A, A')$, and so u and u' differ by an element of t_F, by (Ex. 5.7) and (10.2). But u and u' define the same element $X' \in F(A')$, lying over X, so by condition (H_4), this element of t_F must be zero. Hence u and u' are equal as elements of $\mathrm{Def}(X/A, A')$; in other words there exists an isomorphism $\tau : X' \to X'$ over X_0 such that $u' = \tau \circ u$. Restricting to X we obtain $\sigma = \tau \mid X$. Thus $\tau \in \mathrm{Aut}(X'/X)$ lifts σ, and the map is surjective.

Satisfying as it may be to have a necessary and sufficient condition for pro-representability, this condition is difficult to apply in practice, so we will give a corollary and some examples.

Corollary 18.3. *Let X_0/k be a projective scheme with $H^0(X_0, \mathcal{T}_{X_0}) = 0$ (in which case we say "X_0 has no infinitesimal automorphisms"). Then the functor of deformations of X_0/k is pro-representable.*

Proof. We will show, by induction on the length of A, that for any deformation X of X_0 over A, $\mathrm{Aut}(X/X_0) = \{\mathrm{id}\}$. Then obviously the condition of (18.2) will be satisfied.

We start the induction by noting that $\mathrm{Aut}(X_0/X_0) = \{\mathrm{id}\}$. And here it is important that we are using the functor F, and not the other functor F_1 mentioned at the beginning of this section! Thus it does not matter whether X_0 has automorphisms as a scheme over k.

Inductively, assume that $\mathrm{Aut}(X/X_0) = \{\mathrm{id}\}$, where X is a deformation over A. Consider a small extension $A' \to A$ and any $X' \in F(A')$ restricting to X. Choose a map $X \hookrightarrow X'$ inducing the isomorphism $X \overset{\sim}{\to} X' \otimes_{A'} A$. Any automorphism of X' restricts to the identity on X, by the induction hypothesis, so it is an automorphism of the deformation $X \hookrightarrow X'$. Since these are classified by $H^0(X_0, \mathcal{T}_{X_0}) = 0$, (10.2.2) this automorphism is the identity.

Example 18.3.1. Let X_0 be a nonsingular projective curve over the algebraically closed field k. If the genus g is zero, $X_0 = \mathbb{P}_k^1$. In this case $H^0(X_0, \mathcal{T}_{X_0})$ has dimension 3, but still, as we have seen (15.2.2), the functor is pro-representable, represented by a single point. Thus the condition of (18.3) is not necessary for the functor to be pro-representable.

If $g = 1$, we have an elliptic curve. This case is discussed below, (18.4.2) and (Ex. 18.2).

If $g \geq 2$, then the tangent sheaf $\mathcal{T}$ has degree $2-2g < 0$, so $H^0(X_0, \mathcal{T})$ is 0. Thus the functor of deformations of X_0 is pro-representable. Since there are no obstructions (Ex. 10.4), the formal moduli space is smooth, of dimension $3g-3$ (5.3.2). Note that it does not matter whether X_0 has a finite group of automorphisms. What counts here is that it has no infinitesimal automorphisms, such as might arise for example from a continuous group of automorphisms of X_0.

18.3.2. What can we say about algebraic surfaces? We have seen that $\mathbb{P}^2$ is rigid, so its deformations are pro-represented by a point, even though it has infinitesimal automorphisms.

For surfaces of degree $d \geq 2$ in $\mathbb{P}^3$, if $d = 2$, the quadric surface is rigid (20.2.2), so its deformations are pro-representable. For $d \geq 3$ we obtain $H^0(T^0) = 0$ (20.2.2), so (18.3) applies, and the functor is pro-representable.

There are rational ruled surfaces for which the functor is not pro-representable (Ex. 19.2).

An abelian surface has infinitesimal automorphisms, but still its deformation functor is pro-representable (Ex. 18.3).

Now let us return to the question of comparing the functors F and F_1 mentioned at the beginning of this section. Recall that given a scheme X_0/k, F is the functor of deformations of X_0 over A, that is, pairs (X, i) where X is flat over A, and $i : X_0 \hookrightarrow X$ is a morphism such that $i \otimes k : X_0 \to X \otimes k$ is an isomorphism, while F_1 is the functor of flat families X/A such that there exists an isomorphism $X \otimes k \cong X_0$. There is a natural "forgetful" functor from F to F_1.

The following result is proved using the same kind of arguments as in the earlier part of this section.

Theorem 18.4. *Suppose the hypotheses of* (18.1) *satisfied. Then:*

(a) *The crude local functor F_1 has a versal family.*
(b) *F_1 has a miniversal family if and only if in addition, $\mathrm{Aut}\,X \to \mathrm{Aut}\,X_0$ is surjective for each flat family X over the dual numbers D. In this case $t_{F_1} = t_F$.*
(c) *The following conditions are equivalent:*
  (i) *$F_1 = F$ and F is pro-representable.*
  (ii) *F_1 is pro-representable.*
  (iii) *$\mathrm{Aut}\,X' \to \mathrm{Aut}\,X$ is surjective for every small extension $A' \to A$, where X' is a flat family over A' and $X = X' \otimes_{A'} A$.*

Proof. (a) The map $F \to F_1$ is strongly surjective, so a miniversal family for F gives a versal family for F_1.

(b) and (c) are proved by arguments similar to those above (Ex. 18.1).

Example 18.4.1. Let us take X_0 to be the affine scheme $\mathrm{Spec}\,k[x, y]/[xy]$. This is the node that was discussed previously (14.0.1).

(a) It is easy to check that the automorphisms of X_0 are of two types:

$$(1) \quad \begin{cases} x' = ax, \\ y' = by, \end{cases} \qquad (2) \quad \begin{cases} x' = ay, \\ y' = bx, \end{cases}$$

where $a, b \in k^*$. If we attempt to lift an automorphism of type (1) to the family $xy - t$ over the dual numbers $D = \operatorname{Spec} k[t]/t^2$, we will need

$$\begin{cases} x' = ax + tf, \\ y' = by + tg, \end{cases}$$

for some $f, g \in k[x, y]$ satisfying $u(xy - t) = x'y' - t$, where u is a unit $u = \lambda + th$ in $D[x, y]$. To satisfy this equation, we find that $ab = 1$, $f = xf_1$, $g = yg_1$, and $h = ag_1 + bf_1$. Thus the lifted automorphism is of the form

$$x' = (a + tf_1)x,$$
$$y' = (b + tg_1)y,$$

subject to the condition $ab = 1$, and with f_1, g_1 arbitrary elements of $k[x, y]$.

In particular, if we consider an automorphism of X_0 with $ab \neq 1$, it does not lift. Thus $\operatorname{Aut} X \to \operatorname{Aut} X_0$ is not surjective, and F_1 has a versal family, but does not have a miniversal family.

Another way to interpret this is to let $\operatorname{Aut} X_0$ act on the set $F(D) = t_F$ of deformations of X_0 over D. Any element of $F(D)$ is defined by $xy - ct = 0$ for some $c \in k$. We let $\operatorname{Aut} X_0$ act on this set by replacing $i : X_0 \hookrightarrow X$ by $i \circ \sigma : X_0 \hookrightarrow X$ for any $\sigma \in \operatorname{Aut} X_0$. The calculation above shows that this action is nontrivial. In fact, it has two orbits, corresponding to $c = 0$ and $c \neq 0$, and the set $F_1(D) = t_{F_1}$ is the quotient space consisting of two elements, the trivial deformation and the nontrivial deformation. Thus t_{F_1} is not even a vector space over k.

(b) Now let us consider lifting automorphisms of the deformation X given by $xy - t$ over the dual numbers to the deformation X' given by $xy - t$ over the ring $A' = k[t]/(t^3)$.

Automorphisms of X/X_0 are given by

$$x' = (1 + tf)x,$$
$$y' = (1 + tg)y,$$

with $f, g \in k[x, y]$ arbitrary. To lift it to an automorphism of X' we need

$$x' = (1 + tf)x + f't^2,$$
$$y' = (1 + tg)y + g't^2,$$

for some $f', g' \in k[x, y]$. A calculation similar to the one above shows that for this to be possible, there must exist a polynomial $h \in k[x, y]$ for which

$$hxy = f + g + ay + bx + fgxy.$$

In particular, $f + g \in (x, y)$. So if we take $f = 1$, $g = 0$, for example, the automorphism does not lift. This confirms, by (18.2), that F is not pro-representable, as we have noted earlier (14.0.4).

Example 18.4.2 (Pointed elliptic curves). Let X_0 be a nonsingular projective curve of genus 1 over k, and let P_0 be a fixed point. Assume char $k \neq 2, 3$. We consider two functors associated to the pair (X_0, P_0). One, $F(A)$, consists of isomorphism classes of deformations of the pointed curve (X_0, P_0) over A, that is, flat families X, together with a section $P : \operatorname{Spec} A \to X$, and an inclusion $i : X_0 \to X$ such that $i \otimes k$ is an isomorphism of (X_0, P_0) to $(X, P) \otimes k$. The other is the crude functor $F_1(A)$, which is just isomorphism classes of flat families X over A, with a section P, such that $(X, P) \otimes k \cong (X_0, P_0)$.

Repeating the analysis of [57, IV, 4.7] we find that any family of pointed curves (X, P) over the dual numbers D has an equation

$$y^2 = x(x - 1)(x - \lambda)$$

with $\lambda \in D$, and that the group of automorphisms of (X, P) has order

$$\begin{cases} 6 & \text{if } \lambda = -\omega, -\omega^2 \ (j = 0), \\ 4 & \text{if } \lambda = -1, \frac{1}{2}, 2 \ (j = 12^3), \\ 2 & \text{otherwise.} \end{cases}$$

If we take X_0 to be a curve with $j = 12^3$, and X to be the family over $D = k[t]/t^2$ defined by $\lambda = -1 + t$, then the group of automorphisms of (X, P) has order 2, while $\operatorname{Aut}(X_0, P_0)$ has order 4. In particular, $\operatorname{Aut}(X, P) \to \operatorname{Aut}(X_0, P_0)$ is not surjective, so F_1 does not have a miniversal family (18.4).

On the other hand, even though X_0 has infinitesimal automorphisms, since $H^0(X_0, \mathcal{T}) = H^0(X_0, \mathcal{O}_{X_0}) \neq 0$, there are none leaving P_0 fixed, and so the method of (18.3) shows that F is pro-representable.

In this case $t_F = H^1(\mathcal{T})$ has dimension 1, and the deformations over D are given by the equation above with $\lambda = -1 + at$ for any $a \in k$. The action of $\operatorname{Aut}(X_0, P_0)$ on this space sends a to $-a$, so $t_{F_1} = k/\{\pm 1\}$, which is not a k-vector space.

Suppose now we take X_0 to be an elliptic curve with $j \neq 0, 12^3$. Then $\operatorname{Aut}(X_0, P_0)$ has order 2, corresponding to the automorphisms $y \mapsto \pm y$, and these lift to any deformation. So in this case $F_1 = F$ is pro-representable.

Because of the form of the equations above, we can think of the ring pro-representing F as the completion of the λ-line at the corresponding point. In the case $j \neq 0, 12^3$, the λ-line is étale over the j-line, so this is also equal to the completion of the j-line at that point.

References for this section. For a description of versal deformation spaces of germs of singularities in the complex analytic category, with Grauert's theorem on the existence of a versal deformation of an isolated singularity

[38], and a detailed study of deformations of plane curve singularities in the analytic case, see the recent book [42]. While we have shown the existence of miniversal deformation spaces (18.1) in certain cases, the actual computation of examples is difficult, but can be done using a computer algebra package [159], [160].

Exercises.

18.1. Complete the proof of (18.4).

18.2. Here is another approach to deformations of an elliptic curve. Let X_0/k be a nonsingular projective curve of genus 1, over an algebraically closed field k, of char $k \neq 2, 3$. Let $P_0 \in X_0$ be a fixed point. Recall [57, V, 1.3.7, 4.8] that the assignment $Q_0 \in X_0$ goes to $\mathcal{O}_{X_0}(Q_0 - P_0)$ is a bijection from the set of closed points of X_0 to the group $\mathrm{Pic}^0(X_0)$ of invertible sheaves of degree 0. This makes X_0 into a group variety.

(a) If X is a flat deformation of X_0 over A, show that there is a section $P : \mathrm{Spec}\, A \to X$ restricting to $P_0 \in X_0$. Then show that the assignment that to each section Q of X/A gives $\mathcal{O}_X(Q - P)$ is a bijection from sections of X to invertible sheaves on X whose restrictions to X_0 have degree 0.
(b) If Q is another section of X then the operation $\otimes \mathcal{O}_X(Q - P)$ on $\mathrm{Pic}^0 X$ gives, via the bijection of (a), an automorphism of X over A sending P to Q. Show also that an automorphism of X leaving P fixed and restricting to the identity on X_0 must be the identity.
(c) Using (b), show that for any extension X' of X to an A', where $A' \to A$ is a small extension, $\mathrm{Aut}\, X'/X_0 \to \mathrm{Aut}\, X/X_0$ is surjective.
(d) Conclude from (18.2) that the functor of deformations of X_0 is pro-representable, even though (18.3) does not apply.
(e) Show that the forgetful functor sending deformations of pointed elliptic curves (18.4.2) to deformations of (unpointed) elliptic curves is actually an isomorphism.

18.3. Generalize the methods of (18.4.2) and (Ex. 18.2) to show that deformations of pointed abelian surfaces are pro-representable, and that this functor is isomorphic to deformations of (plain) abelian surfaces.

18.4. Taking a hint from the example of a jump phenomenon for plane cuspidal curves (26.6.7), show that the local deformation functor of the cuspidal curve $y^2 = x^3$ is not pro-representable, as follows. We consider the deformation $y^2 = x^3 + t^2 ax + t^3 b$ over the Artin ring $k[t]/t^4$.

(a) Show that $x' = x(1 + 4t)$, $y' = y(1 + 6t + 6t^2)$ gives an automorphism of this family over the ring $k[t]/t^3$, restricting to the identity over k.
(b) Show that the automorphism of (a) does not lift to an automorphism of this family over $k[t]/t^4$.

18.5. Let C be a nonsingular curve in $\mathbb{P}^3$, and let X be the threefold obtained by blowing up C.

(a) Using the exact sequence of (13.1.1), show that if the genus of C is ≥ 2, and C is not contained in a plane, then $H^0(\mathcal{T}_X) = 0$, so the functor of deformations of X is pro-representable.

(b) Now let C be the curve of Sernesi's example (Ex. 13.2–13.4), and show that the scheme Spec R pro-representing the functor of deformations of X is reduced, with two irreducible components of dimension 57, meeting along a subscheme of codimension 1.

18.6. Let X_0 and X_0' be affine schemes over k each having a single isolated singularity at points P, P'. Assume that the singularities at P and P' are analytically isomorphic. Show that there is an isomorphism of the deformation functors $\mathrm{Def}(X_0)$ and $\mathrm{Def}(X_0')$ with the property that for any Artin ring C, if X and X' are the corresponding deformations of X_0 and X_0' over C, then the singularities of X and X' at P and P' are analytically isomorphic. In particular, if (R, ξ) and (R', ξ') are miniversal families of deformations of X_0 and X_0', then R and R' are isomorphic complete local rings. *Hint:* Reviewing the proofs of (10.1) and (18.1), show that obstructions and extensions of deformations can all be computed over the completed local rings of X_0 and X_0' at the singular points, where the T^i functors are isomorphic (Ex. 4.4).

18.7. Let X_0 and Y_0 be linked by a complete intersection scheme Z_0 inside P_0, the spectrum of a regular local ring over k, as in (Ex. 9.4). Let $H = \mathrm{Def}(X_0, Z_0)$ be the functor of deformations of the pair $X_0 \subseteq Z_0$, which is a local analogue of the Hilbert-flag scheme (Ex. 6.8).

(a) Use (Ex. 9.4) to show that the forgetful morphism $H \to \mathrm{Def}(X_0)$ is a strongly surjective morphism of functors.
(b) Show that H is also equal to $\mathrm{Def}(Y_0, Z_0)$, so that we have a strongly surjective morphism $H \to \mathrm{Def}(Y_0)$.
(c) Now assume that X_0 and Y_0 are local Cohen–Macaulay schemes with isolated singularities, so they have miniversal families with complete local rings R for $\mathrm{Def}(X_0)$ and S for $\mathrm{Def}(Y_0)$. We wish to compare R and S using H, but H may not have a versal family because in general its tangent space t_H will not be finite-dimensional. So let H' be the image of H in $\mathrm{Def}(X_0) \times \mathrm{Def}(Y_0)$. That is, for any Artin ring C, $H'(C)$ is the set of pairs (X, Y), where X and Y are deformations of X_0 and Y_0 over C, such that there exists a complete intersection Z linking X to Y. Show then that $t_{H'}$ is finite-dimensional and that H' has a miniversal family.
(d) Now use (Ex. 15.7) to show that there is an isomorphism $R[[t_1, \ldots, t_r]] \cong S[[u_1, \ldots, u_s]]$ of formal power series rings over R and S for some r and s. We say that the miniversal deformation spaces of X_0 and Y_0 are "equivalent up to power series rings." (This is a theorem of Buchweitz [13, 6.4.4].)

18.8. If the affine scheme X_0 over k is rigid, then every infinitesimal deformation is trivial (Ex. 10.3), so the one-point space will be a universal deformation space for X_0. We cannot expect global deformations of X_0 to be trivial (Ex. 4.9), (5.3.1). However, if X_0 has an isolated singularity and is rigid, we can expect that nearby singularities of a global family will be analytically isomorphic to the singularity of X_0.

(a) Over the complex numbers $\mathbb{C}$, Grauert and Kerner [39] have shown that if a germ of a local analytic space has an isolated singularity that is rigid (in the complex analytic sense), then any germ of deformations is locally trivial, so that nearby fibers have isomorphic complex analytic singularities. Show that if X_0 is a reduced affine scheme over $\mathbb{C}$ with a rigid singularity (in the algebraic sense), then the associated complex analytic space is also rigid in the analytic sense. Thus the theorem of Grauert and Kerner will apply to the associated complex analytic space of any global algebraic deformation of X_0. Conclude that the completions of the local rings at singularities of nearby fibers are isomorphic to the completion of the local ring at the singular point of X_0.

(b) For X_0 an affine scheme with a rigid isolated singularity over any field k, give a proof of the same result using Artin's approximation theorem (21.4) applied to a global deformation X_1 of X_0 compared to the trivial deformation X_2 of X_0 over the same base scheme S.

(c) Can you find a purely algebraic proof of this result using only the methods of this book (without the methods of (a) and (b) above)?

19. Versal Families of Sheaves

Suppose we are given a scheme X_0 over k and a coherent sheaf $\mathcal{F}_0$ on X_0. For each local Artin k-algebra A, let $X = X_0 \times_k A$ be the trivial deformation of X_0. We consider the functor F that to each A assigns the set of deformations of $\mathcal{F}_0$ over A, namely $\mathcal{F}$ coherent on X, flat over A, together with a map $\mathcal{F} \to \mathcal{F}_0$ inducing an isomorphism $\mathcal{F} \otimes_A k \cong \mathcal{F}_0$, modulo isomorphisms of $\mathcal{F}$ over $\mathcal{F}_0$.

Theorem 19.1. *In the above situation, assume that X_0 is projective. Then the functor F has a miniversal family.*

Proof. We apply Schlessinger's criterion (16.2), the proof being similar to the case of deformations of schemes (18.1).

(H_0) $F(k)$ has just one element $\mathcal{F}_0 \xrightarrow{\text{id}} \mathcal{F}_0$.

(H_1) Given $\mathcal{F}'/X'$ and $\mathcal{F}''/X''$ restricting to the same $\mathcal{F}/X$, we can choose maps $\mathcal{F}' \to \mathcal{F}$ and $\mathcal{F}'' \to \mathcal{F}$, compatible with the given maps to $\mathcal{F}_0$, inducing isomorphisms $\mathcal{F}' \otimes A \to \mathcal{F}$ and $\mathcal{F}'' \otimes A \to \mathcal{F}$. We now take $\mathcal{F}^*$ to be the fibered product sheaf $\mathcal{F}' \times_\mathcal{F} \mathcal{F}''$, which will be flat over $A^* = A' \times_A A''$ by (16.4).

(H_2) In case $A = k$, the maps to $\mathcal{F} = \mathcal{F}_0$ are already specified, so $\mathcal{F}^*$ is uniquely determined.

(H_3) Since X_0 is projective, by (2.7) we have $t_F = \text{Ext}^1_{X_0}(\mathcal{F}_0, \mathcal{F}_0)$, which is finite-dimensional.

Hence F has a miniversal family.

Remark 19.1.1. We could also consider the crude functor $F_1(A)$, which is the set of isomorphism classes of $\mathcal{F}$ flat over X such that $\mathcal{F} \otimes k \cong \mathcal{F}_0$, but without specifying the map $\mathcal{F} \to \mathcal{F}_0$. As in the case of deformations of schemes (18.4), if $\text{Aut}\,\mathcal{F} \to \text{Aut}\,\mathcal{F}_0$ is surjective for every such $\mathcal{F}$ over the ring of dual numbers D, then F_1 will also have a miniversal family, and $t_{F_1} = t_F$.

Theorem 19.2. *Assume X_0 projective as above, but now assume in addition that $\mathcal{F}_0$ is simple, i.e., $H^0(\mathcal{E}nd\,\mathcal{F}_0) = k$. Then the functors F and F_1 are equal and pro-representable.*

Proof. As in the case of deformations of schemes (18.2) and (18.4), it is merely a matter of showing that $\operatorname{Aut}\mathcal{F}' \to \operatorname{Aut}\mathcal{F}$ is surjective for any $\mathcal{F}' \to \mathcal{F}$.

We have assumed $\mathcal{F}_0$ simple, so $H^0(\mathcal{E}nd\,\mathcal{F}_0) = k$, and $\operatorname{Aut}\mathcal{F}_0 = k^*$. For any deformation $\mathcal{F}$ over A, there is a natural map $A^* \to \operatorname{Aut}\mathcal{F}$, where A^* denotes the group of units in A. If $\mathcal{F}'$ is an extension of $\mathcal{F}$ over A' where $A' \to A$ is a small extension, then $\operatorname{Aut}(\mathcal{F}'/\mathcal{F}) = \operatorname{End}(\mathcal{F}_0) = k^*$ by (7.1a), whose proof does not need the hypothesis $\mathcal{F}_0$ locally free. Then we see, by induction on the length of A, that for any $\mathcal{F}$ over A, $\operatorname{Aut}\mathcal{F} \cong A^*$. Now clearly $\operatorname{Aut}\mathcal{F}' \to \operatorname{Aut}\mathcal{F}$ is surjective and so F is pro-representable. In particular, $\operatorname{Aut}\mathcal{F} \to \operatorname{Aut}\mathcal{F}_0$ is surjective, and so the two functors F and F_1 are equal.

Example 19.2.1. Let X_0 be a nonsingular projective curve of genus g and let $\mathcal{F}_0$ be a simple vector bundle of degree d and rank r. Then the deformations of $\mathcal{F}_0$ are pro-represented by a regular local ring of dimension $r^2(g-1)+1$ (Ex. 7.2).

Theorem 19.3. *Let X_0 be a projective scheme over k, and let $\mathcal{E}_0 \to \mathcal{F}_0 \to 0$ be a surjective map of coherent sheaves. For any local Artin k-algebra A, let $X = X_0 \times_k A$, and let $\mathcal{E} = \mathcal{E}_0 \times_k A$. Then the Quot functor F of quotients $\mathcal{E} \to \mathcal{F} \to 0$ with $\mathcal{F}$ flat over A and $\mathcal{F} \otimes_A k = \mathcal{F}_0$ is pro-representable.*

Proof. Conditions $(H_0), (H_1), (H_2)$ of Schlessinger's criterion are verified as in the previous proof. The tangent space t_F is $H^0(X_0, \mathcal{H}om(Q_0, \mathcal{F}_0))$, which is finite-dimensional, since X_0 is projective. Note that statement (b) of (7.2) does not make use of the hypotheses $\mathcal{E}_0$ locally free and $\operatorname{hd}\mathcal{F}_0 \le 1$. Since there are no automorphisms of a quotient of a fixed sheaf $\mathcal{E}$, the criterion (b) of (7.2) allows us to verify (H_4), and so the functor is pro-representable. Note also that in this case the functors F and F_1 are the same, since there are no automorphisms.

Remark 19.3.1. In fact, Grothendieck [45, exp. 221] has shown that given X_0/k and $\mathcal{E}_0$ on X_0, the global Quot functor of quotients $\mathcal{E}_0 \times S \to \mathcal{F} \to 0$ on $X = X_0 \times S$, flat over S, for any base scheme S, with given Hilbert polynomial P, is representable by a scheme, projective over k.

Example 19.3.2 (Deformations of $\mathcal{O}(-1) \oplus \mathcal{O}(1)$ on $\mathbb{P}^1_k$). Over any Artin ring A, we can construct a coherent sheaf $\mathcal{F}$ on $\mathbb{P}^1_A$ as an extension

$$0 \to \mathcal{O}_{\mathbb{P}^1_A}(-1) \to \mathcal{F} \to \mathcal{O}_{\mathbb{P}^1_A}(1) \to 0.$$

These extensions are classified by $\operatorname{Ext}^1_{\mathbb{P}^1_A}(\mathcal{O}(1), \mathcal{O}(-1)) = H^1(\mathcal{O}_{\mathbb{P}^1_A}(-2)) = A$. If we take a sheaf $\mathcal{F}$ defined by an element $f \in A$ that is contained in the maximal ideal $\mathfrak{m}_A$, then the image of f in k is 0, and so the sheaf $\mathcal{F} \otimes_A k = \mathcal{F}_0$ will be the trivial extension $\mathcal{O}(-1) \oplus \mathcal{O}(1)$ on $\mathbb{P}^1_k$.

Taking Hom of the above sequence into $\mathcal{F}$, we obtain an exact sequence

$$0 \to H^0(\mathcal{F}(-1)) \to \text{Hom}(\mathcal{F}, \mathcal{F}) \to H^0(\mathcal{F}(1)) \to \cdots .$$

The group on the right is a free A-module of rank 4. The group on the left depends on the choice of $f \in A$. Tensoring the original sequence with $\mathcal{O}(-1)$ and taking cohomology, we get

$$0 = H^0(\mathcal{O}(-2)) \to H^0(\mathcal{F}(-1)) \to H^0(\mathcal{O}) \xrightarrow{\delta} H^1(\mathcal{O}(-2)) \to \cdots ,$$

and the image $\delta(1)$ is the element $f \in A$ determining the extension.

(a) First we take $A = D = k[t]/(t^2)$ the dual numbers, and let the sheaf $\mathcal{F}$ be defined by $f = t$. Then $\mathcal{F} \otimes_D k$ is the trivial extension $\mathcal{F}_0$. Furthermore, since $\delta : A \to A$ is multiplication by t, $H^0(\mathcal{F}(-1)) \cong kt$ and the map $H^0(\mathcal{F}(-1)) \to H^0(\mathcal{F}_0(-1))$ is zero. Hence $\text{End}\,\mathcal{F} \to \text{End}\,\mathcal{F}_0$ is not surjective, so $\text{Aut}\,\mathcal{F} \to \text{Aut}\,\mathcal{F}_0$ is not surjective, and we conclude that the functor F_1 does not have a miniversal family.

(b) Next, take $A' = k[t]/t^3 \to A = D$, and let $\mathcal{F}'$ over A' be defined by $f = t^2 \in A'$. Then, by considering the automorphism of $\mathcal{F}$ defined by t in $H^0(\mathcal{F}(-1))$, the same reasoning shows that $\text{Aut}(\mathcal{F}'/\mathcal{F}_0) \to \text{Aut}(\mathcal{F}/\mathcal{F}_0)$ is not surjective, and so we see that the functor F is not pro-representable. All of this is related to the fact that the global family over $S = \text{Spec}\,k[t]$ defined by $f = t$ exhibits a jump phenomenon: the fiber over $t = 0$ is $\mathcal{O}(-1) \oplus \mathcal{O}(1)$, while the fiber over any point $t \neq 0$ is isomorphic to $\mathcal{O} \oplus \mathcal{O}$. So this study of the automorphisms of extensions over Artin rings is the infinitesimal analogue of a global jump phenomenon.

Exercises.

19.1. Let C be a closed subscheme of a scheme X. We consider deformations of the sheaf $\mathcal{O}_C$ as a sheaf of $\mathcal{O}_X$-modules. If we assume that C is integral and projective, then $H^0(\mathcal{E}nd\,\mathcal{O}_C) = k$, so the deformations of C are pro-representable, with tangent space given by $\text{Ext}^1_X(\mathcal{O}_C, \mathcal{O}_C)$. Show that there is an exact sequence

$$0 \to H^0(\mathcal{N}_C) \xrightarrow{\alpha} \text{Ext}^1_X(\mathcal{O}_C, \mathcal{O}_C) \xrightarrow{\beta} H^1(\mathcal{O}_C) \to \cdots .$$

Here, of course, $H^0(\mathcal{N}_C)$ represents deformations of C as a closed subscheme of X. How do you interpret the map β? Can you give an example of a deformation of the sheaf $\mathcal{O}_C$ that is not in the image of α?

19.2. Use (19.3.2) to give an example of a nonsingular projective surface whose deformation functor is not pro-representable. Let $Y_0 = \mathbb{P}^1_k$, let $\mathcal{F}_0$ be $\mathcal{O}(-1) \oplus \mathcal{O}(1)$, and let $X_0 = \mathbb{P}_{Y_0}(\mathcal{F}_0)$, the associated projective space bundle. Then X_0 is a rational ruled surface with a morphism $\pi_0 : X_0 \to Y_0$ [57, V, §2].

(a) For any local Artin ring A, let $Y = \mathbb{P}^1_A$, let $\mathcal{F}$ be a deformation of $\mathcal{F}_0$, and let $X = \mathbb{P}_Y(\mathcal{F})$. Show that X is a deformation of X_0, so that we have a morphism of the functor F of deformation of $\mathcal{F}_0$ to the functor G of deformation of X_0.

(b) Show that $t_F \to t_G$ is an isomorphism.

(c) If X is any deformation of X_0, show that the sheaf $\pi_0^* \mathcal{O}_{Y_0}(1)$ lifts uniquely to X, and that its sections also lift. Hence the map $\pi_0 : X_0 \to Y_0$ extends to a map $\pi : X \to Y$.

(d) If σ is an automorphism of the surface X, then σ acts on $H^0(\pi^* \mathcal{O}_Y(1)) = H^0(\mathcal{O}_Y(1))$ and hence determines an automorphism τ of Y.

(e) Show that any automorphism of X can be factored into the product of an automorphism of Y and an automorphism of X compatible with π, which then arises from an automorphism of $\mathcal{F}$.

(f) Conclude from (19.3.2) that there are deformations X and extensions X' for which $\mathrm{Aut}(X'/X_0) \to \mathrm{Aut}(X/X_0)$ is not surjective, and hence the functor of deformation of X_0 is not pro-representable.

20. Comparison of Embedded and Abstract Deformations

There are many situations in which it is profitable to compare one deformation problem to another. In this section we will compare deformations of a scheme X_0 as a closed subscheme of $\mathbb{P}^n$ to its deformations as an abstract scheme. If F_1 is the functor of Artin rings of embedded deformations, and F_2 is the functor of abstract deformations, then we have a "forgetful morphism" from F_1 to F_2, which for every Artin ring A maps $F_1(A) \to F_2(A)$ by forgetting the embedding.

As an application we compare embedded and abstract deformations of surfaces in $\mathbb{P}^3$ and prove Noether's theorem that a general surface of degree ≥ 4 in $\mathbb{P}^3$ contains only complete intersection curves.

We begin with a result on morphisms of functors.

Proposition 20.1. *Let $f : F_1 \to F_2$ be a morphism of functors on Artin rings. Assume that F_1 and F_2 both have versal families corresponding to complete local rings R_1, R_2. Then there is a morphism of schemes $\bar{f} : \mathrm{Spec}\, R_1 \to \mathrm{Spec}\, R_2$ corresponding to a homomorphism of rings $\varphi : R_2 \to R_1$ such that for each Artin ring A the following diagram is commutative:*

$$
\begin{array}{ccc}
\mathrm{Hom}(R_1, A) & \xrightarrow{\varphi^*} & \mathrm{Hom}(R_2, A) \\
\downarrow & & \downarrow \\
F_1(A) & \xrightarrow{f} & F_2(A)
\end{array}
$$

where the vertical arrows are the maps expressing the versal families. Furthermore, if R_1 and R_2 are miniversal families, then the map induced by $\bar{f}$ on Zariski tangent spaces $t_{R_1} \to t_{R_2}$ is just $t_{F_1} \to t_{F_2}$ given by $F_1(D) \to F_2(D)$, where D is the dual numbers.

Proof. Consider the inverse system $(R_1/\mathfrak{m}^n)$. The natural maps $R_1 \to R_1/\mathfrak{m}^n$ induce elements $\xi_n \in F_1(R_1/\mathfrak{m}^n)$ forming a compatible sequence. By f we get a compatible sequence $f(\xi_n) \in F_2(R_1/\mathfrak{m}^n)$. By the versal property of R_2 we get compatible maps $R_2 \to R_1/\mathfrak{m}^n$ and hence a homomorphism of $R_2 \to \varprojlim R_1/\mathfrak{m}^n = R_1$. The rest is straightforward.

Now we will consider the case that F_1 is the functor of embedded deformations of a projective scheme $X_0 \subseteq \mathbb{P}_k^n$, F_2 is the functor of abstract deformations of X_0, and $f : F_1 \to F_2$ is the forgetful morphism. We know that F_1 is pro-representable (17.1) by a ring R_1 and that F_2 has a miniversal family (18.1) given by a ring R_2, and so we have an associated morphism $\bar{f} : \operatorname{Spec} R_1 \to \operatorname{Spec} R_2$.

Proposition 20.2. *Suppose that $X = X_0$ is a nonsingular subscheme of $\mathbb{P}^n$. Then the exact sequence*

$$0 \to \mathcal{T}_X \to \mathcal{T}_{\mathbb{P}^n}|_X \to \mathcal{N}_{X/\mathbb{P}^n} \to 0$$

gives rise to an exact sequence of cohomology

$$0 \to H^0(\mathcal{T}_X) \to H^0(\mathcal{T}_{\mathbb{P}^n}|_X) \to H^0(\mathcal{N}_X) \xrightarrow{\delta^0} H^1(\mathcal{T}_X)$$

$$\to H^1(\mathcal{T}_{\mathbb{P}^n}|_X) \to H^1(\mathcal{N}_X) \xrightarrow{\delta^1} H^2(\mathcal{T}_X) \to H^2(\mathcal{T}_{\mathbb{P}^n}|_X) \to \cdots$$

in which the boundary map $\delta^0 : H^0(\mathcal{N}_X) \to H^1(\mathcal{T}_X)$ is just the induced map on tangent spaces $t_{F_1} \to t_{F_2}$ of the deformation functors, and $\delta^1 : H^1(\mathcal{N}_X) \to H^2(\mathcal{T}_X)$ maps the obstruction space of F_1 to the obstruction space of F_2.

Proof. The only thing to prove is the identification of δ^0 and δ^1 with the corresponding properties of the functors F_1 and F_2, and this we leave to the reader.

Remark 20.2.1. Because of this exact sequence, we can interpret $H^1(\mathcal{T}_{\mathbb{P}^n}|_X)$ as the obstructions to lifting an abstract deformation of X to an embedded deformation of X. We can also interpret the image of $H^0(\mathcal{T}_{\mathbb{P}^n}|_X)$ in $H^0(\mathcal{N}_X)$ as those deformations of X_0 induced by automorphisms of $\mathbb{P}^n$.

Example 20.2.2. Let us apply this proposition to the case of a nonsingular surface X of degree $d \geq 2$ in $\mathbb{P}^3$.

Restricting the Euler sequence on $\mathbb{P}^3$ to X we obtain

$$0 \to \mathcal{O}_X \to \mathcal{O}_X(1)^4 \to \mathcal{T}_{\mathbb{P}^3}|_X \to 0.$$

From the cohomology of this sequence we obtain $h^0(\mathcal{T}_{\mathbb{P}^3}|_X) = 15$, $h^1(\mathcal{T}_{\mathbb{P}^3}|_X) = 0$ except for the case $d = 4$, in which case it is 1; and $h^2(\mathcal{T}_{\mathbb{P}^3}|_X) = 0$ for $d \leq 5$, but $\neq 0$ for $d \geq 6$.

Next, we observe that the map $H^0(\mathcal{T}_{\mathbb{P}^3}|_X) \to H^0(\mathcal{N}_X)$ is surjective for $d = 2$ and injective for $d \geq 3$. Noting that $\mathcal{N}_X \cong \mathcal{O}_X(d)$ and using the sequence above, this is a consequence of the following lemma.

Lemma 20.3. *Let* $f \in k[x_0, \ldots, x_n]$ *be a homogeneous polynomial of degree* $d \geq 3$ *whose zero scheme is a nonsingular hypersurface in* $\mathbb{P}^n$ *and assume that* char $k \nmid d$. *Let* f_i, $i = 0, \ldots, n$, *be the partial derivatives of* f. *Then the forms* $x_i f_j$, *for* $i, j = 0, \ldots, n$, *are linearly independent forms of degree* d.

Proof. Since the zero scheme of f is nonsingular, the subset of $\mathbb{P}^n$ defined by $(f, f_0, \ldots, f_n)$ is empty. The Euler relation $d \cdot f = \sum x_i f_i$ shows that this ideal is the same as the ideal $(f_0, \ldots, f_n)$. Therefore it is primary for the maximal ideal $(x_0, \ldots, x_n)$, and the f_i form a regular sequence. Now the exactness of the Koszul complex shows that the relations among the f_i are generated by the relations $f_i f_j - f_j f_i = 0$. Since $d \geq 3$, there are no relations with linear coefficients.

Example 20.2.2 (continued). Now, using the fact that $H^1(\mathcal{N}_X) = H^1(\mathcal{O}_X(d)) = 0$ for any surface in $\mathbb{P}^3$, we can construct the following table for the dimensions of the groups of (20.2):

d	$h^0(T_X)$	$h^0(T_{\mathbb{P}^3}|_X)$	$h^0(\mathcal{N}_X)$	$h^1(T_X)$	$h^1(T_{\mathbb{P}^3}|_X)$
2	6	15	9	0	0
3	0	15	19	4	0
4	0	15	34	20	1
≥ 5	0	15	large	large	0

For $d = 2$, the quadric surface X has no abstract deformations, i.e., it is rigid (5.3.1). On the other hand, it has a 6-dimensional family of automorphisms, since $X \cong \mathbb{P}^1 \times \mathbb{P}^1$. The chart shows a 9-dimensional family of surfaces in $\mathbb{P}^3$, any two related by an automorphism of $\mathbb{P}^3$.

For $d \geq 3$ we obtain $h^0(T_X) = 0$. There are no infinitesimal automorphisms, and so the functor F_2 of abstract deformations is also pro-representable.

Excepting the case $d = 4$ (for which see below), every abstract deformation of a surface of degree d in $\mathbb{P}^3$ is realized as a deformation inside $\mathbb{P}^3$. Indeed, since both functors are pro-representable, F_1 has no obstructions, and the map $t_{F_1} \to t_{F_2}$ is surjective, it follows that the morphism of functors $F_1 \to F_2$ is strongly surjective (Ex. 15.8).

The same exact sequences as above show also that $h^2(T_X) = 0$ for $d \leq 5$, but $h^2(T_X) \neq 0$ for $d \geq 6$. Thus at least for $2 \leq d \leq 5$ the abstract deformations are unobstructed. For $d \geq 6$ see (Ex. 20.1).

Example 20.3.1. We examine more closely the case of a nonsingular surface of degree 4 in $\mathbb{P}^3$, which is a $K3$ surface. The functor of embedded deformations is unobstructed, since $h^1(\mathcal{N}_X) = 0$ as noted above. Also the functor F_1 is pro-representable, so the universal family is defined by a complete regular local ring R_1 of dimension 34. The abstract deformations are also unobstructed, as noted above, so the functor F_2 is pro-represented by a complete regular local ring R_2 of dimension 20. The induced map on the Zariski tangent spaces of the morphism $\operatorname{Spec} R_1 \to \operatorname{Spec} R_2$ is not surjective, however, as we see from the table above: its image has only dimension 19. Computing the image step by step, we see that the image, which corresponds to abstract deformations that lift to embedded deformations, is a smooth subspace of $\operatorname{Spec} R_2$ of dimension 19. In particular, there are abstract deformations of X_0 that cannot be realized as embedded deformations in $\mathbb{P}^3$. (Over the complex numbers, this corresponds to the fact that there are complex manifold $K3$ surfaces that are not algebraic.)

Example 20.3.2. Using (20.3.1), we can give an example of an obstructed deformation of a line bundle. Let X_0 be a nonsingular quartic surface in $\mathbb{P}^3$. Let X be a deformation over the dual numbers D that does not lift to $\mathbb{P}^3$. Let $\mathcal{L}_0$ be the invertible sheaf $\mathcal{O}_{X_0}(1)$. I claim that $\mathcal{L}_0$ does not lift to X. For suppose it did lift to an invertible sheaf $\mathcal{L}$ on X. Then the exact sequence

$$0 \to \mathcal{L}_0 \to \mathcal{L} \to \mathcal{L}_0 \to 0$$

and $H^1(\mathcal{O}_{X_0}(1)) = 0$ would show that the sections $x_0, x_1, x_2, x_3 \in H^0(\mathcal{L}_0)$ that define the embedding $X_0 \subseteq \mathbb{P}^3$ lift to $\mathcal{L}$. Using these sections we would obtain a morphism of X to $\mathbb{P}^3_D$, which must be a closed immersion by flatness. Thus X lifts to $\mathbb{P}^3_D$, a contradiction. So this is an example in which the obstruction in $H^2(\mathcal{O}_{X_0})$ to lifting $\mathcal{L}_0$ is nonzero (6.4).

For another approach to the deformations of the quartic surface in $\mathbb{P}^3$, we use the theory of deformations of a scheme together with a line bundle (Ex. 10.6).

Proposition 20.4. *Let X be a nonsingular quartic surface in $\mathbb{P}^3$ over a field k of characteristic 0. Then for any nontrivial line bundle $\mathcal{L}$ on X, there is an abstract deformation X' of X over the dual numbers to which $\mathcal{L}$ does not lift.*

Proof. Recall (Ex. 10.6) that deformations of the pair $(X, \mathcal{L})$ are given by $H^1(X, \mathcal{P}_\mathcal{L})$ and that there is an exact sequence

$$\cdots \to H^1(\mathcal{O}_X) \to H^1(\mathcal{P}_\mathcal{L}) \to H^1(\mathcal{T}_X) \overset{\delta}{\to} H^2(\mathcal{O}_X) \to \cdots,$$

where for any deformation $\tau \in H^1(\mathcal{T}_X)$ of X, the obstruction to extending $\mathcal{L}$ over that deformation of X is given by $\delta(\tau) \in H^2(\mathcal{O}_X)$. Furthermore, $\delta(\tau)$ is the cup product of τ with the cohomology class $c(\mathcal{L}) \in H^1(\Omega_X^1)$ via the pairing $\mathcal{T}_X \otimes \Omega_X^1 \to \mathcal{O}_X$.

Thus to prove the proposition, we must show that there is a $\tau \in H^1(\mathcal{T}_X)$ with $\delta(\tau) \neq 0$. Since $H^2(\mathcal{O}_X)$ is one-dimensional, we need to show that the map δ is nonzero. Since the canonical class of X is zero, this map δ is dual to a map $H^0(\mathcal{O}_X) \to H^1(\Omega_X^1)$, which, because of how the exact sequence $0 \to \mathcal{O}_X \to \mathcal{P}_\mathcal{L} \to \mathcal{T}_X \to 0$ was defined, is none other than the map sending 1 to $c(\mathcal{L})$, the cohomology class of $\mathcal{L}$. Thus we need only show that $c(\mathcal{L})$ is not zero, and this is a consequence of the following lemma.

Lemma 20.5. *Let X be a nonsingular surface in $\mathbb{P}^3$ over a field k of characteristic 0. Let $\mathcal{L}$ be a nontrivial line bundle. Then $c(\mathcal{L}) \in H^1(X, \Omega^1_X)$ is not zero.*

Proof. The formation of the cohomology class is compatible with intersection theory on the surface [57, V, Ex. 1.8], in the sense that for any two divisor classes D, E on X, the cup product of the cohomology classes $c(\mathcal{O}_X(D))$ and $c(\mathcal{O}_X(E))$ from $H^1(\Omega^1_X) \times H^1(\Omega^1_X)$ to $H^2(\omega_X) = k$ is just $(D.E) \cdot 1$ in the field, where $(D.E)$ is the intersection number.

Let H be an ample divisor on X. If D is any divisor of degree $\neq 0$, then $(H.D) \neq 0$, and since char $k = 0$, we obtain $c(\mathcal{O}_X(D)) \neq 0$ in $H^1(\Omega^1_X)$. On the other hand, if $(H.D) = 0$, but $D \neq 0$, then by the Hodge index theorem [57, V, 1.9], it follows that $D^2 < 0$, since Pic X is discrete and has no torsion. Then $c(\mathcal{O}_X(D)) \cup c(\mathcal{O}_X(D)) = (D^2) \cdot 1 \neq 0$, so again $c(\mathcal{O}_X(D)) \neq 0$.

Taking the same ideas a step farther, we can prove the theorem of Noether for quartic surfaces.

Theorem 20.6 (Noether). *A very general quartic surface X in $\mathbb{P}^3$ over an uncountable algebraically closed field k of characteristic 0 contains only curves that are complete intersections with other surfaces in $\mathbb{P}^3$. (Here very general means that we must avoid a countable union of proper closed subsets of the parameter space.)*

Proof. Let H denote the hyperplane class on X. Since X is projectively normal, any curve C linearly equivalent to mH for some m is a complete intersection. So if X contains a curve C that is not a complete intersection, then C and H are linearly independent in Pic X. (Here we use the fact that Pic $X/\mathbb{Z}H$ is torsion-free (Ex. 20.7).)

The first step is to show that there is a deformation of X in $\mathbb{P}^3$ over the dual numbers to which C does not lift. Let $d = \deg C$ and consider the divisor $D = 4C - dH$. Then $D.H = 0$, and $D \neq 0$. Let $\mathcal{L} = \mathcal{O}_X(D)$, and let X' be a deformation of X as an abstract surface to which $\mathcal{L}$ does not lift (20.4). If $\tau \in H^1(\mathcal{T}_X)$ corresponds to X', then as in (20.4), $\delta(\tau) = \tau \cup c(\mathcal{L}) \neq 0$. Now on the quartic surface, $\mathcal{T}_X \cong \Omega^1_X$ and $\omega_X \cong \mathcal{O}_X$, so the pairing $H^1(\mathcal{T}_X) \times H^1(\Omega^1_X) \to H^2(\mathcal{O}_X)$ is the same as the pairing $H^1(\Omega_X) \times H^1(\Omega_X) \to H^2(\omega_X)$, which is compatible with intersection pairing. Since $c(\mathcal{L}) \cup c(\mathcal{O}_X(1)) = 0$, we can choose τ in such a way that $\delta(\tau) \neq 0$ and $\tau \cup c(\mathcal{O}_X(H)) = 0$, so that $\mathcal{O}_X(H) = \mathcal{O}_X(1)$ lifts to X'. Using this lifting we can embed X' in $\mathbb{P}^3$, as in (20.3.2). So we have found a deformation X' of X as a closed subscheme of $\mathbb{P}^3$ to which $\mathcal{L}$ does not lift. It follows that the curve C does not lift to X', for if it did, the corresponding invertible sheaf $\mathcal{O}_X(C)$ would also lift (Ex. 6.7) and then so would $\mathcal{L} \cong \mathcal{O}_X(4C) \otimes \mathcal{O}_X(-d)$.

Let d, g be the degree and genus of C. The next step is to consider the Hilbert-flag scheme $H\{C, X\}$ of curves of degree d and genus g in quartic surfaces X in $\mathbb{P}^3$. By the lemma (20.7) below, there are no complete intersection curves on X with this degree and genus. We will show that the forgetful morphism from $H\{C, X\}$ to the Hilbert scheme $H\{X\} \cong \mathbb{P}^{34}$ of quartic surfaces is not surjective. Suppose to the contrary. Then some reduced irreducible component $Z \subseteq H\{C, X\}$ would map surjectively to $H\{X\}$. Let U be an open dense subset of Z that is nonsingular, and let $f : U \to H\{X\}$ be the induced map. By the theorem of generic smoothness [57, III, 10.7] there is an open subset $V \subseteq H\{X\}$ such that $f : f^{-1}(V) \to V$ is smooth. Take a pair $C \subseteq X$ corresponding to a point $u \in f^{-1}(V)$, and let v be its image in V. We apply the earlier argument to C and X. The deformation X' to which C does not lift gives a map of $\operatorname{Spec} D \to V$ with image centered at v, where D is the dual numbers. By the infinitesimal lifting property of smooth morphisms (Ex. 4.7), this lifts to a map of $\operatorname{Spec} D \to f^{-1}(v)$ at the point u. Pulling back the universal family over $H\{C, X\}$ we get a lifting of C to X', a contradiction. We conclude that the image of $H\{C, X\}$ in $H\{X\}$, that is, the set of quartic surfaces containing some curve of degree d and genus g, is a proper closed subset of $H\{X\}$.

Now, as the pair (d, g) ranges over all possible pairs of integers different from those of a complete intersection on a quartic surface, we find that the set of quartic surfaces in $\mathbb{P}^3$ that contain some non-complete-intersection curve is a countable union of proper closed subsets of the parameter space $\mathbb{P}^{34}$. Over the uncountable field k, a very general point of $\mathbb{P}^{34}$ will therefore correspond to a nonsingular quartic surface containing only complete intersection curves.

Lemma 20.7. *Let X be a nonsingular surface in $\mathbb{P}^3$ (of any degree), and let C be an effective Cartier divisor on X having the same degree and genus as a complete intersection of X with another surface. Then C itself is a complete intersection of X with another surface.*

Proof. Let H be a hyperplane class, and let D be a complete intersection curve having the same degree and genus as C. Since C and D have the same degree, $C.H = D.H$. Furthermore, since $D \sim mH$ for some H, this implies $C.D = D^2$. Also the canonical class K_X is a multiple of H, so $C.K_X = D.K_X$. Now since C and D have the same genus, using the adjunction formula $2g_C - 2 = C.(C + K_X)$, and ditto for D, we find that $C^2 = D^2$. Now consider the divisor $C - D$. Note that $(C - D).H = 0$ and $(C - D)^2 = C^2 - 2C.D + D^2 = 0$. Hence by the Hodge index theorem [57, V, 1.9], we conclude that $C - D$ is numerically equivalent to zero. But since $\operatorname{Pic} X$ is discrete and has no torsion, it follows that C is linearly equivalent to D. Then since X is projectively normal, C is the intersection of X with some other surface.

References for this section. What we call Noether's theorem appears in his treatise on space curves [125, §11, p. 58] as a fact mentioned in a subordinate clause that he uses in his study of the dimensions of families of algebraic space curves. It appears again [ibid, p. 64] with reference to its earlier mention, but in neither case is there any justification or hint of a proof. Fano

[30] says "it has been observed in the researches of M. Noether, based essentially on a computation of constants, that a general surface of any degree ≥ 4 contains no other curves than complete intersections." Lefschetz [94] claims to give "the first complete proof of a theorem first considered by Noether." Later [ibid, p. 359] he says, "For ordinary space this theorem has been stated and proved many years ago by Noether, but his proof, based on enumeration of constants, has long been considered unsatisfactory." The German encyclopedia article [143, p. 1329] mentions this result, saying that "the proof of M. Noether is based on counting constants. A rigorous proof has been given by S. Lefschetz." So we see that what Noether simply mentioned as a fact, with the passage of time is attributed to him as a theorem with a proof.

The only classical proof that I could find is in a paper of Rohn [142]. He refers to Noether and Halphen in his introduction, but does not refer to Noether on pp. 641, 642 when he states the theorem. His proof by counting constants starts out well enough, along the lines of (Ex. 20.6), but gets more difficult to justify as he deals with families of space curves of larger degree.

Lefschetz's proof [94], [95] uses topological methods for surfaces defined over $\mathbb{C}$. He considers a pencil of general nonsingular surfaces containing special fibers with nodes as singularities (a "Lefschetz pencil") and studies the monodromy of cycles for a small loop in the parameter space around a singular fiber. Deligne [18] brings Lefschetz's proof up to date by reinterpreting it in l-adic cohomology over an arbitrary base field. He proves the theorem in all characteristics, where now "general" means a surface defined by an equation with indeterminate coefficients. Besides these, there have been many papers giving new proofs and refinements of the theorem [34], [108], [16], [22], [43], [40], [41], [168], [169], [170], [98], [28], [29], [75], [107]. The proof given here is my own, though I am sure all the ideas have been used before.

Exercises.

20.1.   For $d \leq 5$ we have seen that the abstract deformations of a nonsingular surface X of degree d in $\mathbb{P}^3$ are unobstructed. Show that this holds also for $d \geq 6$, even though $H^2(\mathcal{T}_X) \neq 0$. Do this by lifting an abstract deformation to an embedded deformation and then deforming the embedded deformation.

20.2.   Make a similar analysis of nonsingular curves in $\mathbb{P}^2$, comparing their embedded deformations to their abstract deformations. Show that for $d \leq 4$, every abstract deformation of the curve lifts to an embedded deformation, but that for $d \geq 5$ this no longer holds.

20.3.   In the proof of (20.6), to show that the map of schemes $H\{C, X\}$ to $H\{X\}$ is not surjective, it is not sufficient to say merely that the induced map on Zariski tangent spaces is not surjective.

(a) Give some examples of surjective morphisms of schemes $f : X \to Y$ where the induced map on tangent spaces is not surjective.
(b) Show, however, that if f is smooth and surjective, then the induced map on tangent spaces is surjective.

20.4. Extend the proof of Noether's theorem (20.6) to surfaces X of any degree $d \geq 5$ in $\mathbb{P}^3$.

(a) Show that the problem reduces to the following: If $\delta_{\mathcal{L}} : H^1(\mathcal{T}_X) \to H^2(\mathcal{O}_X)$ is the map coming from the exact sequence of the sheaf of principal parts of $\mathcal{L}$, then $\delta_{\mathcal{L}} = 0$ if $\mathcal{L} \cong \mathcal{O}_X(1)$, while $\delta_{\mathcal{L}} \neq 0$ for any line bundle $\mathcal{L}$ linearly independent of $\mathcal{O}_X(1)$ in $\operatorname{Pic} X$.

(b) By duality, remembering that $K_X = \mathcal{O}_X(d - 4)$, show that this is equivalent to saying that multiplication by any polynomial f of degree $d - 4$ from $H^1(\Omega_X^1) \to H^1(\Omega_X^1(d - 4))$ kills $c(\mathcal{O}_X(1))$, but that for $c(M)$, where M is linearly independent of L, there exists some f with $f.c(M) \neq 0$.

(c) To prove (b), use the cohomology of the sequence

$$0 \to \mathcal{O}_X(-d) \to \Omega_{\mathbb{P}^3}^1 \otimes \mathcal{O}_X \to \Omega_X^1 \to 0$$

and the same sequence twisted by $d - 4$, and note that $c(\mathcal{O}_X(1))$ comes from $c(\mathcal{O}_{\mathbb{P}^3}(1))$, hence is in the image of $H^1(\Omega_{\mathbb{P}^3}^1 \otimes \mathcal{O}_X)$.

20.5. Use the method of proof of (20.6) to prove a result of Mori [110]: if there exists a nonsingular quartic surface X_0 containing a nonsingular curve C_0 of degree d and genus g, then there also exists a nonsingular quartic surface X_1 containing a nonsingular curve C_1 of the same degree and genus, and with the property that $\operatorname{Pic} X_1$ is generated by C_1 and H, the hyperplane class.

20.6. Quartic surfaces containing various curves.

(a) Show that the family of those quartic surfaces that contain a line has dimension 33 inside the parameter space $\mathbb{P}^{34}$ of all quartic surfaces in $\mathbb{P}^3$. Do this by counting for each fixed line the quartic surfaces containing that line, and adding the dimension of the family of lines in $\mathbb{P}^3$.

(b) Make a similar argument for nonsingular, nonspecial, ACM curves of any degree ≤ 8 in $\mathbb{P}^3$. In this case you need to take into account the dimension of the linear system of those curves on a fixed quartic surface. In each case the family of quartic surfaces containing curves of this type has dimension 33.

(c) For surfaces of degree $d \geq 5$ in $\mathbb{P}^3$, show that the codimension of the family of surfaces containing lines in the family of all surfaces of degree d is $d - 3$.

(d) For surfaces of degree $d \geq 4$ in $\mathbb{P}^3$, the codimension of the family of surfaces containing conics is $2d - 7$.

20.7. If X is a complete intersection surface in $\mathbb{P}^n$ over a field k of characteristic 0, show that $\operatorname{Pic} X / \mathbb{Z}H$ has no torsion, where H is the hyperplane class. To do this, follow Grothendieck's proof of the fact that a complete intersection scheme X of dimension ≥ 3 in $\mathbb{P}^n$ has $\operatorname{Pic} X \cong \mathbb{Z}$, generated by H [55, Ch IV, 3.1]. The difference, in case X is a surface, is that $H^2(X, \mathcal{I}^n/\mathcal{I}^{n+1})$ may no longer be zero. But over a field k of characteristic zero it is a torsion-free abelian group. Use this to show that any torsion class (mod H) would lift to $\mathbb{P}^n$ and so be equal to zero.

21. Algebraization of Formal Moduli

The question is, once we have a formal family of deformations of some object, either a versal, a miniversal, or a universal family in the sense of this chapter, can we extend it to an actual family of deformations, defined over a scheme of finite type over k? And this family, if it exists, does it inherit the good properties of the formal family (versal, miniversal, universal)? What we are asking, in effect, is, can we leverage our way from the formal family to a global moduli space?

To fix the ideas, suppose that X_0 is a given scheme over k and that we have a versal family of deformations of X_0. This means we have a complete local k-algebra $R, \mathfrak{m}$ with residue field k, and we have a compatible collection of schemes $X_0 \subseteq X_1 \subseteq X_2 \subseteq \cdots$, where for each n, X_n is a deformation of X_0 over $R_n = R/\mathfrak{m}^{n+1}$, with the versal property for arbitrary deformations of X_0 over Artin rings, as described in §15. The first step is to create a formal scheme as limit of the schemes X_n.

Proposition 21.1. *Let $R, \mathfrak{m}$ be a complete local ring with residue field k, and suppose we are given a formal family of deformations of X_0 over R, that is, for each n, schemes X_n flat and of finite type over $R_n = R/\mathfrak{m}^{n+1}$ and maps $X_n \to X_{n+1}$ inducing isomorphisms $X_n \xrightarrow{\sim} X_{n+1} \otimes_{R_{n+1}} R_n$. Then there is a noetherian formal scheme $\mathcal{X}$, flat over $\operatorname{Spf} R$, the formal spectrum of R, such that for each n, $X_n \cong \mathcal{X} \times_R R_n$.*

Proof. We define $\mathcal{X}$ to be the locally ringed space formed by taking the topological space X_0, together with the sheaf of rings $\mathcal{O}_\mathcal{X} = \varprojlim \mathcal{O}_{X_n}$ [57, II, 9.2]. To show that $\mathcal{X}$ is a noetherian formal scheme, we must show that $\mathcal{X}$ has an open cover $\mathcal{U}_i$, such that on each $\mathcal{U}_i$, the induced ringed space is obtained as the formal completion of a scheme U_i along a closed subset Z_i.

Let U be an open affine subset of X_0, with $U = \operatorname{Spec} B_0$. Then for each n the restriction of X_n to U will be $\operatorname{Spec} B_n$ for a suitable ring B_n. Furthermore, the rings B_n form a surjective inverse system with $\varprojlim B_n = B_\infty$, and $H^0(U, \mathcal{O}_\mathcal{X}) = B_\infty$.

Take a polynomial ring $A_0 = k[x_1, \ldots, x_n]$ together with a surjective map $A_0 \to B_0$. For each n, let $A_n = R_n[x_1, \ldots, x_n]$. Lifting the images of x_i we get a surjective map $A_n \to B_n$, with kernel I_n. Because of the flatness of B_n over R_n, we find that the inverse system $\{I_n\}$ is also surjective, and hence [57, II, 9.1] the map of inverse limits $\varprojlim A_n \to B_\infty$ is also surjective. Now $\varprojlim A_n$ is equal to $R\{x_1, \ldots, x_n\}$, the convergent power series in $x_1, \ldots, x_n$ over R, which is a noetherian ring, so B_∞ is a noetherian ring also, complete with respect to the $\mathfrak{m}B_\infty$-adic topology, and each $B_n = B_\infty/\mathfrak{m}^n B_\infty$. Thus we see that the ringed space $(U, \mathcal{O}_\mathcal{X}|_U)$ is just the formal completion of $\operatorname{Spec} B_\infty$ along the closed subset U. Such open sets U cover X_0, so by definition $(\mathcal{X}, \mathcal{O}_\mathcal{X})$ is a noetherian formal scheme.

Remark 21.1.1. If in addition we are given a collection of coherent sheaves $\mathcal{F}_n$ on X_n (resp. locally free, resp. invertible), flat over R_n, and maps $\mathcal{F}_n \cong \mathcal{F}_{n+1} \otimes_{R_{n+1}} R_n$, then $\mathcal{F} = \varprojlim \mathcal{F}_n$ will be a coherent (resp. locally free, resp. invertible) sheaf on $\mathcal{X}$ [57, II, 9.6].

The next step, once we have a formal scheme $\mathcal{X}$ over Spf R, is to ask whether there exists a scheme X, flat and of finite type over R, whose formal completion along the closed fiber is $\mathcal{X}$. In this case, following Artin, we say that the formal scheme $\mathcal{X}$ is *effective* (in [57, II, 9.3.2] this property was called *algebraizable*). This is not always possible (21.2.1), and in that case we can go no further. However, there is a good case in which it is possible, namely when $\mathcal{X}$ is projective, thanks to an existence theorem of Grothendieck.

Theorem 21.2 (Grothendieck). *Let $\mathcal{X}$ be a formal scheme, proper over Spf R, where $R, \mathfrak{m}$ is a complete local ring, and suppose there exists an invertible sheaf $\mathcal{L}$ on $\mathcal{X}$ such that $\mathcal{L}_0 = \mathcal{L} \otimes_R k$ is ample on $X_0 = \mathcal{X} \otimes_R k$. Then there exists a scheme X over R, together with an ample line bundle L, such that $\mathcal{X} = \hat{X}$ and $\mathcal{L} = \hat{L}$, taking completions along the closed fiber over R. In particular, $\mathcal{X}$ is effective.*

Proof. We refer to [48, III, 5.4.5] for the proof.

Example 21.2.1 (A noneffective formal deformation). This example shows that in the theorem, it is not enough to assume X_0 projective. It must admit an ample invertible sheaf that lifts to the formal scheme $\mathcal{X}$.

Let X_0 be a nonsingular quartic surface in $\mathbb{P}^3$ over a field of characteristic zero. Then we have seen (20.4) that for any nonzero divisor D on X_0, there is some deformation of X, already over the dual numbers, to which $L = \mathcal{O}_X(D)$ does not lift. We know from (18.1) that there is a miniversal formal family of deformations of X_0 and hence a formal scheme $\mathcal{X}$ as in (21.1). However, (20.4) shows that no ample line bundle on X_0 lifts to $\mathcal{X}$, and so (21.2) does not apply.

I claim in this case that the formal family $\mathcal{X}$ is actually not effective. For suppose there were a scheme X over Spec R with $\hat{X} \cong \mathcal{X}$. Then X would be a family of smooth surfaces over R. Since $H^2(\mathcal{T}_{X_0}) = 0$ (20.2.2), there are no obstructions to deforming X_0, so R will be a power series ring. Thus the total space X is a regular scheme. Take an open affine piece U that meets X_0, and take a hyperplane section $Y \subseteq U$ that also meets X_0. Then the closure $\bar{Y}$ of Y in X is a subscheme of codimension one. Since the local rings of X are regular, $\bar{Y}$ is a Cartier divisor, whose intersection with X_0 is a nonzero effective divisor. Then the corresponding invertible sheaf L lifts to $\mathcal{X}$ by construction, a contradiction. Thus $\mathcal{X}$ is not effective.

Let us see how the theorem applies to our standard situations.

Example 21.2.2. In Situation A we start with a closed subscheme $X_0 \subseteq \mathbb{P}^n_k$. The Hilbert functor of deformations of X_0 as a closed subscheme of $\mathbb{P}^n$ is pro-representable (17.1). Then by (21.1) we obtain a formal scheme $\mathcal{X} \subseteq \hat{\mathbb{P}}^n_R$. It is projective by construction, so (21.2) applies, and the formal family is effective.

Example 21.2.3. In Situation B we start with an invertible sheaf $\mathcal{L}_0$ on a scheme X_0. Assuming X_0 projective and $H^0(\mathcal{O}_{X_0}) = k$, the local Picard functor of deformations of $\mathcal{L}_0$ on X_0 is pro-representable (17.2), so by (21.1.1) we obtain an invertible sheaf $\hat{\mathcal{L}}$ on $(X_0 \times R)^\wedge$. We use the theorem of Grothendieck [48, III, 5.1.4], which says that if X is a scheme, projective over a complete local ring R, and if $\hat{X}$ is the completion along the closed fiber, then the functor $\mathcal{F} \mapsto \hat{\mathcal{F}}$ is an equivalence of the category of coherent sheaves on X with the category of coherent sheaves on $\hat{X}$. In our case, this guarantees that there is an invertible sheaf $\mathcal{L}$ on $X_0 \times R$ whose completion is isomorphic to $\hat{\mathcal{L}}$. Thus the formal family of invertible sheaves is effective.

Example 21.2.4. In Situation C we start with a projective scheme X_0 and a locally free sheaf $\mathcal{E}_0$ on X_0. The local deformation functor then has a miniversal family (19.1), which is universal if $\mathcal{E}_0$ is simple (19.2). For the same reason as in the previous example, we conclude that the formal family is effective.

Example 21.2.5. In Situation D, suppose we start with an (abstract) projective scheme X_0. Then the local deformation functor has a miniversal family (18.1), which is universal if $H^0(X_0, \mathcal{T}_{X_0}) = 0$ (18.3). Since X_0 is projective, there is an ample invertible sheaf L_0 on X_0. But in order to apply (21.2), we must be able to extend L_0 to an invertible sheaf $\mathcal{L}$ on the formal scheme $\mathcal{X}$. If $H^2(\mathcal{O}_{X_0}) = 0$, then by (6.4) there are no obstructions to deformations of L_0, and we obtain the desired $\mathcal{L}$. In this case the formal family of deformations of X_0 is effective, as in the case of (21.2.1). This applies in particular to any projective curve, since then there is no H^2. If X_0 is a nonsingular projective surface, then $H^2(\mathcal{O}_{X_0})$ is dual to $H^0(K_{X_0})$, whose dimension is the *geometric genus* p_g of X_0. So for surfaces with $p_g = 0$, the formal family of deformations is effective. This applies in particular to rational, ruled, and Enriques surfaces. However, for other surfaces, such as $K3$ surfaces and abelian surfaces, this method fails, and in fact the formal family of deformations is not effective, as in the case of (21.2.1). This is a serious impediment to the construction of moduli, and is the principal reason why one considers *polarized varieties* instead of "bare" varieties with no polarization [119, Ch 5, §1].

Example 21.2.6. As a special case of the previous example, Artin [7, 5.5] shows that the formal deformations of a surface of general type X_0 are effective, using the fact that a sufficiently high multiple of the canonical bundle defines a birational morphism to some $\mathbb{P}^n$, and this map extends to infinitesimal deformations.

The third and most difficult step is to descend from a family over a complete local ring R to a family defined over a scheme of finite type over k. More specifically, we ask for a scheme S of finite type over k, a scheme X flat and of finite type over S, and a point $s_0 \in S$, such that the fiber of X over s_0 is X_0, and the formal completion of X along X_0 is isomorphic to the formal scheme $\mathcal{X}$ above. This problem was addressed in a series of deep papers by

Michael Artin [4], [6], [7]. To explain his work in detail would carry us too far afield, so we will merely describe his results briefly. His main result, stated in the special case of deformations of a scheme X_0 as above, is the following:

Theorem 21.3 ([4, 1.6, 1.7]). *Let X_0 be a projective scheme over k, and assume that X_0 admits an effective formal versal deformation $\bar{X}$ over the complete local ring R. Then $\bar{X}$ is algebraizable in the following sense: there exists a scheme S of finite type over k, a point $s_0 \in S$, and a flat finite-type family X over S, with fiber X_0 over s_0, such that $R \cong \hat{\mathcal{O}}_{S,s_0}$ and $\bar{X}$ is isomorphic to $X \times_S \operatorname{Spec} R$. Furthermore, the triple (X, S, s_0) is unique locally around s_0 in the étale topology, meaning that if (X', S', s_0') is another such, then there exist an S'' with a point s_0'' and an étale morphism $S'' \to S$, $S'' \to S'$ sending s_0'' to s_0, s_0' respectively such that $X \times_S S'' \cong X' \times_{S'} S''$.*

Note that the resulting family X/S is unique only to within étale coverings, so we do not get an actual moduli space by this method. However, as so often happens in mathematics, when you don't get what you want, make a definition! In the absence of automorphisms, the notion of *algebraic space* [3], [85] works well; in the general case one needs the notion of *stack* [21]. In both cases, the definition is crafted in such a way that having these algebraic families up to local étale isomorphism is enough to determine the structure. In this connection, see the discussion of *modular families* in §§26, 27. Much of the language of schemes carries over to the categories of algebraic spaces and of stacks, so that one can think of these notions as generalizations of the concept of scheme and can work with moduli problems in this more general context. The question whether the moduli space is actually a scheme can be deferred until later, or even ignored. The only drawback of this approach is the considerable technical baggage necessary to work in these larger categories, and for this reason we have avoided them in this book (but see (27.7.1) for some comments on stacks).

Example 21.3.1. If X_0 is a projective curve, the miniversal formal family $\bar{X}$ of local deformations is effective (21.2.5), so by the theorem it is algebraizable. In the case of nonsingular curves of genus $g \geq 2$, we will find another proof of this result using a modular family (27.2).

Remark 21.3.2. For another example illustrating the ideas of this section, consider the problem, given a flat family X/T of schemes and an invertible sheaf $\mathcal{L}_0$ on a special fiber X_0, of lifting $\mathcal{L}_0$ to an invertible sheaf on the whole family. Of course if $\dim X_0 \geq 2$, there may be obstructions to lifting $\mathcal{L}_0$ (20.3.2). But if $\dim X_0 = 1$, then $H^2(\mathcal{O}_{X_0}) = 0$ and there are no obstructions to infinitesimal liftings of $\mathcal{L}_0$ (6.4). Thus we obtain a sheaf $\hat{\mathcal{L}}$ on the formal completion of X along X_0, and as in (21.2.3) we obtain an invertible sheaf $\bar{\mathcal{L}}$ on the scheme $X \times_{\mathcal{O}_{T,0}} \operatorname{Spec} \hat{\mathcal{O}}_{T,0}$. By analogy with Artin's theorem (21.3), we may then expect to find an $\mathcal{L}$ on X after some étale base extension $T' \to T$. In fact, this is true, and we can give a direct proof without using Artin approximation (Ex. 21.4).

While the above discussion concerned projective varieties, another of Artin's approximation results is purely local: it tells us that a formal isomorphism can be approximated by étale morphisms.

Theorem 21.4 ([5, 2.6]). *Let S be a scheme of finite type over a field k, let X_1 and X_2 be schemes of finite type over S, let $x_i \in X_i$ be points, and suppose that there is an isomorphism of completions of the local rings $\hat{O}_{X_1,x_1} \cong \hat{O}_{X_2,x_2}$ over S. Then X_1 and X_2 are locally isomorphic in the étale topology, that is, there is another scheme X' of finite type over S together with a point $x' \in X'$, and there are étale morphisms $X' \to X_1$ and $X' \to X_2$ sending x' to x_1 and x_2, respectively, and inducing isomorphisms on the residue fields at x', x_1, x_2.*

References for this section. Grothendieck's study of formal schemes and comparison with algebraic schemes first appeared in [44] and was later incorporated into [48, III, §5]. For a recent survey of this work, see [74]. Artin's work on algebraization occurs in a series of papers [3], [4], [5], [6], [7]. For an analogous situation comparing algebraic varieties over $\mathbb{C}$ with complex analytic spaces, see [153].

Exercises.

21.1. Let X_0 be a nonsingular surface of degree $d \geq 5$ in $\mathbb{P}^3$.
(a) Show that the functor of local deformation of X_0 is pro-representable.
(b) Show that the formal deformation space $\mathcal{X}$ of X_0 is effective.
(c) Show that another surface $X_1 \subseteq \mathbb{P}^3$ is isomorphic to X_0 as an abstract surface if and only if X_1 is obtained from X_0 by an automorphism of $\mathbb{P}^3$. Conclude that the subset of the Hilbert scheme H consisting of surfaces isomorphic to X_0 is a smooth subvariety V of dimension 15.
(d) Note that H is isomorphic to $\mathbb{P}^N$ with $N = \binom{d+3}{3} - 1$. Let $S \subseteq H$ be a linear space of dimension $N-15$ passing through the point $x_0 \in H$ corresponding to X_0, and transversal to the subvariety V at x_0. Show that S with the induced family X is an algebraic family whose formal completion along the fiber above x_0 is isomorphic to $\mathcal{X}$. Thus (X, S, x_0) satisfies the conclusion of Artin's algebraization theorem (21.3).

21.2. In the situation of (21.2), show that the scheme X over R is unique up to isomorphism. *Hint:* Use the theorem of Grothendieck mentioned in (21.2.3). Apply this to the graph of the isomorphism $\hat{X}_1 \cong \hat{X}_2$ for two choices X_1, X_2, and deduce an isomorphism of X_1 and X_2.

21.3. Using (Ex. 10.7), adapt the argument of (21.2.1) to show that if X_0 is an abelian surface, the functor of local deformations is unobstructed and pro-representable, giving rise to a formal scheme $\mathcal{X}$ over a power series ring R of dimension 4. However, the formal family $\mathcal{X}$ is not effective.

21.4. Let X/T be a flat family of reduced projective curves over a nonsingular curve T. Let $0 \in T$ be a point, and suppose we are given an invertible sheaf $\mathcal{L}_0$ on the fiber X_0 over 0. Show that there is an étale morphism $T' \to T$ whose image contains 0 and there is an invertible sheaf $\mathcal{L}'$ on $X' = X \times_T T'$ restricting to $\mathcal{L}_0$ on X_0.

(a) Since X/T is projective, it has a relatively ample sheaf. Tensoring by this one, we reduce to the case that $\mathcal{L}_0$ is very ample on X_0, hence can be represented by a divisor D_0 consisting of distinct nonsingular points of X_0.

(b) If $P \in X_0$ is a nonsingular point, the total space X is a nonsingular surface in a neighborhood of P, so we can find a curve C on X intersecting X_0 transversally at P (and at other points). Replacing C by an open neighborhood of P in C, we may assume that the projection map $C \to T$ is étale. Then making the base extension $C \to T$, we obtain a section of $X \times_T C/C$ that meets X_0 in P.

(c) Applying (b) successively to each of the points of D_0, we obtain an étale map $T' \to T$ having sections corresponding to all the points of D_0. Their union is a divisor on $X' = X \times_T T'$ that defines an invertible sheaf $\mathcal{L}'$ on X' restricting to $\mathcal{L}_0$ on X_0.

21.5. Again let X/T be a flat family of reduced projective curves, let $0 \in T$ be a point, and suppose we are given a projective embedding $X_0 \hookrightarrow \mathbb{P}^n$ with the property that $h^1(\mathcal{O}_{X_0}(1)) = 0$. Show that there is an étale map $T' \to T$ whose image contains 0 and there is a closed immersion of $X' = X \times_T T' \hookrightarrow \mathbb{P}^n_{T'}$, restricting to the given embedding of X_0.

(a) Use the previous exercise to extend $\mathcal{O}_{X_0}(1)$ to an invertible sheaf $\mathcal{L}'$ on X' for a suitable étale base extension $T' \to T$.

(b) Use the hypothesis $h^1(\mathcal{O}_{X_0}(1)) = 0$ to show that the sections of $\mathcal{L}_0 = \mathcal{O}_X(1)$ defining the embedding lift to sections of $\mathcal{L}$ (after shrinking T' if necessary), and that these define a morphism of X' to $\mathbb{P}^n_{T'}$, which will be a closed immersion after possibly shrinking T' a little more.

(c) Give an example to show that the conclusion of this problem is false without the hypothesis $h^1(\mathcal{O}_{X_0}(1)) = 0$. For example, let X_0 be a plane quintic curve, and let X/T be a family of general curves of genus 6 having X_0 as a limit.

22. Lifting from Characteristic p to Characteristic 0

If we have a scheme X flat over $\operatorname{Spec} R$, where R is a ring of characteristic zero, but with residue fields of finite characteristic (for example R could be the ring of integers in an algebraic number field or the ring of Witt vectors over a field of characteristic p), then the generic fiber X_η of X over R will be a scheme over a field of characteristic zero, while a special fiber X_0 will be a scheme over a field of characteristic $p > 0$. In this case we can think of X_0 as a specialization of X_η. The lifting problem is the reverse question: Given a scheme X_0 over a field k of characteristic $p > 0$, does there exist a flat family X over an integral domain R whose special fiber is X_0 and whose general fiber is a scheme over a field of characteristic zero?

To fix the ideas, let us suppose that X_0 is a nonsingular projective variety over a perfect field k of characteristic $p > 0$. Let us fix a complete discrete valuation ring $(R, \mathfrak{m})$ of characteristic 0 with residue field k. Such a ring always exists, for example the ring of Witt vectors over k. Then we ask whether there is a scheme X, flat over R, with closed fiber X_0.

For each $n \geq 1$, let $R_n = R/\mathfrak{m}^{n+1}$. Then we have exact sequences

$$0 \to \mathfrak{m}^n/\mathfrak{m}^{n+1} \to R_n \to R_{n-1} \to 0$$

and we can use our study of infinitesimal liftings to try to lift X_0 successively to a scheme X_n flat over R_n. Note that in contrast to the equicharacteristic situation of a local ring R containing its residue field k, there is no "trivial" deformation of X_0: already the step from X_0 to X_1 may be obstructed. But we know (10.3) that the obstructions to lifting at each step lie in $H^2(X_0, T_{X_0})$, and when an extension exists, the set of all such is a torsor under $H^1(X_0, T_{X_0})$. Suppose we can find a family of liftings X_n flat over R_n for each n, with $X_n \otimes R_{n-1} = X_{n-1}$. Then by (21.1) the limit $\mathcal{X} = \varprojlim X_n$ will be a noetherian formal scheme, flat over Spf R, and restricting to X_n over each R_n.

The next problem is that while $\mathcal{X}$, as a noetherian formal scheme, is locally isomorphic to the completion of a (usual) scheme along a closed subset, it may not be globally so, in other words, it may not be effective (in the sense of §21). In general, the effectivity is a difficult question, but we can deal with it in the projective case using the theorem of Grothendieck (21.2).

Putting these results together, we can prove the following lifting theorem.

Theorem 22.1. Let X_0 be a nonsingular projective variety over a perfect field k of characteristic $p > 0$. Assume that $H^2(X_0, \mathcal{O}_{X_0}) = 0$ and $H^2(X_0, T_{X_0}) = 0$. Let $(R, \mathfrak{m})$ be a complete discrete valuation ring with residue field k. Then X_0 can be lifted to a scheme X, flat over R, with closed fiber isomorphic to X_0.

Proof. Since $H^2(X_0, T_{X_0}) = 0$, the obstructions to infinitesimal lifting are zero, so we obtain a compatible sequence of liftings X_n flat over R_n. Their limit gives a noetherian formal scheme $\mathcal{X}$ by (21.1). Since X_0 is assumed to be projective, it has an ample invertible sheaf $\mathcal{L}_0$. The obstruction to lifting an invertible sheaf lies in $H^2(X_0, \mathcal{O}_{X_0})$. Since this is zero, we may lift $\mathcal{L}_0$ to a compatible sequence of invertible sheaves $\mathcal{L}_n$ on each X_n. Then $\mathcal{L} = \varprojlim \mathcal{L}_n$ is an invertible sheaf on $\mathcal{X}$ (21.1.1) restricting to $\mathcal{L}_0$ on X_0. Now we use (21.2) to conclude that $\mathcal{X}$ is effective, hence comes from a scheme X flat over R with closed fiber X_0.

Corollary 22.2. Any nonsingular projective curve over a perfect field k_0 of characteristic $p > 0$ is liftable to characteristic zero.

We can apply the same techniques to the embedded lifting problem.

Theorem 22.3. Let X_0 be a closed subscheme of $\mathbb{P}^r_{k_0}$, with k_0 a perfect field of characteristic p. Assume that X_0 is locally unobstructed (e.g., X_0 is a locally complete intersection (9.2), or X_0 is locally Cohen–Macaulay of codimension 2 (8.5)). Assume also that $H^1(X_0, N_{X_0}) = 0$. Let R be a complete discrete valuation ring with residue field k. Then X_0 lifts to R, as a subscheme of $\mathbb{P}^r$, i.e., there exists a closed subscheme X of $\mathbb{P}^r_R$, flat over R, with $X \times_R k = X_0$.

Proof. The method is already contained in the proof of the previous theorem. Because of $H^1(X_0, \mathcal{N}_0) = 0$, one can lift X_0 stepwise to a sequence of closed subschemes X_n of $\mathbb{P}^r$, flat over R_n. The limit of these is a projective formal subscheme of $\hat{\mathbb{P}}^r_R$, which is effective by (21.2).

Remark 22.3.1. Without assuming $H^1(\mathcal{N}) = 0$, it seems to be an open question whether any nonsingular curve in $\mathbb{P}^3_k$ (or any other $\mathbb{P}^r_k$) lifts to characteristic zero as an embedded curve (cf. [26]).

Remark 22.3.2. Using (22.3) we can give a different proof of (22.1), by converting it to a problem of embedded deformations. Under the hypotheses of (22.1), take a projective embedding of X_0. Replacing this by a d-uple embedding, if necessary, we may assume that $H^1(X_0, \mathcal{O}_{X_0}(1)) = 0$. Combined with the hypotheses of (22.1), using the exact sequence of the normal bundle and the Euler sequence on projective space, this implies that $H^1(X_0, \mathcal{N}_{X_0}) = 0$. Therefore by (22.3), X_0 lifts as an embedded scheme, and a fortiori as an abstract scheme.

In theorems (22.1) and (22.3) above, the problem of lifting from characteristic p to characteristic 0 is solved in the *strong* sense, namely, given the object X_0 over k and the valuation ring R with residue field k, the lifting is possible over that ring R. One can also ask the *weak* lifting problem: given X_0 over k, does there exist a local integral domain R of characteristic 0 and residue field k_0 over which X_0 lifts to an X flat over R? Oort has given an example [130] of a curve together with an automorphism of that curve that is not liftable over the Witt vectors, but is liftable over a ramified extension of the Witt vectors. Thus the strong and the weak lifting problems are in general not equivalent. Next we will give Serre's example that even the weak lifting problem is not always possible for nonsingular projective varieties.

Theorem 22.4 (Serre). *Over an algebraically closed field k of characteristic $p \geq 5$, there is a nonsingular projective 3-fold Z that cannot be lifted to characteristic 0, even in the weak sense.*

Proof. Let k be algebraically closed of characteristic $p \geq 5$. Let $r \geq 5$, and let $G = (\mathbb{Z}/p)^r$. Then G is a finite abelian group, and by choosing elements $e_1, \ldots, e_r \in k$ that are linearly independent over $\mathbb{F}_p$, we can find an additive subgroup $G \subseteq k^+$ isomorphic to the abstract group G.

Now let N be the 5×5 matrix $(a_{ij})_{i,j=0,\ldots,4}$ defined by $a_{i,i+1} = 1$ for $i = 0, 1, 2, 3$ and $a_{ij} = 0$ otherwise. Then N is a nilpotent matrix with $N^5 = 0$. For each $t \in G \subseteq k^+$, consider the matrix $e^{tN} = I + tN + \frac{1}{2}t^2 N^2 + \frac{1}{6}t^3 N^3 + \frac{1}{24}t^4 N^4$ in $\mathrm{SL}(5, k)$. The fractions are well-defined because we assumed characteristic $k = p \geq 5$. This gives a homomorphism of the additive group G to the multiplicative group $\mathrm{SL}(5, k)$, and hence an action of G on $\mathbb{P}^4_k$. It is easy to check that the only fixed point of this action is $P_0 = (1, 0, 0, 0, 0)$.

Now $\mathbb{P}^4/G$ is a (singular) projective variety. Taking a suitable projective embedding we can find a smooth 3-dimensional hyperplane section Z. This is the required example.

To prove that Z is not liftable, we proceed as follows. First of all, let $Y \subseteq \mathbb{P}^4$ be the inverse image of Z under the quotient map $\mathbb{P}^4 \to \mathbb{P}^4/G$. Then Y is a hypersurface, stable under the action of G, and $Y/G \cong Z$. Since Z is smooth, it does not contain the image of the fixed point P_0, so G acts freely on Y, and the map $Y \to Z$ makes Y into an étale Galois cover of Z with group G.

Now suppose there is a local integral domain R of characteristic 0 with residue field k, and a scheme Z', flat over R, with $Z' \times_R k = Z$. First we show that the étale cover Y lifts.

Proposition 22.5. *Let Z be a scheme over a field k, let $Y \to Z$ be a finite étale cover, let R be a complete local ring with residue field k, and suppose there exists a scheme Z', flat over R, with $Z' \times_R k = Z$. Then there is a finite étale cover $Y' \to Z'$ (necessarily flat over R) with $Y' \times_R k = Y$.*

Proof. By definition $Y \to Z$ is a finite, affine, smooth morphism of relative dimension zero. As we showed in (4.11), for any smooth ring extension $A \to B$, the functors $T^i(B/A, M)$ are 0 for $i = 1, 2$ and for all M. If $A \to B$ is étale, then we also have $\Omega_{B/A} = 0$ and so $T^0(B/A, M) = 0$.

Thus, for a finite étale morphism, over each open affine subset of the base, the obstructions in T^2 to lifting vanish. A lifting exists, and because of $T^1 = 0$, it is unique. Thus the liftings patch together, and we get a unique lifting of the entire étale cover over each $R_n = R/\mathfrak{m}^{n+1}$. In the limit, these give an étale cover of the formal scheme $\hat{Z}'$. Since the morphism $Y \to Z$ is projective, the effectivity theorem (21.2) gives the cover Y' of Z' desired.

Proof of (22.4), continued. Using (22.5), we obtain a finite étale cover Y' of Z' that reduces to Y over Z. Because of the uniqueness in each step of lifting the étale cover, the group action G extends to Y' and makes Y' a Galois covering of Z' with group G.

Next, since Y is a hypersurface in $\mathbb{P}^4$ we have $H^i(Y, \mathcal{O}_Y) = 0$ for $i = 1, 2$. Now $H^2(Y, \mathcal{O}_Y)$ contains the obstructions to lifting an invertible sheaf, and $H^1(Y, \mathcal{O}_Y)$ tells the number of ways to lift an invertible sheaf. Since both of these are zero, the invertible sheaf $\mathcal{L} = \mathcal{O}_Y(1)$ on Y lifts uniquely to an invertible sheaf $\mathcal{L}'$ on Y'. Furthermore, since $H^1(Y, \mathcal{O}_Y(1)) = 0$, the sections of $H^0(\mathcal{O}_Y(1))$ defining the projective embedding also lift, and we find that $H^0(\mathcal{L}')$ is a free R-module of rank 5.

Since the group acts on $\mathbb{P}^4_k$ sending Y to itself, G also acts on $H^0(Y, \mathcal{O}_Y(1)) = k^5$, and this action also lifts to $H^0(Y', \mathcal{L}')$. Let K be the quotient field of R. Then we get an embedding of Y'_K in $\mathbb{P}^4_K$, and the group action G extends to an action on $\mathbb{P}^4_K$. In other words, we get a homomorphism $\varphi : G \to \mathrm{PGL}(5, K)$ compatible with the original action on $\mathbb{P}^4_k$. In particular,

φ is injective. Thus $\text{PGL}(5, K)$ contains a subgroup isomorphic to G, which is impossible as long as the rank of G is $r \geq 5$.

Hence Z cannot be lifted, as was to be shown.

References and further results. As a general reference for this section, see §8.5 of Illusie's article [74]. Theorems (21.2) and (22.1) and Corollary (22.2) first appeared in Grothendieck's Bourbaki Seminar [44] of May 1959. There he also raised the problem of liftability of smooth projective varieties, which was answered by Serre's example (22.4), in 1961 [154]. Mumford modified Serre's method to give an example of a nonliftable surface [74, 8.6.7].

Since then, many lifting problems have been studied. The two articles of Oort [129], [130] are extremely useful. He shows that finite commutative group schemes can be lifted [120], but gives an example of a noncommutative finite group scheme that cannot be lifted (as a group scheme).

Mumford [116] showed that one can lift any principally polarized abelian variety (as an abelian variety).

Deligne [19] showed that any $K3$ surface can be lifted. This is an interesting case, because the lifting of the abstract surface to a formal scheme may not be effective: the ample invertible sheaf need not lift. It requires some extra subtlety to show that there is an effective lifting. This is analogous to the complex analytic theory, where the deformation space, as complex manifolds, has dimension 20, but the algebraic $K3$ surfaces form only 19-dimensional subfamilies. This lifting result has been sharpened by Ogus [127, 2.3].

Several authors have studied the problem of lifting a curve along with some of its automorphisms. One cannot expect to lift a curve with its entire group of automorphisms, because the order of that group in characteristic $p > 0$ can exceed $84(g - 1)$, which is impossible in characteristic 0 [57, IV, Ex. 2.5]. However, one can lift a curve C together with a cyclic group H of automorphisms, provided that p^2 does not divide the order of H.

Raynaud [140] has given examples of surfaces, the "false ruled surfaces," that cannot be lifted, and W. E. Lang [89] has generalized these.

Hirokado [68] and Schröer [149] have given examples of nonliftable Calabi–Yau threefolds, and Ekedahl [25] has shown that these examples have all their deformations limited to characteristic p. In particular, they cannot be lifted even to the Witt vectors $\bmod p^2$.

Vakil [164] shows that there are smooth varieties of each dimension ≥ 2 that cannot be lifted in an arbitrarily complicated way, e.g., liftable $\bmod p^5$ but not $\bmod p^6$.

It seems to be unknown whether there are nonliftable singular curves or nonliftable zero-schemes.

4

Global Questions

In this chapter we apply the methods of infinitesimal and formal deformations from Chapters 1, 2, and 3 to global questions. The foremost question in every situation is whether there exists a global "moduli space" parametrizing isomorphism classes of the objects in question. To make this question precise, we introduce the functorial language in Section 23, we define the notions of coarse moduli space and fine moduli space, and mention various properties of a functor that help determine whether it may be representable. The "easy" cases—that is, easy to state, though the proofs are not easy—are the cases of closed subschemes and invertible sheaves, Situations A and B, where the functor is representable respectively by the Hilbert scheme and the Picard scheme (Section 24).

Our first example of a coarse moduli space in Situation D is the one-point moduli space of curves of genus 0 (Section 25). The detailed study of families of curves of genus 0 gives a good illustration, in a case in which everything can be made explicit, of the functorial language introduced in Section 23. For curves of genus 1, we show in Section 26 that the j-line is a coarse moduli space and study the structure of families of elliptic curves.

For curves of genus $g \geq 2$, one knows [119] that there is a coarse moduli space. We do not prove this but instead in Section 27 develop Mumford's concept of a "modular family," which is a precursor of the theory of stacks.

In Section 28 we discuss the problem of moduli of vector bundles, Situation C, and explain the importance of simple and stable vector bundles. A final Section 29 discusses the problem of smoothing singularities, an interesting interaction between local deformation theory and existence of global families. In particular, we introduce the notion of a formally smoothable scheme, which has infinitesimal deformations tending toward a smooth scheme.

R. Hartshorne, *Deformation Theory*, Graduate Texts in Mathematics 257,
DOI 10.1007/978-1-4419-1596-2_5, © Robin Hartshorne 2010

23. Introduction to Moduli Questions

What is a variety of moduli or a moduli scheme? In this section we will consider the general question and make some definitions. Then in subsequent sections we will give some examples to illustrate the various issues that often arise in dealing with moduli questions.

To fix the ideas, let us work over an algebraically closed base field k (though everything that follows can be generalized to work over a fixed base scheme). Suppose we have identified a certain class of objects $\mathcal{M}$ over k that we wish to classify. You can think of closed subschemes with fixed Hilbert polynomial of $\mathbb{P}_k^n$, or curves of genus g over k, or vector bundles of given rank and Chern classes on a fixed scheme X over k, and so on. We will deal with specific cases later. But for the moment, let us just say we have focused our attention on a set of objects $\mathcal{M}$, and we have given a rule for saying when two of them are the same (usually isomorphism). We wish to classify the objects of $\mathcal{M}$.

The first step is to list the possible elements of $\mathcal{M}$ up to isomorphism. This determines $\mathcal{M}$ as a set. To go further, we wish to put a structure of algebraic variety or scheme on the set $\mathcal{M}$ that should be natural in some sense. So we look for a scheme M of finite type over k whose closed points are in one-to-one correspondence with the elements of the set $\mathcal{M}$ and whose scheme structure reflects the possible variations of elements in $\mathcal{M}$: how they behave in families.

To make this precise, we must say what we mean by a family of elements in $\mathcal{M}$. For a parameter scheme S of finite type over k, this will usually mean a scheme X, flat over S, with an extra structure whose fibers at closed points are elements of $\mathcal{M}$. Then for the scheme M to be a variety of moduli for the class $\mathcal{M}$, we require that for every family X/S there be a morphism $f : S \to M$ such that for each closed point $s \in S$, the image $f(s) \in M$ corresponds to the isomorphism class of the fiber X_s in $\mathcal{M}$.

But that is not enough. We want the assignment of the morphism $f : S \to M$ to the family X/S to be functorial. To explain what this means, for every scheme S/k, let $\mathcal{F}(S)$ be the set of all families X/S of elements of $\mathcal{M}$ parametrized by S. If $S' \to S$ is a morphism, then by base extension, a family X/S will give rise to a family X'/S' (note here that we should define our notion of family in each situation so that it does extend by base extension). Thus the morphism $S' \to S$ gives rise to a map of sets $\mathcal{F}(S) \to \mathcal{F}(S')$. In this way $\mathcal{F}$ becomes a contravariant functor from (Sch $/k$) to (Sets). (In the above discussion we spoke only of schemes S of finite type over k, but to make the functorial language work well, we should extend the definition of $\mathcal{F}$ to include all schemes over k.) So what we are asking for is a morphism of functors $\varphi : \mathcal{F} \to \mathrm{Hom}(\cdot, M)$ that for each scheme S/k and each element $X/S \in \mathcal{F}(S)$ assigns the associated morphism $f_{X/S} : S \to M$. If we denote the functor $\mathrm{Hom}(\cdot, M)$ from (Sch $/k$) to (Sets) by h_M, then we can say that φ is a morphism of functors $\varphi : \mathcal{F} \to h_M$.

What we have said so far still does not determine the scheme structure on M uniquely. To make M unique, it should be the "largest possible" with the above properties. So we require that if N is any other scheme, and $\psi : \mathcal{F} \to h_N$ a morphism of functors, then there should exist a unique morphism $e : M \to N$ such that $\psi = h_e \circ \varphi$, where $h_e : h_M \to h_N$ is the induced map on associated functors.

Summing up, we come to the following definition.

Definition. We consider a certain class $\mathcal{M}$ of objects over an algebraically closed field k. Suppose we have defined what we mean by families of elements of $\mathcal{M}$ parametrized by a scheme S, and we have said when two families are the same. We consider the functor $\mathcal{F} : (\mathrm{Sch}\,/k) \to (\mathrm{Sets})$ that to each S/k assigns the set $\mathcal{F}(S)$ of equivalence classes of families of elements of $\mathcal{M}$ over S. Then we define a *coarse moduli scheme* for the family $\mathcal{M}$ (or the functor $\mathcal{F}$) to be a scheme M/k together with a morphism of functors $\varphi : \mathcal{F} \to h_M$ such that

(a) the induced map $\mathcal{F}(k) \to h_M(k)$ is bijective, and
(b) φ is universal in the sense that if $\psi : \mathcal{F} \to h_N$ is any other morphism from $\mathcal{F}$ to a functor of the form h_N, then there is a unique morphism of schemes $e : M \to N$ such that $\psi = h_e \circ \varphi$.

Remark 23.0.1. The condition (a) tells us that the elements of $\mathcal{M}$, considered as trivial families over k, are in one-to-one correspondence with the closed points of M, which are just morphisms of $\operatorname{Spec} k$ into M. The fact that φ is a morphism of functors tells us that for any family $X/S \in \mathcal{F}(S)$, its image $\varphi_X \in \operatorname{Hom}(S, M)$ has the property that for each closed point $s \in S$, the point $\varphi_X(s) \in M$ corresponds to the equivalence class of the fiber X_s in $\mathcal{M}$. Just consider the functoriality for a morphism of $\operatorname{Spec} k$ to S sending the point to s. The condition (b) implies that M is unique, up to unique isomorphism, if it exists.

Definition. If M is a coarse moduli scheme for the moduli problem $\mathcal{M}$, we define a *tautological family* for $\mathcal{M}$ to be a family X/M such that for each closed point $m \in M$, the fiber X_m is the element of $\mathcal{M}$ corresponding to m by the bijection $\mathcal{F}(k) \to h_M(k)$ above.

Example 23.0.2. Let $\mathcal{M}$ be the set of nonsingular projective curves of genus 0 over k, up to isomorphism. Then $\mathcal{M}$ has a single element, namely $\mathbb{P}^1$. We will see (25.1) that $M = \operatorname{Spec} k$ is a coarse moduli space, and the trivial family $\mathbb{P}^1/k$ becomes a tautological family.

Example 23.0.3. On the other hand, we will see (26.3.1) that the j-line is a coarse moduli scheme for curves of genus 1, but that it has no tautological family.

Remark 23.0.4. A coarse moduli scheme may fail to exist. A *jump phenomenon* for a moduli problem $\mathcal{M}$ is a family X/S of elements of $\mathcal{M}$, where S is an integral scheme of dimension at least one, of finite type over k, such that all fibers X_s for $s \in S$ are isomorphic except for one X_{s_0} that is different. In this case there can be no coarse moduli space, because it is not possible to have a morphism of S to a scheme M sending s_0 to one point and all other closed points of S to another single point.

Here are some examples of moduli problems with jump phenomena:

(a) Vector bundles of rank 2 and degree 0 over $\mathbb{P}^1$. We have seen (19.3.2) that there is a family parametrized by $S = \operatorname{Spec} k[t]$ whose fiber at $t = 0$ is $\mathcal{O}(-1) \oplus \mathcal{O}(1)$, but whose fiber for every $t \neq 0$ is $\mathcal{O} \oplus \mathcal{O}$.
(b) Plane curve singularities up to analytic isomorphism. For example there are families whose fiber is a cusp at one point, but is a node at all other points (14.2.2).
(c) Integral projective curves of arithmetic genus one. There are families (26.6.7) where the special curve has a cusp, but all the other fibers are isomorphic nonsingular curves.

Definition. If the functor $\mathcal{F}$ associated to a moduli problem $\mathcal{M}$ is isomorphic to a functor of the form $h_M = \operatorname{Hom}(\cdot, M)$, then we say that $\mathcal{F}$ is a *representable functor*, or that M *represents* the functor $\mathcal{F}$, and we call M a *fine moduli space* for $\mathcal{M}$ (or for $\mathcal{F}$).

Remark 23.0.5. If $\mathcal{F} \to h_M$ is an isomorphism, then in particular $\mathcal{F}(M) \to h_M(M)$ is bijective, and there is a family X_u/M corresponding to the identity map $1_M \in \operatorname{Hom}(M, M)$. We call X_u the *universal family* associated with the fine moduli space M. It has the additional property that for any family X/S, there is a unique morphism $S \to M$ such that the family X/S is obtained by base extension from the universal family X_u/M. Conversely, if there is a scheme M and a family X_u with this latter property, then the functor $\mathcal{F}$ is represented by M.

Example 23.0.6. The Hilbert scheme. Let $\mathcal{M}$ be the set of closed subschemes Y of $X = \mathbb{P}^n_k$ with a given Hilbert polynomial P. A family of elements of $\mathcal{M}$ will be a closed subscheme $\mathcal{Y} \subseteq X_S = \mathbb{P}^n_S$, flat over a scheme S, all of whose closed fibers are elements of $\mathcal{M}$. The associated functor $\mathcal{F}(S)$ is called the *Hilbert functor*. This functor is represented by the Hilbert scheme (1.1a), so it is a representable functor.

Proposition 23.1. *If the functor $\mathcal{F}$ associated to a moduli problem $\mathcal{M}$ is representable by a scheme M, then M is also a coarse moduli scheme for $\mathcal{F}$, and the universal family X_u/M is a tautological family.*

Proof. Since $\mathcal{F} \cong h_M$, we take the isomorphism as our morphism of functors $\varphi : \mathcal{F} \to h_M$. Property (a) of the definition, namely $\mathcal{F}(k) \to h_M(k)$

bijective, follows from the isomorphism. We have only to check the universal property (b), namely, if $\psi : \mathcal{F} \to h_N$ is another morphism, then there exists a unique morphism $e : M \to N$ such that $\psi = h_e \cdot \varphi$. By hypothesis there is a morphism $h_M = \mathcal{F} \to h_N$, so it remains to show that this comes from a morphism $e : M \to N$, and this is left as an exercise (Ex. 23.1).

Remark 23.1.1. A family X/S for a moduli problem $\mathcal{M}$ is *trivial* if it is obtained by base extension from the family consisting of one element of $\mathcal{M}$ over a point. A family X/S with S of finite type over k is *fiberwise trivial* if the fibers X_s are isomorphic for all closed points $s \in S$. If $\mathcal{M}$ has a fine moduli space M, then every fiberwise trivial family must be trivial, because it is obtained by pulling back the universal family at a single point. Because of this property, we will see (25.2.1) that the one-point coarse moduli space for curves of genus 0 is not a fine moduli space.

Remark 23.1.2. Since there are nonreduced fine moduli spaces (for example Mumford's curves in $\mathbb{P}^3$ (13.1)), it follows that there are nonreduced coarse moduli spaces, even though it seems in the definition of a coarse moduli space that we dealt only with closed points and hence apparently cannot distinguish a scheme from its associated reduced scheme.

One of the great benefits of having a fine moduli space is that we can study it using infinitesimal methods.

Proposition 23.2. *Let M be a fine moduli scheme for the moduli problem $\mathcal{M}$, and let $X_0 \in \mathcal{M}$ correspond to a point $x_0 \in M$. Then the Zariski tangent space to M at x_0 is in one-to-one correspondence with the set of families X over the dual numbers D whose closed fibers are isomorphic to X_0.*

Proof. Indeed, the Zariski tangent space to M at x_0 can be identified with $\mathrm{Hom}_{x_0}(D, M)$, [57, II, Ex. 2.8], and this in turn corresponds to the subset of those elements of $\mathcal{F}(D)$ restricting to X_0 over k.

Extending the same argument to higher-order infinitesimal neighborhoods we obtain the following.

Proposition 23.3. *Let $\mathcal{F}$ be the functor associated to a moduli problem $\mathcal{M}$, let $X_0 \in \mathcal{M}$, and consider the functor on Artin rings $\mathcal{F}_0$ that to each local Artin ring A over k assigns the set of families of elements of $\mathcal{M}$ over $\mathrm{Spec}\,A$ whose closed fiber is isomorphic to X_0. If $\mathcal{M}$ has a fine moduli scheme, then the functor $\mathcal{F}_0$ is pro-representable (§15).*

Proof. Let M be a fine moduli scheme for $\mathcal{M}$, let $x_0 \in M$ correspond to $X_0 \in \mathcal{M}$, and let R be the completion of the local ring of x_0 on M. Since M is a fine moduli space, each element of $\mathcal{F}_0(A)$ corresponds to a unique morphism $\mathrm{Spec}\,A \to M$ whose image lands at the point x_0. Such morphisms correspond to ring homomorphisms $R \to A$. Thus the functor $\mathcal{F}_0$ is pro-representable, and its formal family arises from the base extension of the universal family X_u/M to $\operatorname{Spec} R$, hence is effective (§21).

Warning 23.3.1. The local functor $\mathcal{F}_0$ considered here consists of families over $\operatorname{Spec} A$ whose closed fiber is isomorphic to X_0. This is what we called the crude local functor in §18, and in general is not the same as the local functor of deformations of X_0, which require a fixed isomorphism of X_0 with the closed fiber; cf. (18.4) and the discussion following, which contrasts these two different local functors.

Corollary 23.4. *Let $\mathcal{F}$ be a representable functor, represented by a scheme M, and let $x_0 \in M$ be a point. If we have an obstruction theory for the local functor $\mathcal{F}_0$, then knowing its tangent space t_0 and its obstruction space V_0, the dimension of M at x_0 is bounded:*

$$\dim_{x_0}(M) \geq \dim t_0 - \dim V_0.$$

Proof. Just apply (11.2) to the local ring of x_0 on M.

Definition. We say that a contravariant functor $\mathcal{F}$ from (Sch/k) to (Sets) is a *sheaf for the Zariski topology* if for every scheme S and every covering of S by open subsets $\{U_i\}$, the diagram

$$\mathcal{F}(S) \to \prod \mathcal{F}(U_i) \rightrightarrows \prod \mathcal{F}(U_i \cap U_j)$$

is exact. Spelled out, this means two things:

(a) given elements $x, x' \in \mathcal{F}(S)$ whose restrictions to $\mathcal{F}(U_i)$ are equal for all i, then $x = x'$, and
(b) given a collection of elements $x_i \in \mathcal{F}(U_i)$ for each i such that for each i, j, the restrictions of x_i and x_j to $U_i \cap U_j$ are equal, then there exists an element $x \in \mathcal{F}(S)$ whose restriction to each $\mathcal{F}(U_i)$ is x_i.

Proposition 23.5. *If the moduli problem $\mathcal{M}$ has a fine moduli space, then the associated functor $\mathcal{F}$ is a sheaf for the Zariski topology.*

Proof. Indeed, if $\mathcal{F} = h_M$, then for any scheme S, $\mathcal{F}(S) = \operatorname{Hom}(S, M)$, and one knows that morphisms from one scheme to another are determined locally, and can be glued together if they are given locally and are compatible on overlaps [57, II.3.3, Step 3].

Example 23.5.1. We will see that the functor of families of curves of genus 0 is not a sheaf in the Zariski topology (25.2.1).

Remark 23.5.2. Grothendieck's theory of *descent* [45, exposé 190] shows more generally that a representable functor is a sheaf for the fpqc (faithfully flat quasi-compact) topology, and hence also for the étale topology.

References for this section. There are many general discussions of moduli problems in the literature, for example [119], [157], [137], just to name a few. Grothendieck's Séminaire Bourbaki talks [45] develop his method of representable functors, and in particular their application to the Hilbert and Picard schemes.

Exercises.

23.1. (a) For each scheme M we have defined the contravariant functor $h_M = \text{Hom}(\cdot, M)$ from (Sch $/k$) to (Sets). Show that h is a covariant functor from (Sch $/k$) to the category (Funct) of contravariant functors from (Sch $/k$) to (Sets), and that it is fully faithful, meaning that the natural map

$$\text{Hom}_{(\text{Sch})}(M, N) \to \text{Hom}_{(\text{Funct})}(h_M, h_N)$$

is bijective.

(b) Use part (a) to show that if $\mathcal{F}$ is a representable functor, then $\mathcal{F}$ has the universal property (b) of the definition of a coarse moduli space, and so the fine moduli space M is also a coarse moduli space.

23.2. Let $\mathcal{M}$ be a moduli problem and suppose that some element $X_0 \in \mathcal{M}$ has a finite group of nontrivial automorphisms (for example, a curve of genus 2). Then $\mathcal{M}$ cannot have a fine moduli space. Consider a triangle S made of three lines. On each line put the trivial family X_0; at two corners, glue by the identity, and at the third corner, glue by a nontrivial automorphism σ. Show that this makes a fine moduli scheme impossible. (One could also use a base scheme S made of two curves meeting at two distinct points.)

23.3. Let $\mathcal{F}$ be the Hilbert functor of closed subschemes with a given Hilbert polynomial P of $X = \mathbb{P}_k^n$. Show that $\mathcal{F}$ is a sheaf for the Zariski topology (without using the fact that $\mathcal{F}$ is representable!).

23.4. Let X/k be a projective scheme, and consider the functor $\mathcal{F}(S) = \text{Pic}(X \times S)$ for each scheme S/k. Here Pic denotes the group of invertible sheaves modulo isomorphism. Give an example to show that $\mathcal{F}$ is not a sheaf for the Zariski topology.

23.5. Let $\mathcal{M}$ be the family of nonsingular projective curves of genus $g \geq 3$ having no nontrivial automorphisms. Show that the functor $\mathcal{F}$ of flat families of elements of $\mathcal{M}$ is a sheaf for the Zariski topology.

23.6. If $X \subseteq \mathbb{P}_T^n$ is a flat family of closed subschemes of $\mathbb{P}_k^n$, with T integral and of finite type over k, show that a jump phenomenon of the Hilbert functor is impossible. Here we are considering closed subschemes, so this means that the fibers X_t are equal for all $t \neq 0$. You have to show that X_0 is equal to the others. You think it is obvious? Do not use representability of the Hilbert functor.

23.7. For families of invertible sheaves on a fixed nonsingular projective variety, show that jump phenomena are impossible. Do not use representability of the Pic functor.

24. Some Representable Functors

In this section we mention some representable functors, in particular Hilb and Pic. We will not give complete proofs, which can be found elsewhere, and which extend beyond the aims of this book. However, we will discuss some aspects of the proofs, since these shed light on the nature of representable functors and help in understanding why some other functors are not representable.

Theorem 24.1. *The Hilbert functor, which to every scheme S/k associates the set of subschemes $Y \subseteq \mathbb{P}_S^N$, flat over S, whose fibers all have a given Hilbert polynomial P, is representable by a scheme M, projective over k.*

We have already stated this theorem, in different words, as $(1.1a)$. A complete proof can be found in the article of Nitsure [124]. In the course of the proof, one has to establish certain properties of the functor. These are useful to consider for any functor.

Definition. A contravariant functor $\mathcal{F} : (\text{Sch } /k) \to (\text{Sets})$ is *bounded* if there exists a scheme S of finite type over k and a family $X \in \mathcal{F}(S)$ such that every $X_0 \in \mathcal{F}(k)$ is isomorphic to the fiber X_s for some closed point $s \in S$. (Here by fiber, of course, we mean the image of X in $\mathcal{F}(k)$ corresponding to the morphism $\text{Spec } k \to S$ that sends the point to s.)

We say that $\mathcal{F}$ is *separated* if for any nonsingular curve S/k and a point $s_0 \in S$, if X and X' are two elements of $\mathcal{F}(S)$ whose fibers X_s, X_s' are isomorphic for all $s \in S$, $s \neq s_0$, then also $X_{s_0} \cong X_{s_0}'$.

We say that $\mathcal{F}$ is *complete* if for any nonsingular curve S/k and point $s_0 \in S$, given an element $X \in \mathcal{F}(S - \{s_0\})$, then there exists an element $X' \in \mathcal{F}(S)$ such that the fibers X_s and X_s' are isomorphic for all $s \neq s_0$.

Proposition 24.2. *If the functor $\mathcal{F}$ is represented by a scheme M of finite type over k, then $\mathcal{F}$ is bounded. In that case the scheme M is separated (resp., proper over k) if and only if $\mathcal{F}$ is separated (resp., separated and complete).*

Proof. Left to reader as (Ex. 24.1).

To show that the Hilbert functor is bounded, one uses Castelnuovo–Mumford regularity.

Definition. A coherent sheaf $\mathcal{F}$ on a projective scheme X is *m-regular* if $H^i(\mathcal{F}(m - i)) = 0$ for each $i > 0$.

Proposition 24.3 (Mumford). *If $\mathcal{F}$ is m-regular, then $\mathcal{F}$ is also m'-regular for all $m' \geq m$. Furthermore, $\mathcal{F}(m)$ is generated by global sections.*

Proof. [115] or [124, 5.1].

Proposition 24.4. *A family $\mathcal{M}$ of coherent sheaves on a projective scheme X/k, all having the same Hilbert polynomial, is bounded (meaning the functor of flat families of sheaves in $\mathcal{M}$ is bounded) if and only if there is a uniform m_0 such that all members of $\mathcal{M}$ are m_0-regular.*

Proof (in outline). One direction is easy. If $\mathcal{M}$ is a bounded family, then there is a scheme S of finite type over k together with a coherent sheaf $\mathcal{F}$ on $X \times S$, flat over S, containing among its fibers at closed points of S all elements of $\mathcal{M}$. For any i, n, the function $h^i(\mathcal{F}_s(n))$ is semicontinuous for $s \in S$. By Serre's vanishing theorem it is zero for $i > 0$, $n \gg 0$ for each s. Looking at the generic points of all irreducible components of S, we find a uniform n_0 to make all the $h^i(\mathcal{F}_s(n)) = 0$ for all $n \geq n_0$ at those generic points. Then by semicontinuity, the h^i are 0 on a dense open subset U of S. Let $S_1 = S - U$ and find an n_1 such that $h^i(\mathcal{F}_s(n)) = 0$ for all $n \geq n_1$ at the generic points of S_1. Then also they are zero on an open dense subset $U_1 \subseteq S_1$. Let $S_2 = S_1 - U_1$, and repeat, until the h^i are zero everywhere. Then each $\mathcal{F}_s$ is $(\max\{n_i\} + \dim X)$-regular.

Now suppose conversely that there is a uniform m_0 such that all the elements of $\mathcal{M}$ are m_0-regular. In that case each $\mathcal{F}(m_0)$ will be generated by global sections and $h^0(\mathcal{F}(m_0)) = P(m_0)$, the value of the associated Hilbert polynomial, since $h^i = 0$ for $i > 0$. Thus we can find a surjective map $\mathcal{O}_X^N \to \mathcal{F}(m_0)$ for each $\mathcal{F}$, where $N = h^0(\mathcal{F}(m_0))$. Let $\mathcal{G}$ be the kernel. Then a diagram chase of cohomology shows that there is another uniform m_0' such that all the $\mathcal{G}$'s are m_0'-regular. Hence $\mathcal{G}(m_0')$ is generated by global sections. Let $h^0(\mathcal{G}(m_0')) = M$. Then $\mathcal{F}$ is completely determined by the M-dimensional subspace $H^0(\mathcal{G}(m_0'))$ of $H^0(\mathcal{O}_X^N(m_0'))$. These vector subspaces are parametrized by a finite-dimensional Grassmann variety, and over a suitable subspace of that Grassmann variety we obtain a family containing all of our initial sheaves $\mathcal{F}$.

An important step in the proof of existence of the Hilbert scheme is the following.

Proposition 24.5. *The set of subschemes Y of $\mathbb{P}_k^n$ with Hilbert polynomial P forms a bounded family.*

Proof. Using the previous proposition it is enough to show that there is a uniform m_0 such that the ideal sheaves $\mathcal{I}_Y$ of all such Y in $\mathbb{P}_k^n$ are m_0-regular. This is accomplished by induction on $\dim Y$, using a generic hyperplane section. See [124, 5.3] for details.

Remark 24.5.1. The fact that the Hilbert functor is separated and complete is immediate, because if S is a curve and U an open subset, and if $Y \subseteq \mathbb{P}_U^n$ is a closed subscheme flat over U, then there is a unique $\bar{Y} \subseteq \mathbb{P}_S^n$ flat over S restricting to Y, namely the scheme-theoretic closure of Y in $\mathbb{P}_S^n$.

Remark 24.5.2. If to each closed subscheme $Y \subseteq X = \mathbb{P}_k^n$ we associate its structure sheaf $\mathcal{O}_Y$, then we can regard the Hilbert scheme as parametrizing all quotients $\mathcal{O}_X \to \mathcal{O}_Y \to 0$ with Hilbert polynomial P. A generalization of the Hilbert scheme is the *Quot scheme*, which parametrizes all quotients $\mathcal{E} \to \mathcal{F} \to 0$ where $\mathcal{E}$ is a fixed coherent sheaf on $X = \mathbb{P}_k^n$, and $\mathcal{F}$ runs through all quotients with a given Hilbert polynomial. Cf. (19.3), where we showed that the associated local deformation functor is pro-representable. Grothendieck shows that the Quot functor is representable by a projective scheme using the same techniques as for the Hilbert functor.

Next we consider the Picard scheme, which should parametrize invertible sheaves. Fix a scheme X/k. For any base scheme S/k we can consider the group $\mathrm{Pic}(X \times S)$ of invertible sheaves on $X \times S$. This is a contravariant functor in S, but in general it is not a sheaf in the Zariski topology (Ex. 23.4), so it cannot be representable. Besides, what we really care about is the family of the invertible sheaves on the fibers $X \times \{s\}$, and not the invertible sheaf on $X \times S$. If $\mathcal{L}$ is an invertible sheaf on $X \times S$, and if $\mathcal{N}$ is an invertible sheaf on S, then $\mathcal{L}$ and $\mathcal{L}' = \mathcal{L} \otimes p_2^* \mathcal{N}$ have the same fibers, so represent the same family. Therefore we consider the modified functor $\mathrm{Pic}(X \times S/S) = \mathrm{Pic}(X \times S)/p_2^* \mathrm{Pic}\, S$. But this one may also fail to be a sheaf for the Zariski topology. In that case we can take the associated sheaf and consider this new functor. But this one may still fail to be representable, because a representable functor is always a sheaf for the étale topology according to descent theory, and this functor, which is a sheaf for the Zariski topology, may fail to be a sheaf for the étale topology. See Kleiman's article [79, §9.2] for a detailed discussion of these subtleties.

We can avoid all these difficulties by trivializing families along a section. So let X be a scheme of finite type over the algebraically closed field k, and let P be a fixed point. We consider the functor $\mathrm{Pic}_{X/k,P}$, which to each base scheme S assigns the group of invertible sheaves $\mathcal{L}$ on $X \times S$, together with a fixed isomorphism of $\mathcal{L} \mid \{P\} \times S \cong \mathcal{O}_S$.

Theorem 24.6. *With the above hypotheses, assume furthermore that X is integral and projective. Then the functor $\mathrm{Pic}_{X/k,P}$ is represented by a separated scheme, locally of finite type over k, which we call the* Picard scheme *of X/k.*

Proof. See [79, 9.4.8]. ∎

As a variant of the Hilbert scheme, we can consider the *Hilbert-flag scheme* parametrizing nested sets of closed subschemes (cf. (Ex. 6.8)). For a flag of length 2, fix $X = \mathbb{P}_k^n$ and let P, Q be two Hilbert polynomials. We consider the functor $\mathcal{F}$ that to each base scheme S assigns a pair of closed subschemes $Y \subseteq Z \subseteq X \times S$, both flat over S, and where the fibers of Y (resp., Z) have Hilbert polynomial P (resp., Q).

Theorem 24.7. *The functor $\mathcal{F}$ is represented by a scheme, projective over k, which we call the* Hilbert-flag scheme.

Proof. One can deduce this from the existence of the relative Hilbert scheme: First let H be the Hilbert scheme associated to the Hilbert polynomial Q, with universal family Z_u/H. Then take the Hilbert scheme of relative subschemes $Y \subseteq Z_u \times S/H \times S$ with Hilbert polynomial P. For details see [152, §4.5] (cf. (Ex. 6.8)).

It is often useful to consider deforming not only schemes, but also morphisms of schemes. Given a morphism $f : X \to Y$ of schemes over k, a *deformation of f* (keeping X, Y fixed) over an Artin ring A is a morphism $f' : X \times A \to Y \times A$ such that $f' \otimes k = f$.

Lemma 24.8. *To give a deformation of a morphism $f : X \to Y$ (keeping X and Y fixed) it is equivalent to give a deformation of the graph Γ_f as a closed subscheme of $X \times Y$.*

Proof. To any deformation f' of f we associate its graph $\Gamma_{f'}$, which will be a closed subscheme of $X \times Y \times A$. It is a deformation of Γ_f. Conversely, given a deformation Z of Γ_f over A, we need only verify that it is a graph of some morphism. The projection $p_1 : Z \to X \times A$ gives an isomorphism when tensored with k. From flatness of Z over A it follows that p_1 is an isomorphism, and so Z is the graph of $f' = p_2 \circ p_1^{-1}$.

Proposition 24.9. *Assume that Y is nonsingular. Then the tangent space to the deformation functor of $f : X \to Y$ (keeping X and Y fixed) is $H^0(X, f^*T_Y)$, and the obstructions to deforming f lie in $H^1(X, f^*T_Y)$. If X and Y are also projective, the deformation functor of f is pro-representable.*

Proof. From (24.8) we must consider the deformations of Γ_f as a closed subscheme of $X \times Y$. Note that $\Gamma_f = (f \times \mathrm{id})^{-1} \Delta_Y$, where $\Delta_Y \subseteq Y \times Y$ is the diagonal. Since Y is nonsingular, Δ_Y is a local complete intersection in $Y \times Y$, and $\mathcal{I}_\Delta / \mathcal{I}_\Delta^2 = \Omega^1_{Y/k}$. It follows that Γ_f is a local complete intersection in $X \times Y$, and that its normal bundle is f^*T_Y. Now our result follows from the corresponding discussion for the Hilbert scheme (6.2), (17.1).

Theorem 24.10. *Given X, Y projective schemes over k, the global functor of families of morphisms $f : X \times S \to Y \times S$ over a scheme S is represented by a union of quasi-projective schemes over k.*

Proof. This follows from the existence of the Hilbert scheme of closed subschemes of $X \times Y$ (24.1), and the observation that the set of subschemes Z representing graphs of morphisms is an open subset of the Hilbert scheme. There will be different quasi-projective components depending on the Hilbert polynomial of Z.

Remark 24.10.1. If $f : X \to Y$ is a closed immersion, there is a natural morphism of functors $\mathrm{Def}(f) \to \mathrm{Hilb}(Y)$ obtained by assigning to f the closed subscheme image. If Y is nonsingular, the exact sequence

$$\mathcal{I}/\mathcal{I}^2 \to \Omega_Y^1|_X \to \Omega_X^1 \to 0$$

dualizes to give a cohomology sequence

$$0 \to H^0(T_X^0) \to H^0(f^*T_Y) \to H^0(\mathcal{N}_{X/Y}).$$

The middle group represents infinitesimal deformations of f. The image on the right is the corresponding deformation of the subscheme. If the subscheme is unchanged, then f comes from an infinitesimal automorphism of X. See also (20.2).

Remark 24.10.2. A special case of the Hom functor is the functor of isomorphisms. Given X, Y schemes over a base scheme S, for any base extension $T \to S$, we denote by $\mathcal{F}(T)$ the set of isomorphisms $\varphi : X \times_S T \xrightarrow{\sim} Y \times_S T$, as schemes over T. Giving φ is equivalent to giving its graph $\Gamma_\varphi \subseteq X \times Y \times T$. Thus, if X and Y are projective over S, the representability of the Hilbert scheme shows that $\mathcal{F}$ is globally represented by a scheme $S' = \operatorname{Isom}_S(X, Y)$, quasi-projective over S. What this means, in more detail, is that S' comes together with a universal isomorphism $\varphi : X \times_S S' \xrightarrow{\sim} Y \times_S S'$, and for any other base scheme T and any isomorphism $\psi : X \times_S T \to Y \times_S T$, there is a unique morphism $f : T \to S'$ such that $\psi = \varphi \times_{S'} T$.

Remark 24.10.3. A generalization of the above discussion allows us to treat deformations of $f : X \to Y$, keeping Y fixed, but allowing both f and X to vary. We consider the functor $\mathcal{F} = \operatorname{Def}(X, f)$ that to each Artin ring assigns a deformation X'/A, together with its closed immersion $X \hookrightarrow X'$, and a morphism $f' : X' \to Y \times A$, restricting to f on X. If X and Y are projective over k, one can apply Schlessinger's criterion as before to see that $\mathcal{F}$ has a miniversal family. If X and Y are both nonsingular, there is an exact sequence of tangent spaces

$$H^0(f^*T_Y) \to t_{\mathcal{F}} \to H^1(T_X),$$

where the right-hand arrow is the forgetful functor $\operatorname{Def}(X, f) \to \operatorname{Def}(X)$, and the kernel comes from those deformations of f that leave X fixed, which we studied above.

A spectacular application of the deformation theory of a morphism was Mori's proof that a nonsingular projective variety with ample tangent bundle is isomorphic to $\mathbb{P}^n$ ("Hartshorne's conjecture") [109]. We will not describe how he deduced the existence of rational curves on such a variety in characteristic 0 from their existence in characteristic $p > 0$; nor will we trace the steps leading from the existence of rational curves to the final result. We will only prove the key step, using the technique that is now called "bend and break," which is the following criterion for the existence of a rational curve on a manifold in characteristic $p > 0$.

Theorem 24.11 (Mori). *Let X be a nonsingular projective variety over an algebraically closed field k of characteristic $p > 0$. Assume that the canonical divisor K_X is not numerically effective, i.e., there exists an irreducible curve C with $C.K_X < 0$. Then X contains a rational curve, i.e., an integral curve whose normalization is isomorphic to $\mathbb{P}^1$.*

Proof. Let $C_0 \subseteq X$ be an integral curve with $(C_0.K_X) < 0$. Let $C_1 \to C_0$ be the normalization of C_0, and let $g = $ genus of C_1. If $g = 0$ there is nothing to prove, so we suppose $g > 0$. Since $C_0.K_X < 0$, we can find $q = p^r$ for $r \gg 0$ such that

$$-q(C_0.K_X) \geq ng + 1,$$

where $n = \dim X$.

Let $f : C \to C_1$ be the qth k-linear Frobenius morphism, i.e., C is the same abstract curve as C_1, but with structural morphism to k modified by qth powers in k, so that f is a purely inseparable k-morphism of degree q. Note that the genus of C is still g. We denote also by f the composed map $C \to C_1 \to C_0 \subseteq X$.

Fix a point $P \in C$. We will consider the deformation theory of the morphism $f : C \to X$, keeping C and X fixed, and also keeping fixed the image $f(P) = P_0 \in C_0$. As in (24.10), the corresponding deformation functor is represented by a scheme $\mathrm{Hom}_P(C, X)$, quasi-projective over k; its tangent space is $H^0((f^*T_X)(-P))$ and its obstructions lie in $H^1((f^*T_X)(-P))$.

Now the dimension estimate for representable functors (23.4) tells us that

$$\dim \mathrm{Hom}_P(C, X) \geq h^0((f^*T_X)(-P)) - h^1((f^*T_X)(-P)).$$

To compute this, note that T_X is locally free of rank n; the restriction $T_X|_{C_0}$ has degree $-(C_0.K_X)$, and so $f^*(T_X)$ has degree $-q(C_0.K_X)$. The twist $-P$ subtracts n from the degree. Then by Riemann–Roch on C we have

$$\chi((f^*T_X)(-P)) = -q(C_0.K_X) - n + n(1 - g),$$

and by our choice of q this number is at least 1, so $\dim \mathrm{Hom}_P(C, X) \geq 1$.

Thus there exist a nonsingular curve D, not necessarily complete, and a morphism $F : C \times D \to X$ representing a nonconstant family of morphisms of C to X, parametrized by D, all sending P to P_0.

I claim, in fact, that D is not complete. For suppose D were complete. Then $C \times D$ would be a nonsingular projective surface. For any point $Q \in C$, the curve $Q \times D$ is algebraically equivalent to $P \times D$. Now let $\mathcal{L}$ be a very ample invertible sheaf on X, corresponding to a projective embedding of X in some projective space. The degree of the image curve $F(Q \times D)$ is then measured by $(Q \times D).F^*\mathcal{L}$. Since $Q \times D \sim P \times D$, and $F(P \times D) = P_0$ is a point, this degree is zero. So $F(Q \times D)$ is also a point, and this implies that F is a constant family $f : C \to X$, contrary to hypothesis. Thus D cannot be complete.

Now let $D \subseteq \bar{D}$ be a completion to a projective nonsingular curve $\bar{D}$, and let $\bar{F} : C \times \bar{D} \dashrightarrow X$ be the corresponding rational map, which by the previous argument cannot be a morphism. The undefined points of $\bar{F}$ can be resolved after a finite number of blowings up of points $\pi : Y \to C \times \bar{D}$ into a morphism $F' : Y \to X$. Let $E \subseteq Y$ be the exceptional curve of the last blowing up that was needed to get the morphism F'. Then F' does not collapse E to a point, and the image $F'(E)$ is the required rational curve in X.

References for this section. Mori's theorem occurs in his paper [109], which is also where the proof of the dimension estimate for representable functors (23.4) was proved. The Isom scheme is used by Mumford in his discussion of Picard groups of moduli problems [114]; cf. §§26, 27. The general theory of the Hom and Isom schemes as representable functors appears in the same exposé of Grothendieck's as the Hilbert scheme [45]. For the existence of the Hilbert scheme, see also [161] and [152]. The existence of the Picard scheme is in [45, exposé 232]. See also [79].

Exercises.

24.1.   Give a proof of (24.2) using the valuative criteria of separatedness and properness.

24.2.   Let $\mathcal{M}$ be the family of invertible sheaves of fixed degree d on an integral projective curve X.

(a) Show that $\mathcal{M}$ is bounded.
(b) Show that $\mathcal{M}$ is separated.
(c) But show that $\mathcal{M}$ may not be complete.

24.3.   Let $\mathcal{M}$ be the family of rank 2 vector bundles of degree 0 on a nonsingular projective curve X of genus g.

(a) Show that the family $\mathcal{M}$ is not bounded.
(b) Show that the family $\mathcal{M}$ is not separated.
(c) But show that the family $\mathcal{M}$ is complete.
   Note: We will see later (§28) that for rank 2 vector bundles of degree d on a curve of genus g:
(d) If one restricts to simple or stable vector bundles, the family is bounded.
(e) The family of stable bundles is separated, but the family of simple bundles is not necessarily separated.
(f) For d even the family of stable bundles is not complete, but for d odd it is complete.

24.4.   For each $g \geq 0$, the family of nonsingular projective curves of genus g is bounded. *Hint:* Take projective embeddings of sufficiently high degree and use boundedness of the Hilb functor.

24.5.   If X is an integral projective curve over k, show that the Picard scheme, which exists by (24.6), is a disjoint union of nonsingular varieties of dimension $g = p_a(X)$. If furthermore X is nonsingular show that each component of the Picard scheme is proper over k.

24.6. Let X_0 be a projective scheme over k with only finitely many automorphisms. Show that a jump phenomenon X/T with special fiber X_0 is impossible, as follows.

Suppose X/T is a flat family of projective schemes with X_t all isomorphic for $t \neq 0$, and fiber X_0 at $t = 0$ not isomorphic to X_t for $t \neq 0$. Consider the two families $X \times T$ and $T \times X$ over $T \times T$, and let $S = \mathrm{Isom}_{T \times T}(X \times T, T \times X)$ (24.10.2). Thus a point of S corresponds to a triple (t_1, t_2, φ), where φ is an isomorphism of $X_{t_1} \xrightarrow{\sim} X_{t_2}$. Consider the projection $\pi : S \to T$ onto the first factor T. Show that the fiber $\pi^{-1}(0)$ is finite, while for each $t \neq 0$, $\pi^{-1}(t)$ has dimension ≥ 1. This will contradict the semicontinuity of dimension of fibers of a morphism [57, II, Ex. 3.22].

24.7. (a) Let X, X' be flat families of projective varieties over a nonsingular curve T of finite type over k algebraically closed. Suppose for every closed point $t \in T$ the fibers X_t and X_t' are isomorphic. Show that there is another nonsingular curve T' and a dominant morphism $T' \to T$ such that the two families $X \times_T T'$ and $X' \times_T T'$ obtained by base extension are isomorphic. *Hint:* Use an Isom scheme.

(b) Let X/T be a flat family of projective varieties over a curve T as in (a), and assume that X/T is a *fiberwise trivial* family, i.e., the fibers X_t for closed points $t \in T$ are all isomorphic to each other. Show that there are another curve T' and a dominant base extension $T' \to T$ such that the family $X' = X \times_T T'$ is trivial over T'.

(c) Let X_0 be a rigid projective scheme over k, and let X/T be a flat family of projective schemes over a curve T as above with fiber over a point $0 \in T$ isomorphic to X_0. Show that this family is fiberwise trivial over some open neighborhood of $0 \in T$, and hence by (b) becomes trivial after a base extension $T' \to T$ as above. *Hint:* Let X'/T be the trivial family $X_0 \times T$, and consider the scheme $\mathrm{Isom}_T(X, X')$. Use the fact that every infinitesimal deformation of X_0 is trivial (Ex. 10.3) to show that the image of the Isom scheme in T must contain an open neighborhood of the point $0 \in T$.

24.8. Give an example to show that the functor of isomorphism classes of nonsingular projective varieties is not separated in general.

24.9. Show, in contrast to the previous exercise, that for any fixed g, the functor of families of nonsingular projective curves of genus g over k is separated. *Hint:* Here is one way to approach the problem. Given a family of curves X/T, use an embedding into projective space and the completeness of the Hilbert scheme to embed this in a larger family, allowing singular fibers, so that we may assume that X is a projective surface mapping to a projective nonsingular curve T. Next, resolve the singularities of X, so that we may assume that X is a nonsingular surface. Now, given two such surfaces over T whose fibers are isomorphic for $t \in U$, an open subset, use (Ex. 24.7) to conclude that X and X' are birationally equivalent. Therefore [57, V, 5.5] there is a third surface X'' together with birational morphisms of X'' to X and to X', each of which can be factored into a sequence of monoidal transformations. Now if X_0 and X_0' are two special fibers that are both nonsingular, but $0 \notin U$, then the inverse images of X_0 and X_0' in X'' are obtained from each by adding rational curves, and are equal. Therefore $X_0 \cong X_0'$.

25. Curves of Genus Zero

Any complete nonsingular curve of genus 0 over an algebraically closed field k is isomorphic to $\mathbb{P}^1_k$ [57, IV, 1.3.5], so you may think that the moduli problem for curves of genus 0 is trivial. But even in this case, there are some interesting aspects to the problem.

So let us consider the moduli problem for nonsingular projective curves of genus 0 over an algebraically closed field k. The set $\mathcal{M}$ of isomorphism classes of such curves has just one element, namely $\mathbb{P}^1_k$. A *family* of curves of genus 0 over a scheme S will be a scheme X, smooth and projective over S, whose geometric fibers are curves of genus 0. That means that for each $s \in S$, if we take the fiber X_s and extend the base field to the algebraic closure $\overline{k(s)}$ of the residue field $k(s)$, then the new curve $X_{\bar{s}} = X_s \times_{k(s)} \overline{k(s)}$ is a nonsingular projective curve of genus 0 over the field $\overline{k(s)}$.

Proposition 25.1. *The one-point space* $M = \operatorname{Spec} k$ *is a coarse moduli scheme for curves of genus 0, and it has a tautological family.*

Proof. The first condition (a) for a coarse moduli scheme (§23) is satisfied because the one point of M corresponds to the one curve $\mathbb{P}^1_k$. We can also see right away that there is a tautological family: just take $\mathbb{P}^1_k / \operatorname{Spec} k$. For any family X/S of curves of genus 0, where S is a scheme over k, there is a unique morphism $S \to M = \operatorname{Spec} k$, so we have the required morphism of functors $\varphi : \mathcal{F} \to h_M$.

For the universal property (b), suppose $\psi : \mathcal{F} \to h_N$ is any morphism of functors, where $\mathcal{F}$ is our functor of families of curves of genus 0. Then in particular, the family $\mathbb{P}^1_k / M$ determines a morphism $e : M \to N$. We need to show that ψ factors through the morphism $\varphi : \mathcal{F} \to h_M$ described above, which maps every scheme S/k to $\operatorname{Spec} k$.

Lemma 25.2. *If C is an Artin ring with residue field k algebraically closed, then any family $X/\operatorname{Spec} C$ of curves of genus 0 is trivial, namely isomorphic to $\mathbb{P}^1_{\operatorname{Spec} C}$.*

Proof. Since k is algebraically closed, the special fiber X_0 is just $\mathbb{P}^1_k$. Then by our infinitesimal study of deformations (10.3) the choices at each step are given by $H^1(X_0, \mathcal{T}_{X_0}) = 0$. Thus at each step there is a unique deformation, which must be equal to $\mathbb{P}^1_{\operatorname{Spec} C}$. Thus $\mathbb{P}^1$ is a rigid scheme (5.3.1), (Ex. 10.3).

Proof of (25.1), continued. Let X be a family of curves of genus 0 over a scheme S of finite type over k. For any closed point $s \in S$, the fiber X_s is just $\mathbb{P}^1_k$, so the point s must go to the same point $n_0 \in N$ as the image of the morphism $e : M \to N$. Thus all closed points of S go to n_0. But we need more. We need to know that the morphism $S \to N$ factors through the reduced point n_0 as a closed subscheme of N. And this follows from the lemma, because the restriction of the family on S to any artinian closed subscheme of S will be trivial, and therefore will factor through the reduced scheme $\operatorname{Spec} k$.

If S is not of finite type over k, a similar argument, making base extensions to geometric points of S and Artin rings over them, shows in that case also that the associated map $S \to N$ factors through the reduced point $n_0 \in N$, and so the morphism ψ factors through φ, as required.

Example 25.2.1. Here we show that the one-point space M is not a fine moduli space for curves of genus 0. Just think of the theory of ruled surfaces. A *ruled surface* is a nonsingular projective surface X together with a morphism π to a nonsingular projective curve C whose fibers are copies of $\mathbb{P}^1$ and that has a section, and therefore is isomorphic to $\mathbb{P}(\mathcal{E})$ for some rank 2 vector bundle $\mathcal{E}$ on C [57, V, 2.2]. In particular, this implies that C can be covered by open subsets U_i over which X is trivial, i.e., $\pi^{-1}(U_i) \cong U_i \times \mathbb{P}^1$. On the other hand, there are many ruled surfaces X that are not trivial. Since a ruled surface is in particular a family of curves of genus 0 parametrized by C according to our definition, the functor $\mathcal{F}$ is not a sheaf for the Zariski topology: the structure of X is not determined by knowing its structure locally on C, so the moduli space cannot be a fine moduli space (23.5).

Another way of putting this is if our space M were a fine moduli space, then every family of curves of genus 0 would be trivial, i.e., a product of the base with $\mathbb{P}^1$, and the ruled surfaces give examples of families that are locally trivial but not globally trivial.

Example 25.2.2. Here we show that families of curves of genus 0 need not even be locally trivial. Let $A = k[t, u]$, and consider the curve in $\mathbb{P}^2_A$ defined by $tx^2 + uy^2 + z^2 = 0$. We take $S = \operatorname{Spec} A - \{tu = 0\}$, and take X to be this family of curves over S. This is a family of curves of genus 0, but it is not even locally trivial. If it were, the generic fiber X_η defined by the same equation over the field $K = k(t, u)$ would be isomorphic to $\mathbb{P}^1_K$. But X_η has no rational points over K. A rational point would be given by taking $x = f(t, u)$, $y = g(t, u)$, $z = h(t, u)$, where f, g, h are rational functions in t and u, not all zero, satisfying the above equation. Clearing denominators, we may assume that f, g, h are polynomials. Then, looking at the terms of highest degree in t, u, we see that they cannot cancel in the equation, which gives a contradiction.

This is an example of a *fiberwise trivial* family (Ex. 24.7), namely a family in which all the geometric fibers are isomorphic to each other. But this family is not trivial, and not even locally trivial. We will see, however, that it is trivial for the étale topology (Ex. 25.1).

This phenomenon comes from the fact that over a non-algebraically closed field, there are curves of genus 0 that have no rational points. For example over $\mathbb{R}$ there is the conic $x^2 + y^2 + z^2 = 0$ in $\mathbb{P}^2$, which has no real points. Over the rational numbers there are many different nonisomorphic curves of genus 0 (25.3.1). This is part of the reason for the subtleties in families. We can improve the situation by changing the moduli problem slightly to consider pointed curves.

Definition. A *pointed* curve of genus 0 over k will be a curve of genus 0 together with a choice of a point, rational over k. So the set of objects $\mathcal{M}$ we are considering still has just one element, namely $\mathbb{P}^1_k$ together with a chosen point P. (The choice of point does not matter, since the automorphisms of $\mathbb{P}^1$ are transitive on rational points.) A *family* of pointed curves will be a flat family X/S, whose geometric fibers are all curves of genus 0, together with section $\sigma : S \to X$ (which some people call an S-point of X). The section σ induces a point on each fiber in a coherent way.

As before, we can show that the one-point space $M = \operatorname{Spec} k$ is a coarse moduli scheme for pointed curves of genus 0, and that it has a tautological family. Also as before it is not a fine moduli scheme, because of the ruled surfaces exhibited in (25.2.1). What is different in this case is that now all families are locally trivial.

Proposition 25.3. *Any family X/S of pointed curves of genus 0 is locally trivial, that is, every point $s \in S$ has an open neighborhood U such that $\pi^{-1}(U) \cong \mathbb{P}^1_U$. In particular, a pointed curve of genus 0 over any field k (not necessarily algebraically closed) is isomorphic to $\mathbb{P}^1_k$.*

Proof. (Cf. [57, V, 2.2] for a special case.) Given the family $\pi : X \to S$ and the section $\sigma : S \to X$, we let D be the scheme-theoretic image of σ. Then D is flat over S, and its restriction to any fiber is one point, so D is a Cartier divisor on X. Let $\mathcal{L}$ be the associated invertible sheaf on X. Then for each point $s \in S$, $H^0(X_s, \mathcal{L}_s)$ is a 2-dimensional vector space, and $H^1(X_s, \mathcal{L}_s) = 0$. Now we apply cohomology and base extension [57, III, 12.11] to the maps

$$\varphi^i(s) : R^i f_*(\mathcal{L}) \otimes k(s) \to H^i(X_s, \mathcal{L}_s).$$

For $i = 1$, since $H^1(X_s, \mathcal{L}_s) = 0$, $\varphi^1(s)$ is surjective, hence an isomorphism, so $R^1 f_*(\mathcal{L}) = 0$. The zero sheaf is locally free, so we find that $\varphi^0(s)$ is surjective, hence also an isomorphism. Since $\varphi^{-1}(s)$ is trivially surjective, we find that $f_*\mathcal{L}$ is locally free of rank 2 on S. Call it $\mathcal{E}$. Then the natural map $\pi^*\mathcal{E} \to \mathcal{L} \to 0$ determines a morphism $X \to \mathbb{P}(\mathcal{E})$, which is an isomorphism on each fiber, hence an isomorphism. If we take $U \subseteq S$ to be an open set over which $\mathcal{E}$ is free, then $\pi^{-1}(U) \cong \mathbb{P}^1_U$ as required.

Remark 25.3.1. A deeper study of families of curves of genus 0 inevitably leads to the Brauer group. In case of a field k, the *Brauer group* $\operatorname{Br}(k)$ is defined as the Galois cohomology $H^2(G, K^*)$ where K is the separable closure of k and G is the Galois group of K/k [155]. A *Brauer–Severi variety* over k is a scheme V/k such that $V \times_k K \cong \mathbb{P}^n_K$ for some n. Then $\operatorname{Br}(k)$ can be described as the union, over all n, of isomorphism classes of Brauer–Severi varieties of dimension n over k. Thus to a curve of genus 0 over k corresponds an element of the Brauer group, which vanishes if and only if that curve is isomorphic to $\mathbb{P}^1_k$.

In the case of a scheme S, Grothendieck [46] defines the *Brauer group* of S as a certain subgroup of $H^2_{\text{ét}}(S, \mathbb{G}_m)$, classifying Brauer–Severi schemes over S, that is, flat schemes X/S whose geometric fibers are all projective spaces. Thus to a flat scheme X/S with geometric fibers $\mathbb{P}^1$ is associated an element of $\mathrm{Br}(S)$ that is zero if and only if X/S is locally trivial for the Zariski topology, i.e., $X \cong \mathbb{P}(\mathcal{E})$ for some rank 2 vector bundle on S.

Exercises.

25.1. For the family X/S described in (25.2.2), let $S' \to S$ be the base extension obtained by adjoining $\sqrt{t}$ and $\sqrt{u}$. Show that S' is a finite surjective étale morphism, and that the extended family X'/S' is trivial. In this case we say that the family X/S is *isotrivial*, meaning that it is trivialized by a finite surjective étale base change.

25.2. Show that any flat family X/S of curves of genus 0 is *locally isotrivial*, or *locally trivial in the étale topology*, meaning that there is a surjective étale morphism $S' \to S$, not necessarily finite, for which the extended family X'/S' becomes trivial. *Hint:* Let $\omega_{X/S}$ be the relative canonical sheaf, and for an open affine subset $U \subseteq S$, let $D \subseteq X_U$ be an effective divisor associated to the invertible sheaf $\omega_{X/S}^{-1}$ that is supported at two distinct points in each fiber. Show that $D \to U$ is étale, and that X_D/D has a section. Then use (25.3).

25.3. Take the functor of families of pointed curves of genus 0, define a new functor that we could call the "sheafification" of the original functor in the Zariski topology, and show that the new functor is representable.

25.4. Another way to get a representable functor of curves of genus 0 is to rigidify the curves by taking three distinct rational points. So a family is a smooth proper morphism X/S whose geometric fibers are curves of genus 0, with the additional data of three sections $\sigma_1, \sigma_2, \sigma_3 : S \to X$, such that at each fiber the three sections have distinct support. Show that the corresponding functor is represented by a one-point space, and the universal family is $\mathbb{P}_k^1$ with three points $0, 1, \infty$.

26. Moduli of Elliptic Curves

In this section we will apply the theory we have developed to elliptic curves. Our provisional definition is that an *elliptic curve* over an algebraically closed field k is a nonsingular projective curve of genus one. We will assume characteristic $k \neq 2, 3$ for simplicity throughout this section.

If one studies one elliptic curve at a time, there is a satisfactory theory, explained in [57, IV, §4]. To each elliptic curve C over k one can assign an element $j(C) \in k$, called the j-invariant, in such a way that two elliptic curves over k are isomorphic if and only if they have the same j-invariant. Furthermore, for any $j \in k$ there is an elliptic curve with j-invariant j. Thus the set of isomorphism classes of elliptic curves over k is in one-to-one correspondence with the set of closed points of the affine line $\mathbb{A}_k^1$.

The problem of moduli is to understand not only individual curves, but also flat families X/S whose geometric fibers are elliptic curves. In particular one can study the formal local problem of deformations over Artin rings of a given elliptic curve C/k. Our general theory tells us that this functor has a miniversal family (18.1), but since $h^0(T_C) \neq 0$, our basic result (18.3) does not guarantee that the local functor is pro-representable. On the other hand, if we consider pointed elliptic curves, namely curves C/k with a fixed point $P \in C$, and consider deformations with a section extending the point, then we have seen (18.4.2), (Ex. 18.2) that the local deformation functor is pro-representable.

We will first prove that the local functor of deformations of a pointed elliptic curve over Artin rings is equivalent to the deformations of the elliptic curve without its point.

Proposition 26.1. *Let C_0 be an elliptic curve over k, and let F be the functor of local deformations of C_0 over local Artin rings $(A, \mathfrak{m})$. Let $P_0 \in C_0$ be a closed point and let F' be the functor of local deformations of the pointed curves C_0, P_0, i.e., an element of $F'(A)$ is a family C/A, flat over A, together with a section $\sigma : \operatorname{Spec} A \to C$, and a closed immersion $C_0 \subseteq C$, so that $\sigma(\mathfrak{m}) = P_0$. Then the "forgetful" morphism $F' \to F$, forgetting the section σ, is an isomorphism of functors.*

Proof. Given a deformation $C_0 \subseteq C$ and given $P_0 \in C_0$, the problem of finding a section σ of C reducing to $P_0 \in C_0$ is a question of the Hilbert scheme of P_0 in C_0. The normal sheaf $\mathcal{N}_{P_0/C_0}$ is a 1-dimensional vector space on the 1-point space P_0, so $h^1(\mathcal{N}_{P_0/C_0}) = 0$, and there are no obstructions (6.2). Hence P_0 deforms to give a section σ. Therefore the map $F'(A) \to F(A)$ is surjective for each A.

To show that $F'(A) \to F(A)$ is injective, we use induction on the length of A. For $A = k$, we note that since k is algebraically closed, every elliptic curve has a closed point, and the choice of closed point does not matter, since the group structure on the curve provides automorphisms that act transitively on the set of closed points.

Now suppose we are given C and a section σ over A, as well as C' over A', where $A' \to A$ is a small extension. Then the ambiguity in extending σ lies in $H^0(\mathcal{N}_{P_0/C_0})$. On the other hand, the automorphisms of C' leaving C fixed are given by $H^0(T_{C_0})$. One checks easily that the natural map $H^0(T_{C_0}) \to H^0(\mathcal{N}_{P_0/C_0})$ is an isomorphism. Hence there is a unique pair (C', σ') up to isomorphism for each C' given, and so $F'(A') \to F(A')$ is bijective.

Remark 26.1.1. Since we know that the functor F' is pro-representable, it follows that the functor F of deformations of (unpointed) elliptic curves is also pro-representable, even though $h^0(T_{C_0}) \neq 0$.

Remark 26.1.2. Even though the formal local functors F and F' are isomorphic, the same does not hold for the global functor of isomorphism classes of families X/S of elliptic curves, because there are families having no section. Consider the family of plane curves defined by $x^3 + ty^3 + t^2z^3 = 0$ in $\mathbb{P}_A^2$, where $A = k[t, t^{-1}]$. This is a flat family of elliptic curves, but has no section, because to give a section would be to give $x = f(t)$, $y = g(t)$, $z = h(t)$, polynomials in t and t^{-1} satisfying this equation, and this is impossible (just consider the terms of highest degree in f, g, h).

For this reason, when studying global moduli we must make a choice whether to consider families of unpointed or pointed elliptic curves. We choose the latter, both because it is easier to handle technically, and also because it gives a better analogy with the case of curves of genus $g \geq 2$, which have only finitely many automorphisms. So for the rest of this section we will use the following definitive definition.

Definition. An *elliptic curve* over a scheme S is a flat morphism $X \to S$ whose geometric fibers are all nonsingular projective curves of genus 1, together with a section $\sigma : S \to X$. In particular, an elliptic curve over any field k is a smooth curve C of genus 1 together with a rational point $P \in C$.

Now we turn to the question of moduli. We fix k algebraically closed, and for any scheme S/k consider the functor $F(S) = \{$isomorphism classes of elliptic curves over $S\}$. We ask what kind of moduli space we can find for F.

Proposition 26.2. *The functor F does not have a fine moduli space.*

Proof. There are several reasons one can give for this. One is that the crude local functor F_1 of local families C/A such that $C \otimes_A k \cong C_0$, but without specifying the inclusion $C_0 \subseteq C$, is not pro-representable (18.4.2). We have seen that this would be a necessary condition for the global functor to be representable (23.3).

A second reason is that if F had a fine moduli space, i.e., if F were representable, then any fiberwise trivial family would be trivial (23.1.1). One way to make a fiberwise trivial family is to take a constant family over $\mathbb{P}^1$, identify the fibers over 0 and 1 by a nonconstant automorphism τ that sends the distinguished point P to itself, and glue to get a nonconstant family over a nodal curve whose fibers are all isomorphic.

Another way to make a fiberwise trivial family is to write an equation like $y^2 = x^3 + t$ over $A = k[t, t^{-1}]$. For each t we get a curve with $j = 0$, but to write an isomorphism between this one and the constant family $y^2 = x^3 + 1$, we need $t^{1/6}$, which is not in the ring A.

Proposition 26.3. *The j-line $\mathbb{A}_j$ is a coarse moduli space for the functor F of families of elliptic curves.*

Proof. Recall (§23) that to be a coarse moduli space for the functor F means several things:

(a) The closed points of $\mathbb{A}_j$ are in one-to-one correspondence with the isomorphism classes of elliptic curves over k. This we know from the basic theory [57, IV, 4.1].

(b) For any family X/S there is a morphism $\varphi : S \to \mathbb{A}_j$ such that for each closed point $s \in S$, $\varphi(s)$ is the j-invariant of the fiber over s. This step is easy. Given X/S together with its section σ, for any open affine $U = \operatorname{Spec} A \subseteq S$, we define an embedding of $X_A \hookrightarrow \mathbb{P}_A^2$ using the divisor 3σ. Then by rational operations over the ring A as in [57, IV, §4] (and here we use the assumption that characteristic $k \neq 2, 3$) we bring the equation of the image into the form $y^2 = x^3 + ax + b$, with $a, b \in A$. Then

$$j = 12^3 \cdot \frac{4a^3}{4a^3 + 27b^2}$$

gives the desired morphism from $\operatorname{Spec} A$ to the j-line. These patch together to give $\varphi : S \to \mathbb{A}_j$.

(c) Lastly, we must show that the j-line is universal with property (b). So let N be some other scheme together with a morphism of the functor F to h_N, i.e., a functorial assignment, for each family X/S of a morphism $S \to N$. We consider in particular the family given by the equation $y^2 = x(x - 1)(x - \lambda)$ over the λ-line $\operatorname{Spec} B$ with $B = k[\lambda, \lambda^{-1}, (\lambda - 1)^{-1}]$. Then there is a morphism $\varphi : \operatorname{Spec} B \to N$. Furthermore, this morphism is compatible with the action of the group G of order 6 acting on the λ-line consisting of the substitutions $\{\lambda, \lambda^{-1}, 1 - \lambda, (1 - \lambda)^{-1}, \lambda(\lambda - 1)^{-1}, (\lambda - 1)\lambda^{-1}\}$, because the transported family X' has fibers isomorphic to those of X. Hence the morphism φ factors through $\operatorname{Spec} B^G$, where B^G is the fixed ring of the action of G on B. All that remains is to identify B^G with $k[j]$. Clearly $j \in B^G$. Considering the function fields $k(j) \subseteq k(B^G) \subseteq k(B)$, the latter is of degree 6 over the two former, so $k(j) = k(B^G)$. Next note that B is integral over $k[j]$: the defining equation of j in terms of λ gives

$$\lambda^2(\lambda - 1)^2 j = 256(\lambda^2 - \lambda + 1)^3.$$

This shows that λ is integral over $k[j]$. Rewriting this equation in terms of λ^{-1} and $(\lambda - 1)^{-1}$ shows that they too are integral over $k[j]$. Therefore B^G is integral over $k[j]$. But these two rings have the same quotient field, and $k[j]$ is integrally closed, so $k[j] = B^G$.

Thus we obtain a morphism $\mathbb{A}_j \to N$, so $\mathbb{A}_j$ has the desired universal property.

Remark 26.3.1. The coarse moduli space $\mathbb{A}_j$ does not have a tautological family. For suppose X/S is a family of elliptic curves, and $s_0 \in S$ is a point whose fiber C_0 has $j = 0$. In an affine neighborhood $\operatorname{Spec} A$ of s_0 we represent the family by $y^2 = x^3 + ax + b$ with $a, b \in A$. At the point s_0, since $j = 12^3 \cdot 4a^3/(4a^3 + 27b^2)$, we must have $a \in \mathfrak{m}$, the maximal ideal of A at the point s_0. Hence $j \in \mathfrak{m}^3$, and the morphism $S \to \mathbb{A}_j$ is ramified at the point s_0. In particular, S cannot be $\mathbb{A}_j$.

To summarize the discussion so far, we consider the functor of families of (pointed) elliptic curves. We have seen that the local deformation functor is pro-representable for each elliptic curve. The global functor does not have a fine moduli space, but it does have a coarse moduli space. The coarse moduli space does not have a tautological family. The global functor is not a sheaf for the Zariski topology (Ex. 26.1).

This is about all we can say within the frame of discourse up to this point. But it is unsatisfactory, since it does not give us, as in the case of a representable functor, a complete description of all possible families of elliptic curves. To go further we must expand the range of concepts, and this leads to the world of Grothendieck topologies, algebraic spaces, and stacks. Without explaining what any of these are, we will rather show explicitly how those theories manifest themselves in the case of elliptic curves.

The main idea is to think of replacing the Zariski topology by the étale topology. A local property will be one that holds after an étale base extension instead of on an open subset. I would like to say that the moduli functor is "representable to within étale morphisms," or that "there is a fine moduli space to within étale morphisms." To be precise, we make a definition and prove a theorem.

Definition. Suppose we have a class of algebrogeometric objects $\mathcal{M}$, defined over our base field k, for which we wish to classify the flat families, up to isomorphism, as in §23. A *modular family* of elements of $\mathcal{M}$ is a flat family X/S, with S a scheme of finite type over k, such that:

(a) For each object $C \in \mathcal{M}$, there is at least one and there are at most finitely many closed points $s \in S$ for which the fiber X_s is isomorphic to C.

(b) For each $s \in S$, the complete local ring $\hat{\mathcal{O}}_{S,s}$, together with the formal family induced from X, pro-represents the functor of local deformations of the fiber X_s.

(c) For any other flat family X'/S' of elements of $\mathcal{M}$, there exists a surjective étale morphism $S'' \to S'$ and a morphism $S'' \to S$ such that $X' \times_{S'} S'' \cong X \times_S S''$ as families over S''.

Remark 26.3.2. It follows from the definition that if X_1/S_1 and X_2/S_2 are two modular families, then there is a third modular family X_3/S_3 and surjective étale maps $S_3 \to S_1$ and $S_3 \to S_2$ such that $X_1 \times_{S_1} S_3 \cong X_3 \cong X_2 \times_{S_2} S_3$ as families over S_3.

Remark 26.3.3. Note that the definition of a modular family cannot be made purely in terms of the associated functor of families up to isomorphism, because the condition (b) concerns the functor of local deformations of a fiber X_s, and these involve a given identification of the special fiber, not just its class up to isomorphism.

Theorem 26.4. *There exists a modular family X/S of elliptic curves over k.*

Proof. (a) We will show that the family of plane cubic curves $y^2 = x(x-1)(x-\lambda)$ over the λ-line $\operatorname{Spec} B$, where $B = k[\lambda, \lambda^{-1}, (\lambda-1)^{-1}]$, is a modular family. First of all, we know that every elliptic curve is isomorphic to one of these for some $\lambda \neq 0, 1$, and that each isomorphism type occurs 2, 3, or 6 times.

(b) Next we need to show that the completion of this family at any point pro-represents the local deformation functor. Since in any case by pro-representability there is a morphism from the formal family over the λ-line to the pro-representing family, and both of these are smooth and one-dimensional (18.4.2), it will be sufficient to show that the induced map on Zariski tangent spaces is nonzero. So let $y^2 = x(x - 1)(x - \lambda - t)$ be the induced family over the dual numbers $D = \operatorname{Spec} k[t]/t^2$ at the point λ. We have only to show that this family is nontrivial over D. Now two curves in $\mathbb{P}_D^2$ with equations of the form above are isomorphic if and only if their λ-values are interchanged by the six-element group G. This group sends any $\lambda \in k$ to another $\lambda \in k$ and never to $\lambda + t$; hence the deformation is nontrivial.

(c) Now let X'/S' be any family of elliptic curves, and let X/S be a family with properties (a), (b) above. Then over $S' \times S$ we have two families $X \times S'$ and $X' \times S$. Let $S'' = \operatorname{Isom}_{S' \times S}(X \times S', X' \times S)$ (24.10.2). Then there is an isomorphism over S'', namely $X \times_S S'' \xrightarrow{\sim} X' \times_{S'} S''$. Furthermore, S'' is universal with this property.

Now I claim that $S'' \to S'$ is surjective and étale. For any point $s' \in S'$, let $C = X'_{s'}$ be the corresponding fiber. Then the isomorphism type of C occurs at least once and at most finitely many times in the family X/S, say at points $s_1, \ldots, s_n \in S$. Furthermore, since the automorphism group G of C as an elliptic curve is finite, for each s_i the scheme $\operatorname{Isom}_k(X'_{s'}, X_{s_i})$ is finite. Thus by the universal property of the Isom scheme, the fiber of S'' over S' is a finite nonempty set. Hence the map $S'' \to S'$ is surjective and quasi-finite.

Finally, consider a point $s'' \in S''$ lying over $s' \in S'$. This fixes the corresponding point $s_i \in S$, and also fixes the isomorphism of $X'_{s'}$ with X_{s_i}. For any Artin ring A, quotient of $\mathcal{O}_{S',s'}$, we get an induced family over $\operatorname{Spec} A$. Since S pro-represents the functor of local deformations, there is a unique morphism of $\operatorname{Spec} A \to S$ at the point s_i inducing an isomorphic family. Furthermore, the isomorphism on the closed fiber having been fixed, there are no further automorphisms of the family over $\operatorname{Spec} A$ (recall the proof of local pro-representability (18.3)). Hence there is a unique morphism of $\operatorname{Spec} A$ to S'' at the point s''. This implies that the induced homomorphism on complete local rings $\hat{\mathcal{O}}_{S',s'} \to \hat{\mathcal{O}}_{S'',s''}$ is an isomorphism, and hence $S'' \to S'$ is étale, as required.

Corollary 26.5. *If Y/T is a fiberwise trivial family of elliptic curves, then there exists a finite étale map $T' \to T$ such that the base extension Y'/T' is isomorphic to the trivial family. In other words, the family Y/T is isotrivial* (Ex. 25.1).

Proof. Indeed, let X/S be a modular family. Then there is a surjective étale morphism $T' \to T$ together with a morphism $T' \to S$ such that the extended families over T' are isomorphic. But Y/T is fiberwise trivial, so the image of T' in S is a single point, so the family over T' is trivial. Furthermore, if we take T' to be given by the Isom scheme as in the proof of (26.4), then since the fibers of Y/T are isomorphic, each point of T will have the same number of points of T' lying over it, so $T' \to T$ will be a finite morphism.

Proposition 26.6. *If X/S is a modular family of elliptic curves, the corresponding map of S to the coarse moduli space A_j is étale over points where $j \neq 0, 12^3$; ramified of order 2 over $j = 12^3$ and ramified of order 3 over $j = 0$.*

Proof. Writing

$$j = 256 \frac{(\lambda + \omega)^3 (\lambda + \omega^2)^3}{\lambda^2 (\lambda - 1)^2},$$

where $\omega^3 = 1$, shows that at $\lambda = -\omega$, corresponding to $j = 0$, the map from the λ-line to the j-line is ramified of order 3. At $\lambda = -1, \frac{1}{2}, 2$, corresponding to $j = 12^3$, there are three roots, and the map is of order 6, so it is ramified of order 2. For $j \neq 0, 12^3$, there are six values of λ, so it is unramified.

Since the modular family is unique up to étale morphisms, the same holds for any modular family.

Remark 26.6.1. Thus we may think of a modular family as "the moduli space," uniquely determined up to étale morphisms, where the universal mapping property holds after an étale morphism. Or we may think of the j-line as "the moduli space," but where we need $\sqrt{j - 12^3}$ and $\sqrt[3]{j}$ as local parameters at the points $j = 12^3$ and $j = 0$. Still this does not tell us everything about the functor F, in contrast to the case of a representable functor, where knowledge of the representing scheme and its universal family is equivalent to knowledge of the functor of families. We can ask, what further data do we need to know the functor entirely? The following remarks will reflect on this question, without, however, giving a complete answer.

Remark 26.6.2. If X/S is a modular family and $S' \to S$ is any surjective étale morphism, then $X' = X \times_S S'/S'$ is another modular family. Thus there are bigger and bigger modular families. This leads us to ask whether there is a smallest modular family. The answer is no. Indeed, there is a family over the j-line minus the points $0, 12^3$, defined by the equation

$$y^2 = x^3 + ax + b, \text{ with } a = b = \frac{27}{4} \cdot \frac{j}{12^3 - j}.$$

A simple calculation shows that for any $j \neq 0, 12^3$, this defines an elliptic curve with the corresponding j-invariant. To get a modular family, we need to take a disjoint union with some patches of families containing curves with $j = 0$ and $j = 12^3$. There is no smallest such choice.

Remark 26.6.3. If we confine our attention to elliptic curves with $j \neq 0, 12^3$, then $\mathbb{A}_j - \{0, 12^3\}$ is a coarse moduli space, and it has a tautological family (§23), given in the previous remark. However, this tautological family is not a universal family. For any family X/S, there is a unique morphism $S \rightarrow \mathbb{A}_j - \{0, 12^3\}$ sending points $s \in S$ to the j-value of the fiber X_s, but the pullback of our tautological family may not be isomorphic to X, so the functor is still not representable, even restricting to $j \neq 0, 12^3$. To see this, note that $y^2 = x^3 + j^2 a x + j^3 b$, with the same a, b as above, is another tautological family over $\mathbb{A}_j - \{0, 12^3\}$, but it does not become isomorphic to the previous one until we take a double covering defined by $\sqrt{j}$. So even in this restricted case, there is no minimal modular family.

Remark 26.6.4. There is one more question one could ask in trying to make sense of the functor of all possible families of elliptic curves. Though there is not a universal family, is there perhaps a small set (say finite) of modular families X_i/S_i such that for any family X/S there is a morphism $S \rightarrow S_i$ for some i such that $X \cong X_i \times_{S_i} S$? No, even this last hope is dashed to the ground by the following examples of incomparable families over subsets of the j-line.

Let X_0/S_0 denote the family described in (26.6.2) over $S_0 = \mathbb{A}_j - \{0, 12^3\}$. For any open set $U \subseteq S_0$, let $T \rightarrow U$ be an étale cover of order 2, and let $X = X_0|_U \times_U T$. Now for each $u \in U$, let t_1, t_2 be the two points lying over u, and identify the fibers X_1, X_2 at t_1, t_2 via the automorphism τ of order 2. Then glue to get a new family X'_T over U. Note that we can recover T as $\mathrm{Isom}_U(X, X'_T)$. The family X'_T is isomorphic to X if and only if T is the trivial cover.

Now if $\pi : C \rightarrow \mathbb{P}^1$ is any hyperelliptic curve, and U is $\mathbb{P}^1$ minus $0, 12^3, \infty$, and the branch points of π, and $T = C|_U$, then we get a family X'_T over U. Two of these for different C, C' are isomorphic on a common open set if and only if the corresponding hyperelliptic curves are isomorphic.

Thus there is not a finite number, there is not even a collection of such families as we desired parametrized by a finite union of finite-dimensional algebraic varieties!

Remark 26.6.5 (Completion of the moduli space). Having once found the coarse moduli space $\mathbb{A}_j$, a natural question is, what extra objects can we consider in order to obtain a complete moduli space? Here we will show that if in addition to elliptic curves as above, one allows irreducible nodal curves with $p_a = 1$, together with a fixed nonsingular point, the whole theory extends. We consider families X/S where the fibers are elliptic curves or pointed nodal curves (the point being chosen as a smooth point of the nodal curve). The projective line $\mathbb{P}^1$ acts as a coarse moduli space, taking $j = \infty$ for the nodal curve. The family $y^2 = x(x-1)(x-\lambda)$ over the whole affine λ-line is a modular family in which the values $\lambda = 0, 1$ correspond to nodal curves. The proofs above all extend without difficulty, once we know the deformation theory of the nodal curve, which we explain in the next remark.

Remark 26.6.6 (Deformation theory of the nodal elliptic curve).
We consider a reduced irreducible curve C over k of arithmetic genus $p_a = 1$
having one node as its singularity (such as the curve $y^2 = x^2(x-1)$ in $\mathbb{P}^2$).
The tangent space $\mathrm{Def}(C)$ to its deformation theory fits in an exact sequence
(Ex. 5.7)

$$0 \to H^1(T_C) \to \mathrm{Def}(C) \to H^0(T_C^1) \to H^2(T_C^0).$$

There are no obstructions to deformations of C (Ex. 10.4), so the local deformation space is smooth.

Since T_C^1 is concentrated at the singular point, we know from the local
discussion of deformations of a node (14.1) that $H^0(T_C^1)$ is a 1-dimensional
k-vector space.

It remains to consider the sheaf T_C. For a nodal cubic curve C in $\mathbb{P}^2$ there
is an exact sequence

$$0 \to T_C \to T_{\mathbb{P}^2}|_C \to \mathcal{N}_{C/\mathbb{P}^2} \to T_C^1 \to 0.$$

One sees easily that $h^0(T_{\mathbb{P}^2}|_C) = 9$, $h^0(\mathcal{N}_{C/\mathbb{P}^2}) = 9$, $h^0(T_C^1) = 1$. Further-
more, the natural map $H^0(\mathcal{N}_{C/\mathbb{P}^2}) \to H^0(T_C^1)$ is surjective because the former
measures deformations of C as a closed subscheme of $\mathbb{P}^2$, the latter measures
abstract deformations of the node, and it is easy to see that there are first-
order deformations of C in $\mathbb{P}^2$ that give a nonzero element of T_C^1. From all
this it follows that $h^0(T_C) \geq 1$.

Now let $s \in H^0(T_C)$ be a nonzero section. Then we get an exact sequence
$0 \to \mathcal{O}_C \xrightarrow{s} T_C \to R \to 0$, where the cokernel R is of finite length. Furthermore, R is not zero, because $T_C \cong \mathcal{H}om(\Omega_C^1, \mathcal{O}_C)$ is not locally free, hence
not isomorphic to $\mathcal{O}_C$. Therefore $(T_C)^\vee$ is properly contained in $\mathcal{O}_C$, and
by Serre duality on C, using the dualizing sheaf $\omega_C \cong \mathcal{O}_C$, we find that
$h^1(T_C) = h^0((T_C)^\vee) = 0$.

Thus $\mathrm{Def}(C)$ is one-dimensional, and the miniversal deformation space of
C is smooth of dimension 1.

Finally, we compare the deformations of C to the deformations of the
pointed curve (C, P), where P is a nonsingular point. We find, as in the case
of a smooth curve (26.1), that the two functors are isomorphic, so we conclude
that the deformations of (C, P) are pro-representable of dimension 1. This is
all we need to complete the argument of (26.6.5).

Remark 26.6.7. One might ask, why do we use the nodal curve, but not the
cuspidal curve or any other connected reduced curve with $p_a = 1$? One reason
is that any other singular curve besides the node has a local deformation
theory of dimension ≥ 2, (14.2.2), (Ex. 14.1), and so would not fit in a modular
family of elliptic curves.

Another reason is the presence of jump phenomena. Consider the family
$y^2 = x^3 + t^2 ax + t^3 b$ over the t-line, for any fixed values of a and b such that
$4a^3 + 27b^2 \neq 0$. Then for $t \neq 0$ we have nonsingular elliptic curves all with the
same j-invariant, while for $t = 0$ we get a cuspidal curve. Thus the cuspidal

curve cannot belong to a deformation theory having a coarse moduli space. Another way of saying this is that if you try to add a point to the j-line representing the cuspidal curve, that point would have to be in the closure of every point on the j-line!

References for this section. I owe a special debt to Mumford's article [114] from the Purdue conference. That is where he first introduced the notion of modular family, a precursor of the notion of a stack. The basic theory of elliptic curves is treated in [57, IV, §4]. Other valuable sources are [21] and [119].

Exercises.

26.1. Show that the functor F of families of (pointed) elliptic curves is not a sheaf for the Zariski topology, by constructing a locally trivial but nontrivial family on a triangle of lines, using an automorphism of order 2, as in (Ex. 23.2). Show, however, that this family becomes trivial after a base extension by a finite étale morphism of a hexagon onto the triangle. Hence this family is isotrivial (Ex. 25.1).

26.2. Show that each of the two fiberwise trivial families mentioned in the proof of (26.2) becomes trivial after a finite étale base extension. Hence they are isotrivial.

26.3. We have seen (26.6.2) that there is a tautological family of elliptic curves over the j-line minus the two points $j = 0, 12^3$. Show, nevertheless, that the functor of families of elliptic curves with $j \neq 0, 12^3$, is not representable.

26.4. Show that the functor of families of pointed elliptic curves is separated, but is not complete (cf. definitions in §24).

26.5. Show that the λ-line with its family of elliptic curves $y^2 = x(x-1)(x-\lambda)$ is a coarse moduli space for the moduli problem of classifying pairs (X, α), where X is an elliptic curve and α is an isomorphism of the group of 2-torsion points on X with the Klein four-group $V = \mathbb{Z}/2\mathbb{Z} \oplus \mathbb{Z}/2\mathbb{Z}$.

26.6. Referring to the classification of curves of genus 2 given in [57, IV, Ex. 2.2], let $U \subseteq \mathbb{A}^3$ be the open affine set of triples $\beta_1, \beta_2, \beta_3$, all distinct and different from $0, 1$, let G be the symmetric group Σ_6 of six letters acting on U as described in [loc. cit.], and let T be the quotient U/G. Show that T is a coarse moduli space for curves of genus 2. According to [139], this moduli space has just one singular point, corresponding to the curve $y^2 = x^6 - x$.

26.7. Show that $\mathbb{P}^1_k / \operatorname{Spec} k$ is a modular family for curves of genus zero.

26.8. Moduli of n points on a line. We consider the moduli problem whose objects are n ordered points $P_1, \ldots, P_n$ in $\mathbb{P}^1_k$, not necessarily distinct, and where two objects are equivalent if there is an automorphism of $\mathbb{P}^1$ sending one to the other.

(a) Show that if we limit our attention to objects containing at least three distinct points, then the local deformation problem is pro-representable, the family is bounded, and it has a modular family.

(b) In the case $n = 4$, the moduli problem of (a) is covered by four open subsets U_i, corresponding to objects where the three points except P_i are distinct. Show that each of these U_i is isomorphic to a $\mathbb{P}^1$, and they are glued together to form a nonseparated scheme that is a $\mathbb{P}^1$ with three doubled points. The doubled points occur because, for example, the object where $P_1 = P_2$ is not separated from the object where $P_3 = P_4$.

(c) In the case $n = 5$, consider the open set U where P_1, P_2, and P_5 are distinct. Then we can normalize points of U to the form 0, 1, x, y, ∞, so that U is isomorphic to $\mathbb{P}^1 \times \mathbb{P}^1$. Show that if we restrict our moduli problem to those objects with no triple point, then the moduli space is a nonsingular projective variety obtained from $\mathbb{P}^1 \times \mathbb{P}^1$ by blowing up the points $(0,0)$, $(1,1)$, and (∞,∞) on the diagonal. This is a surface isomorphic to the Del Pezzo surface of degree 5 in $\mathbb{P}^5$ [57, V, 4.7.1].

(d) Still in the case $n = 5$, show that if we add back the objects with a triple point (there are 10 of them), they correspond in a nonseparated way to the 10 exceptional lines on the Del Pezzo surface. That is, the new nonseparated moduli space can be described by blowing down each exceptional line and gluing the original surface to the blown-down one along the complement of the line and its blown-down point.

(e) Now, if you have courage (see [121] for details), show that for n odd, if we consider only those objects having no point of multiplicity $> \frac{n}{2}$, the corresponding moduli problem is separated and complete, and in fact it is a projective variety. (For n even, the situation is more complicated: there seems to be no natural restriction on the objects that leads to a moduli space that is both separated and complete; cf. the case $n = 4$ above.)

27. Moduli of Curves

It has been understood for a long time that there is some kind of moduli space of curves of genus $g \geq 2$. Riemann gave the dimension as $3g - 3$. Transcendental methods show that it is irreducible over the complex numbers. Fulton extended this result for the coarse moduli scheme to characteristic $p > 0$, with some restrictions on small p, by considering the Hurwitz scheme of branched covers of $\mathbb{P}^1$. Deligne and Mumford proved irreducibility in all characteristics by introducing a compactification of the variety of moduli in which they allowed certain singular "stable" curves. They also hinted at a more sophisticated object, the moduli stack. Mumford, in his article "Picard groups of moduli problems" [114], makes the point that to investigate the more subtle properties of the moduli of curves, the coarse moduli space may not carry enough information, and so one should really work with stacks.

Our purpose here is not to prove all of these results (for which there are ample references); rather it is to disengage the issues involved, to explain why we do things the way we do, and to make some precise statements. We will make some remarks about stacks at the end of this section (27.7.1).

First we state the problem. We fix an algebraically closed field k, and we consider projective nonsingular curves of genus $g \geq 2$ over k. The restriction to $g \geq 2$ is because (a) we have discussed the cases of $g = 0, 1$ separately, and (b) the case of $g \geq 2$ is qualitatively different in that curves of genus $g \geq 2$ can have only finitely many automorphisms [57, IV, Ex. 5.2].

We want to describe isomorphism classes of these curves and families of curves, so we define the *moduli functor* $\mathcal{F}$, which assigns to each scheme S/k the set of isomorphism classes of flat families X/S, proper over S, all of whose geometric fibers are nonsingular curves of genus g. If this functor were representable, we would call the corresponding scheme a fine moduli space. But since there are curves with nontrivial automorphisms, we know (Ex. 23.2) that the functor is not representable. On the other hand here is one of the main results of the theory:

Theorem 27.1. *The moduli functor $\mathcal{F}$ of curves of genus $g \geq 2$ over k algebraically closed has a coarse moduli space M_g, which is a normal quasi-projective variety of dimension $3g - 3$ having at most quotient singularities.*

The existence and the fact that it is quasi-projective are proved in Mumford's book [119]; the irreducibility is proved in the article of Deligne and Mumford [21]. Fulton [35] improved the proof of Deligne and Mumford, making it purely algebraic.

The coarse moduli space is a variety whose closed points are in one-to-one correspondence with the set of isomorphism classes of curves in a natural way. Furthermore, for any flat family X/S, there is a morphism $f : S \to M_g$ with the property that for each k-rational point $s \in S$, the image $f(s)$ corresponds to the isomorphism class of the fiber X_s. However, there is no tautological family over M_g, and knowledge of M_g does not give us full information about all possible flat families and morphisms between them.

To give more information about families of curves, we study modular families, following the ideas of Mumford [114], as in §26. This will help us understand flat families of curves "up to étale base extension."

Theorem 27.2. *For any $g \geq 2$, the class of nonsingular projective curves of genus g over k has a modular family.*

Proof. On a curve of genus g, any divisor of degree $\geq 2g + 1$ is nonspecial and very ample. In particular, if we take the tricanonical divisor $3K$, where K is the canonical divisor, then for any $g \geq 2$, its degree $d = 6g - 6$ is $> 2g + 1$, so we can use it to embed the curve in a projective space $\mathbb{P}^n$, with $n = 5g - 6$, as a nonsingular curve of degree d.

Now we consider the Hilbert scheme H of nonsingular curves of degree d and genus g in $\mathbb{P}^n$. Since the curves are nonspecial, the infinitesimal study of the Hilbert scheme shows that H is smooth (Ex. 1.7) of dimension $h^0(\mathcal{N}) = 25(g - 1)^2 + 4(g - 1)$. Of course H contains curves embedded by any divisor of degree d, not only the tricanonical divisor $3K$. One can show, however, that the subset $H' \subseteq H$ of tricanonically embedded curves, with the reduced induced structure, is also smooth and of dimension $25(g - 1)^2 + 3g - 4$, since the choice of a divisor, up to linear equivalence, is an element of the Picard scheme of C, which has dimension g.

The virtue of using the tricanonical embedding is that if two points of H' correspond to isomorphic curves of genus g, the isomorphism preserves the tricanonical divisor, and so the two embeddings differ only by the choice of basis of $H^0(\mathscr{O}_C(3K))$. Thus the group $G = \mathrm{PGL}(n)$ acts on H', and the orbits of this action are closed subsets of H' in one-to-one correspondence with the isomorphism classes of curves. Note that G has dimension $(n+1)^2 - 1 = 25(g-1)^2 - 1$, so that the "orbit space," if it exists, will have dimension $3g - 3$, as we expect. (At this point one can apply the techniques of geometric invariant theory [119] to show that an orbit space does exist, as a quasi-projective coarse moduli space. We follow a different route to get the existence of the modular family.)

The next step is to consider a particular curve C, and choose a point $P \in H'$ representing it. The orbit of G containing P, being a homogeneous space, is smooth, so we can choose a smooth, locally closed subscheme Z of H' of dimension $3g - 3$ passing through P and transversal to the orbit of G at P. For example, Z could be the intersection of H' with a linear space in some projective embedding, of complementary dimension to the orbit of G. Replacing Z by a smaller open subset still containing P, we may assume that for every orbit of G, whenever Z intersects that orbit, if at all, it intersects in only finitely many points, and that the intersection is transversal at those points. (The idea here is just to throw away points of Z where these properties do not hold.) Since Z is contained in H' and in H, we can restrict the universal family of curves on H to Z and obtain a flat family X/Z.

By construction, the given curve C occurs in the family X/Z, and also by construction, any curve appears at most finitely many times. At a point $z \in Z$, we consider the formal family induced by X over the complete local ring $R = \hat{\mathscr{O}}_{Z,z}$. We know that the local deformation functor of the corresponding curve C is pro-representable and has a smooth universal deformation space of dimension $3g - 3$ with tangent space $H^1(T_C)$ (18.3.1). A standard sequence shows that $H^0(\mathscr{N}_{C/\mathbb{P}^n}) \to H^1(T_C)$ is surjective, and one sees easily that the tangent space to Z maps surjectively also, and hence is isomorphic to $H^1(T_C)$. Since R and the deformation space of C are both smooth, and their tangent spaces are isomorphic, they are isomorphic.

Thus the family X/Z satisfies the property (b) of a modular family (§26). We still have to construct a family containing all curves of genus g. It is easy to show that the image of $G \times Z$ in H' contains an open set. Since H' is quasi-projective, a finite number of such open sets $G \times Z_i$ will cover H'. Thus taking a finite disjoint union of such families X_i/Z_i, we obtain a new family X/Z satisfying also (a), which is the required modular family, and we see into the bargain that Z may be taken to be quasi-projective.

For property (c), given a modular family X/Z and any other family X'/S, we consider the two families $S \times X$ and $X' \times Z$ over $S \times Z$, and let $S' = \mathrm{Isom}_{S \times Z}(S \times X, X' \times Z)$. As in the case of elliptic curves (26.4) it follows that $S' \to S$ is étale surjective, and that the two pulled-back families by the morphisms $S' \to S$ and $S \to Z$ are isomorphic.

Remark 27.2.1. As an application of modular families, we will show the nonexistence of certain g_d^1's on a general curve of genus g. On a nonsingular projective curve C, we denote by g_d^1 a linear system of degree d and dimension 1 without base points. A curve C has a g_d^1 if and only if it admits a morphism $C \to \mathbb{P}^1$ of degree d. A curve with a g_2^1 is called *hyperelliptic*; a curve with a g_3^1 is *trigonal*, and so on. One knows that whenever $2d - 2 \geq g$, every curve of genus g has a g_d^1 (see for example [2, VII, 2.3]). We will show that for $g \geq 3$ a general curve of genus g is not hyperelliptic; for $g \geq 5$, a general curve of genus g is not trigonal, and so on. Of course one can give proofs by construction in special cases [57, IV, 5.5.1ff]. For example, a nonsingular plane quartic curve is a nonhyperelliptic curve of genus 3. But the examples get more difficult as the numbers get bigger, and I think a few hours spent trying to construct as many cases as you can will generate ample appreciation for the general method.

Theorem 27.3. *A general curve of genus $g > 2d - 2$ does not have a g_d^1.*

Proof. First of all, to explain the word "general," what we will show is that if $\pi : X \to S$ is a modular family of curves of genus g, then there is a dense open subset $U \subseteq S$ such that for every $s \in U$, the fiber X_s does not have a g_d^1 for any $d < \frac{1}{2}g + 1$. If $g = 1$, the statement says that an elliptic curve does not have a g_1^1. This is immediate, because a g_1^1 would give a map of degree 1 to $\mathbb{P}^1$, and an elliptic curve is not rational. So we may assume henceforth that $g \geq 2$.

If $s \in S$ and the fiber X_s has a g_d^1, let $f : X_s \to \mathbb{P}^1$ be the corresponding morphism, and let $\Gamma_s \subseteq X_s \times \mathbb{P}^1$ be its graph. Then we consider the Hilbert scheme H parametrizing subschemes $\Gamma \subseteq X \times \mathbb{P}^1_S$, flat over S, that are graphs of morphisms. Since H is a quasi-projective scheme, its image in S is a constructible subset [57, II, Ex. 3.18, 3.19], so that if the theorem is not true, there will be an open subset $V \subseteq S$ contained in the image of H.

Looking at an irreducible component of H whose image contains V, there is a corresponding subscheme $\Gamma \subseteq \pi^{-1}(V) \times \mathbb{P}^1_S$, and hence an invertible sheaf $\mathcal{L}$ on $\pi^{-1}(V)$ and sections t_0, t_1 of $\mathcal{L}$ corresponding to the g_d^1 on each fiber.

Now let C be one of these fibers, and let $f : C \to \mathbb{P}^1_k$ be the morphism given by the g_d^1. We write the sequence of differentials

$$0 \to f^* \Omega^1_{\mathbb{P}^1} \to \Omega^1_C \to R \to 0,$$

where R is the ramification sheaf, which is also the sheaf of relative differentials $\Omega^1_{C/\mathbb{P}^1}$. Dualizing we get

$$0 \to T_C \to f^* T_{\mathbb{P}^1} \to R' \to 0,$$

where R' is the torsion sheaf $\mathcal{E}xt^1(R, \mathcal{O}_C)$, which is also $T^1_{C/\mathbb{P}^1}$ (3.5), (Ex. 3.4). This gives a sequence of cohomology

$$0 \to H^0(T_C) \to H^0(f^*T_{\mathbb{P}^1}) \to H^0(R') \to H^1(T_C) \to H^1(f^*T_{\mathbb{P}^1}) \to 0.$$

Here $H^0(T_C) = 0$, since $g \geq 2$, and one can identify $H^0(R')$ with the tangent space to the deformations of the pair (C, f) described in (24.10.3). Indeed, since C is affine over $\mathbb{P}^1$, the deformations of C as a $\mathbb{P}^1$-scheme are just $T^1_{C/\mathbb{P}^1}$ by (5.1) and (5.2).

Since our g^1_d extends over the whole neighborhood V in S by hypothesis, we conclude that the map $H^0(R') \to H^1(T_C)$ must be surjective; hence $H^1(f^*T_{\mathbb{P}^1}) = 0$. Now $T_{\mathbb{P}^1} = \mathcal{O}(2)$, so $f^*T_{\mathbb{P}^1}$ corresponds to the divisor $2D$, where D is the divisor of the g^1_d. In other words, $H^1(\mathcal{O}_C(2D)) = 0$. Furthermore, since $\dim |D| \geq 1$, it follows that $\dim |2D| \geq 2$. Then by Riemann–Roch,

$$h^0(\mathcal{O}(2D)) = 2d + 1 - g \geq 3,$$

and hence $g \leq 2d - 2$.

Therefore, by contradiction, we find that for $g > 2d - 2$, the general curve of genus g has no g^1_d. This argument justifies the proofs "by counting parameters" used by the ancients.

Remark 27.3.1. What is the relation between the modular family X/Z and the coarse moduli space M? One would like to divide Z by the equivalence relation of having isomorphic fibers. So let $Z' = \mathrm{Isom}_{Z \times Z}(X \times Z, Z \times X)$. Then $Z' \to Z \times Z$ expresses this relation. Unfortunately, since the curves may have automorphisms, Z' is not a subscheme of $Z \times Z$, so this is not a scheme-theoretic equivalence relation. If, however, we restrict to $g \geq 3$ and consider only those curves having no automorphisms, then Z' will be a subscheme of $Z \times Z$, étale over Z by both projections, and the quotient will exist as an *algebraic space* (27.7.1).

To show that the coarse moduli M_g of all curves of genus g is a quasi-projective scheme requires more work, which we do not discuss here.

Remark 27.3.2. To compactify the variety of moduli, Deligne and Mumford [21] introduce *stable* curves. They show then that stable curves behave like nonsingular curves in the theory above. In particular, (27.2) holds also for families of stable curves, and there is a coarse moduli space for stable curves, which they show to be projective and irreducible. The following results give the definition and main properties of stable curves.

Proposition 27.4. *Let X be a reduced, connected, projective curve having at most nodes as singularities. Let $T_X = \mathcal{H}om(\Omega^1_X, \mathcal{O}_X)$ be the tangent sheaf of X. Then the following two conditions are equivalent:*

(i) *$p_a(X) \geq 2$ and if C is any irreducible component of X with $p_a(C) = 0$, then C meets $X - C$ in at least three points.*

(ii) *$H^0(X, T_X) = 0$, i.e., X has no infinitesimal automorphisms.*

Definition. A curve satisfying the equivalent conditions of (27.4) is called *stable*.

Lemma 27.5. *Let X be a reduced curve with at most nodes as singularities, and let Z denote the set of nodes (with the reduced induced scheme structure). Then there are natural exact sequences*

$$0 \to \mathcal{O}_Z \to \Omega_X^1 \to \omega_X \to \mathcal{O}_Z \to 0$$

and

$$0 \to \omega_X^\vee \to T_X \to \mathcal{O}_Z \to 0.$$

Proof. The second sequence follows by dualizing the first, so we have only to prove the first. To construct the natural map $\Omega_X^1 \to \omega_X$, we embed X in a nonsingular variety P (such as projective space), with ideal sheaf $\mathcal{I}$. Then there is an exact sequence

$$\mathcal{I}/\mathcal{I}^2 \xrightarrow{d} \Omega_P^1 \otimes \mathcal{O}_X \xrightarrow{\pi} \Omega_X^1 \to 0$$

[57, II, 8.12]. Say P has dimension n. Then X is a local complete intersection scheme, so $\mathcal{I}/\mathcal{I}^2$ is locally free of rank $n-1$ on X. The map d is injective at nonsingular points of X [57, II, 8.17], which are dense in X, so it is injective everywhere. We define the map

$$\overset{n-1}{\bigwedge} (\mathcal{I}/\mathcal{I}^2) \otimes \Omega_X^1 \to \Omega_P^n \otimes \mathcal{O}_X$$

by sending

$$(f_1 \wedge \cdots \wedge f_{n-1}) \otimes w \mapsto df_1 \wedge \cdots \wedge df_{n-1} \wedge w',$$

where w' is any lifting of w to $\Omega_P^1 \otimes \mathcal{O}_X$. The map is well-defined, since two liftings differ by something in $\mathcal{I}/\mathcal{I}^2$. Now tensoring with the invertible sheaf $\bigwedge^{n-1}(\mathcal{I}/\mathcal{I}^2)^\vee$ we get

$$\Omega_X^1 \to \Omega_P^n \otimes \mathcal{O}_X \otimes \overset{n-1}{\bigwedge} (\mathcal{I}/\mathcal{I}^2)^\vee,$$

and the right-hand side is just the dualizing sheaf ω_X by [57, III, 7.11]. This map is an isomorphism at nonsingular points of X [57, III, 7.12], so its kernel and cokernel will be supported at the singular points.

To find the kernel and cokernel is a local question at each singular point, and since a node is analytically isomorphic to the curve $xy = 0$ in $\mathbb{A}^2$, we can make a local analysis using this curve.

When $X \subseteq \mathbb{A}^2$ is the curve defined by $xy = 0$, we find that Ω_X^1 is generated by dx and dy with the relation $x\,dy + y\,dx = 0$. The sheaf $\mathcal{I}/\mathcal{I}^2$ is generated by $f = xy$, and $\Omega_{\mathbb{A}^2}^2$ is generated by $dx \wedge dy$. The map described above sends

$$f \otimes dx \mapsto df \wedge dx = -x \, dx \wedge dy,$$
$$f \otimes dy \mapsto df \wedge dy = y \, dx \wedge dy.$$

Thus the cokernel is one-dimensional. The kernel will be the torsion submodule of Ω^1_X, which is also one-dimensional, generated by $x \, dy = -y \, dx$. This gives the required sequence of (27.5).

Lemma 27.6. *If X is a reduced curve that is a union $X = C \cup D$ of two curves C and D meeting transversally at a finite set of nodes S, then $T_X \cong (\mathcal{I}_{S,C} \otimes T_C) \oplus (\mathcal{I}_{S,D} \otimes T_D)$.*

Proof. As before, the question is local around each node, so we take X to be the curve $xy = 0$ in $\mathbb{A}^2$, with $C = \operatorname{Spec} k[x]$ and $D = \operatorname{Spec} k[y]$. Now $T_X = \operatorname{Hom}(\Omega^1_X, \mathcal{O}_X)$ is generated by two elements

$$a : \begin{cases} dx \mapsto x \\ dy \mapsto 0 \end{cases} \quad \text{and } b : \begin{cases} dx \to 0 \\ dy \to y \end{cases}$$

and as an $\mathcal{O}_X$-module, it is the direct sum of the submodules generated by a and b. Restricting to C and D, the first generator is x times the generator of T_C that sends $dx \mapsto 1$, and similarly, the second is y times the generator of T_D. Hence $T_X = xT_C \oplus yT_D$, as required.

Proof of (27.4).

Case 1. X is irreducible. Then we have only to show that $p_a(X) \geq 2 \Leftrightarrow H^0(T_X) = 0$.

If $p_a(X) = 0$, then $X \cong \mathbb{P}^1$ and $h^0(T_X) = 3$.

If $p_a(X) = 1$, then X could be a nonsingular elliptic curve, or a rational curve with one node. In the first case $T_X = \mathcal{O}_X$ and $h^0(T_X) = 1$. In the second case, by Lemma 27.5 we have

$$0 \to \mathcal{O}_X \to T_X \to k \to 0$$

and $h^0(T_X) \geq 1$. (In fact, it follows from (26.6.6) that $h^0(T_X) = 1$.)

If $p_a(X) = g \geq 2$, then from the sequence

$$0 \to \omega_X^\vee \to T_X \to \mathcal{O}_Z = 0,$$

where Z is the set of nodes, we see that the degree of T_X is $2 - 2g + z$, where z is the number of nodes. But $z \leq g$, so the degree of T_X is $\leq 2 - g$, with equality only in the case of a rational curve with g nodes. In the case of strict inequality, the degree of T_X is negative, so it cannot have global sections. In the case of equality, $\deg T_X = 0$, so if it had a section, it would be isomorphic to $\mathcal{O}_X$. But this is impossible, because T_X is not locally free at the nodes. Thus in any case $h^0(T_X) = 0$.

Case 2. X reducible. Because of Lemma (27.6), $h^0(T_X) \neq 0$ if and only if for some irreducible component C, we have $h^0(\mathcal{I}_S \otimes T_C) \neq 0$, where $S = C \cap (X - C)$. If $p_a(C) \geq 2$, this cannot happen by Case 1. If $p_a(C) = 1$, then $h^0(T_C) = 1$, but there must be at least one point in S, since $p_a(X) \geq 2$. The nonzero section of T_C is nowhere vanishing on the smooth points of C, so it cannot be in $\mathcal{I}_S \otimes T_C$. If $p_a(C) = 0$, then $C \cong \mathbb{P}^1$ and $h^0(\mathcal{I}_S \otimes T_C) \neq 0$ if and only if S consists of at most two points.

The following result shows that from the point of view of deformation theory, stable curves behave like nonsingular curves.

Proposition 27.7. *Let X be a stable curve of (arithmetic) genus $g \geq 2$. Then the functor of local infinitesimal deformations of X is pro-representable by a regular complete local ring of dimension $3g - 3$.*

Proof. The functor is pro-representable because X is projective and satisfies the critical condition $H^0(X, T_X) = 0 (18.3)$. The local ring representing it is regular because there are no obstructions to deforming X (Ex. 10.4). Since the ring is regular, to find its dimension we need only compute its tangent space $\mathrm{Def}(X/k)$, and by (Ex. 5.7) this belongs to an exact sequence

$$0 \to H^1(X, T_X) \to \mathrm{Def}(X/k) \to H^0(X, T_X^1) \to 0,$$

there being no H^2 on a curve. Let δ be the number of nodes in X. From (Ex. 3.1), we see that T_X^1 is just k at each node, so the right-hand term contributes δ to the dimension. On the other hand, from the exact sequence of (27.5), and the fact that $H^0(T_X) = 0$, we find that $H^0(\omega_X^\vee) = 0$ and we obtain an exact sequence

$$0 \to H^0(\mathcal{O}_Z) \to H^1(\omega_X^\vee) \to H^1(T_X) \to 0.$$

Since $H^0(\omega_X^\vee) = 0$, we obtain $H^1(\omega_X^\vee) = 3g - 3$ by Riemann–Roch. The term $H^0(\mathcal{O}_Z)$ is δ, so $h^1(T_X) = 3g - 3 - \delta$. Adding to the above, we find that $\dim(\mathrm{Def}(X/k)) = 3g - 3$ as required.

Remark 27.7.1 (The idea of stacks). The study of families of elliptic curves that we have described in detail (§26), and its generalization to modular families of curves of genus g for any $g \geq 2$, can be formalized so as to give rise to the theory of stacks. Here we will give a brief introduction to the concepts that make up the theory of stacks. For the technical details, the reader will have to consult another source, such as the article of Vistoli [167], who explains in one hundred pages what I will barely indicate in four or five.

The problem is still to find an appropriate moduli space to explain a class of objects one wishes to classify (e.g., curves of genus g) together with the structure of families of these objects. We have seen how a moduli problem gives rise to a functor: in the case of curves of genus g, the functor $\mathcal{F}$ assigns to each scheme S the set of isomorphism classes of families of curves of genus g over S, that is, smooth proper morphisms $X \to S$ whose geometric fibers are all curves of genus g. If the functor is representable, then one has a *fine moduli space*. Even if it is not representable, one can sometimes approximate it by a *coarse moduli space*, which at least tells us the isomorphism classes of curves over a field, but which does not give a satisfactory answer to the problem of describing all possible families of curves.

Now it often happens, as it does in this particular case, that the functors describing moduli problems we are interested in are not representable, at least not by any scheme. So we search for some object, some structure more general than a scheme that will satisfy our needs. We could, of course, say that the functor itself carries all the information we want. But that is too big an object, too nebulous. We want something more concrete, more geometric, which we can deal with as a sort of "generalized scheme."

A first step in this direction is to realize that we should begin to think of étale maps as generalized open neighborhoods. We have already said that our study seemed to describe a moduli space "up to étale covers." To make this more precise, recall that a scheme is constructed out of affine schemes by gluing together along isomorphisms defined on (Zariski) open subsets. What happens if we try to glue schemes together along étale maps? Imagine a collection of schemes $\{U_i\}$, and for each i, j étale morphisms $V_{ij} \to U_i$ and $V_{ji} \to U_j$ and isomorphisms $\varphi_{ij} : V_{ij} \to V_{ji}$, satisfying a cocycle compatibility condition for each i, j, k. We try to glue the U_i along the isomorphisms φ_{ij}. Another way of saying this is to let U be the disjoint union of the U_i, and let $R \subseteq U \times U$ be the equivalence relation defined by the φ_{ij}. Then we seek a quotient U/R of U by this equivalence relation. This may not exist in the category of schemes, but it is an algebraic space. An *algebraic space* is defined to be the quotient of a scheme U by an equivalence relation $R \subseteq U \times U$, where R is a subscheme étale over U by both projections. The theory of algebraic spaces was developed by Artin and Knutson in the 1970s [85], and they showed that algebraic spaces form a reasonable category, enlarging the category of schemes, to which much of the language and theorems of schemes can be extended. Some functors that were not representable in the category of schemes may become representable in the category of algebraic spaces, and in any case, it is usually easier to show that something is an algebraic space than to show that it is a scheme. For example, those nonalgebraic complex analytic spaces called Moishezon spaces [57, Appendix B] can be realized as algebraic spaces.

The second and essential step is to acknowledge that the functor of sets of isomorphism classes of objects is itself deficient in that it does not keep track of morphisms between families, and in particular the automorphisms that objects or families may have. So instead of the *functor* $\mathcal{F}$ described above, which to each scheme S assigns the *set* of isomorphism classes of families $X \to S$ (of curves of genus g, for example), we consider a new object $\mathcal{F}$, like a functor, but to each scheme S it assigns the *category* $\mathcal{F}(S)$ of families $X \to S$

and isomorphisms between such families. This new object $\mathcal{F}$ is called a *fibered category* over the category of schemes.

In the case of the functor $\mathcal{F}$, we saw that a necessary condition that it should be representable is that it be a *sheaf* (23.5). This means that given a scheme S and an open covering $\{U_i\}$ of S, the map of sets

$$\mathcal{F}(S) \rightarrow \Pi\mathcal{F}(U_i) \rightrightarrows \Pi\mathcal{F}(U_i \cap U_j)$$

is exact: (a) the first arrow is injective, and (b) its image is equal to the set of elements in the middle whose two images on the right are equal.

The sheaf axiom for the functor $\mathcal{F}$ is now replaced by a new analogous property for the fibered category $\mathcal{F}$, the *stack* axiom, which is as follows. For any scheme S, and any étale covering $\{U_i\}$, that is, étale maps $U_i \rightarrow S$ that are collectively surjective, consider the restriction functors

$$\mathcal{F}(S) \rightarrow \Pi\mathcal{F}(U_i) \rightrightarrows \Pi\mathcal{F}(U_i \times_S U_j) \overset{\rightarrow}{\underset{\rightarrow}{\rightarrow}} \Pi\mathcal{F}(U_i \times_S U_j \times_S U_k).$$

(Note that we must write $U_i \times_S U_j$ instead of $U_i \cap U_j$ because the U_i are no longer subsets of S.) The sheaf properties (a) and (b) now look like this.

(a) The first arrow is injective. This means that if $a, b \in \mathcal{F}(S)$, and if we denote their restrictions to $\mathcal{F}(U_i)$ by a_i, b_i, and if for each i there are isomorphisms $\varphi_i : a_i \rightarrow b_i$ such that for each i, j, the isomorphisms φ_i, φ_j restrict to the same isomorphism of a_{ij} and b_{ij} on $U_i \times_S U_j$, then there is a unique isomorphism $\varphi : a \rightarrow b$ inducing φ_i on each U_i. (Note how equality of elements in a set has been replaced by isomorphism of objects in a category.)

(b) The sequence is exact at the first middle term. This now says that if we are given objects $a_i \in \mathcal{F}(U_i)$ for each i and isomorphisms $\varphi_{ij} : a_i \rightarrow a_j$ on U_{ij} $(= U_i \times_S U_j)$ satisfying a cocycle compatibility condition on each U_{ijk}, then there exists a (unique) object $a \in \mathcal{F}(S)$ restricting to each a_i on U_i.

A *Deligne–Mumford stack* is a fibered category $\mathcal{F}$ satisfying the stack axiom above, and such that there exists a scheme X and an étale surjective morphism $h_X \rightarrow \mathcal{F}$ from the functor associated to X to $\mathcal{F}$. An *Artin stack* is the same, but requires only that $h_X \rightarrow \mathcal{F}$ be smooth. (I omit some technical conditions of quasi-compactness, locally of finite presentation, finiteness of the diagonal, etc.)

With this definition, it is not yet obvious that a scheme, or an algebraic space, is a stack. To verify the stack axiom, one needs theorems saying that all kinds of data given on an étale covering $\{U_i\}$ of a scheme S, such as quasi-coherent sheaves, or families of schemes over the U_i, glue together to give similar objects over S. This collection of results, which are well established, is called *descent theory* (23.5.2).

When all this has been done, one shows that the variety of moduli of curves $\mathcal{M}_g$ is a Deligne–Mumford stack, for any $g \geq 2$. The remarkable thing is that although the sheaf axiom fails for the *functor* $\mathcal{F}$, the stack axiom, apparently more complicated, actually holds for the fibered category $\mathcal{F}$ of families of curves. In their fundamental paper [21], Deligne and Mumford introduced stacks for the first time, and showed that one can extend much of the terminology and results of schemes to this larger category. By also compactifying the stack $\mathcal{M}_g$ using stable curves, they were able to prove that it is irreducible in all characteristics.

Even though stacks were not mentioned there, many of the basic ideas appear in Mumford's earlier paper [114], where he discusses the modular families that we have described above. Indeed, the existence of the modular family, when translated into the language of stacks, is just the statement that there is a scheme X together with a surjective étale morphism $h_X \to \mathcal{M}_g$.

References for this section. Mumford [119] contains the proofs of existence of coarse moduli spaces. Deligne and Mumford [21] establish the irreducibility of the compactification of the moduli space and introduce the language of stacks (see also [35]). Mumford [114] explains the motivation behind the theory of stacks. Further introductions to stacks are in [165], [37], [93], and [32]. I would also like to thank Barbara Fantechi for explaining the whole theory of stacks to me in the short space of two hours.

Exercises.

27.1. Using the existence of a modular family, show that any fiberwise trivial family of curves of genus g becomes trivial after a surjective étale base extension. In other words, it is locally isotrivial.

27.2. The coarse moduli space M_3 of nonsingular curves of genus 3 is irreducible. To show this, proceed in the following steps (cf. Ex. 8.8e).

(a) For any g, show that the following conditions are equivalent.
  (i) The coarse moduli space M_g is irreducible.
  (ii) For any modular family $X \to S$ of curves of genus g, and any two nonempty saturated open subsets $U, V \subseteq S$, the intersection $U \cap V$ is nonempty. Here *saturated* means that for every point x (in U or V), that open set also contains all the points equivalent to it under the equivalence relation defined by $x \sim y$ if there is an isomorphism between the geometric fibers over the points x and y.
  (iii) Whenever $X_1 \to T_1$ and $X_2 \to T_2$ are families of curves of genus g, with T_1, T_2 integral schemes of finite type over k, and there exist points $t_1 \in T_1$, $t_2 \in T_2$ for which the maps of the induced families $\hat{X}_i$ over the complete local rings $\hat{\mathcal{O}}_{T_i,t_i}$ give rise to surjective maps of Spec $\hat{\mathcal{O}}_{T_i,t_i}$ to the universal deformation space of the fiber, then there exist points $u_1 \in T_1$ and $u_2 \in T_2$ for which the geometric fibers of X_i over u_i are isomorphic.
  If these conditions hold, we will also say that the moduli problem $\mathcal{M}_g$ is irreducible.
(b) For the special case $g = 3$, show that any curve C is either hyperelliptic or isomorphic to a nonsingular plane quartic curve.

(c) Show that there is a family X_1/T_1 with T_1 irreducible of dimension 5 containing all hyperelliptic curves.

(d) Show that there is a family X_2/T_2 with T_2 irreducible of dimension 14 that parametrizes the nonsingular plane quartic curves.

(e) If $X \to S$ is a modular family, then S is nonsingular of dimension 6.

(f) Now combine the above to show that M_3 is irreducible.

27.3.   Make a list of all possible types of stable curves of genus 2. (I found nine types, including the nonsingular one). Compute the dimension of each family of singular stable curves and thus show that they are all limits of families of nonsingular curves. *Note:* We will see later that every stable curve is a limit of nonsingular curves (29.10.1).

27.4.   If T is a nonsingular curve, $0 \in T$ a point, and X' a family of smooth projective curves of genus $g \geq 2$ defined over $T \backslash \{0\}$, show that one can complete X' to a family X/T whose fiber X_0 is a stable curve of genus g (possibly after a base change $T' \to T$). This will show that the functor of nonsingular and stable curves of genus g is complete. *Hints:* As in (Ex. 24.9), first complete X' to a family X/T, where now T is a projective curve and X is a projective surface. Resolve the singularities of X. Then resolve the singularities of the singular fibers so that they have only normal crossings [57, V, 3.9]. (If there were multiple fibers, one should first make a ramified base change on T to remove them.) Now the only problem is that the fiber X_0 may have smooth rational components meeting the rest of the fiber in only one or two points. Those that meet the rest in only one point have self-intersection -1, so they can be blown down to smooth points of a new X. Those that meet the remainder in 2 points are more troublesome. If there is just one, it blows down to an ordinary normal double point of the surface and the new fiber still has normal crossings there. If there is a chain of these -2 curves it blows down to what is called a rational double point of type A_n for some n. In this way we get a fiber that is a stable curve.

28. Moduli of Vector Bundles

On a fixed scheme X, let us consider the family of all *vector bundles* (i.e., locally free sheaves). For given numerical invariants (rank, degree, Chern classes, ...) we would like to find a moduli space parametrizing isomorphism classes of these vector bundles. We have already seen some obstacles to realizing this hope: the family of vector bundles may not be bounded (Ex. 24.3), and there are jump phenomena (Ex. 2.3). So in general we cannot expect to find even a coarse moduli space.

To remedy this situation, in his talk to the International Congress of Mathematicians in 1962 [113], Mumford introduced the notion of a stable vector bundle on a curve, and announced the theorem that stable bundles of given rank and degree on a nonsingular projective curve are parametrized by a quasi-projective coarse moduli space. Since then, Mumford's definition has been extended to varieties of higher dimension, and one knows that a coarse moduli space for stable vector bundles exists under quite general hypotheses. This space can be compactified by allowing certain equivalence classes of semistable torsion-free sheaves.

Our purpose in this section is not to prove these results, but rather to try to understand the significance of the condition of stability (whose definition itself gives little insight), and at the same time to illustrate the general notions of moduli spaces that we have been discussing in this chapter. Therefore we will focus our discussion on vector bundles on curves.

For the rest of this section X will denote a nonsingular projective curve of genus g over an algebraically closed field k. A *vector bundle* is a locally free sheaf $\mathcal{E}$ on X. We denote its *rank* by r, and its *degree* (which is the degree of the line bundle $\bigwedge^r \mathcal{E}$) by d. We define the *slope* of $\mathcal{E}$ to be $\mu(\mathcal{E}) = d/r$.

Definition. A vector bundle $\mathcal{E}$ on the curve X is *stable* if for every subsheaf $\mathcal{F} \subseteq \mathcal{E}$ with $0 < \operatorname{rank} \mathcal{F} < \operatorname{rank} \mathcal{E}$ we have

$$\mu(\mathcal{F}) < \mu(\mathcal{E}).$$

If the inequality $\leq$ holds for all $\mathcal{F}$, we say that $\mathcal{E}$ is *semistable*. We say that $\mathcal{E}$ is *unstable* if it is not semistable. Recall also that $\mathcal{E}$ is *simple* if the only endomorphisms of $\mathcal{E}$ are constants, i.e., $H^0(X, \mathcal{E}nd(\mathcal{E})) = k$. We say that $\mathcal{E}$ is *indecomposable* if $\mathcal{E}$ cannot be written as a direct sum of bundles of lower rank.

Lemma 28.1. *For a bundle $\mathcal{E}$ on a curve X,*

(a) $\mathcal{E}$ *stable* $\Rightarrow \mathcal{E}$ *semistable.*
(b) $\mathcal{E}$ *stable* $\Rightarrow \mathcal{E}$ *simple.*
(c) $\mathcal{E}$ *simple* $\Rightarrow \mathcal{E}$ *indecomposable.*

Proof. (a) is trivial. To prove (b), assume that $\mathcal{E}$ is stable, and suppose there were an endomorphism $\varphi : \mathcal{E} \to \mathcal{E}$ that was not a scalar multiple of the identity. Choose a point $x \in X$ and consider the action of φ on the finite-dimensional k-vector space $\mathcal{E} \otimes k(x)$. Since k is algebraically closed, there is an eigenvector $v \in \mathcal{E} \otimes k(x)$ with $\varphi(v) = \lambda v$ for some $\lambda \in k$. Then the morphism $\psi = \varphi - \lambda \cdot \mathrm{id}_{\mathcal{E}}$ is an endomorphism of $\mathcal{E}$ whose image $\mathcal{F} = \psi(\mathcal{E})$ is not equal to $\mathcal{E}$. Therefore ψ has a kernel, the rank of ψ must be less than the rank of $\mathcal{E}$, and so $0 < \operatorname{rank} \mathcal{F} < \operatorname{rank} \mathcal{E}$. Now by stability, since $\mathcal{F}$ is a subsheaf of $\mathcal{E}$, it follows that $\mu(\mathcal{F}) < \mu(\mathcal{E})$. But also $\mathcal{F}$ is a quotient of $\mathcal{E}$, so $\mu(\mathcal{F}) > \mu(\mathcal{E})$ (Ex. 28.2), which is a contradiction.

(c) If $\mathcal{E}$ is decomposable, say $\mathcal{E} \cong \mathcal{E}_1 \oplus \mathcal{E}_2$, then the projection onto either factor shows that $\mathcal{E}$ is not simple.

Proposition 28.2. *The set of indecomposable bundles of rank r and degree d on a curve X is a bounded family.*

Proof. We use induction on the rank r.

If $r = 1$, we are dealing with line bundles $\mathcal{L}$ of degree d. To show that they form a bounded family, we apply (24.4): it is sufficient to find a uniform m_0 such that every $\mathcal{L}$ of degree d is m_0-regular with respect to some fixed projective embedding of X. Since X is a curve, we need only show $H^1(\mathcal{L}(m_0 - 1)) = 0$, and this will happen as soon as $\deg \mathcal{L}(m_0 - 1) > 2g - 2$. Since $\deg \mathcal{L}(m_0 - 1) = d + (m_0 - 1) \cdot \deg X$, we see that for m_0 sufficiently large, depending only on $\deg X$, g, and d, all line bundles $\mathcal{L}$ of degree d will be m_0-regular. Hence they form a bounded family.

Now consider a bundle $\mathcal{E}$ of rank 2. We look for sub-line bundles $\mathcal{L} \subseteq \mathcal{E}$. In order for $\mathcal{L}$ to be a sub-line bundle of $\mathcal{E}$ we need $h^0(\mathcal{E} \otimes \mathcal{L}^{-1}) \neq 0$. Let $d = \deg \mathcal{E}$ and $d_1 = \deg \mathcal{L}$. Then $\deg(\mathcal{E} \otimes \mathcal{L}^{-1}) = d - 2d_1$. By Riemann–Roch, $\chi(\mathcal{E} \otimes \mathcal{L}^{-1}) = d - 2d_1 + 2(1 - g)$. As soon as this number is positive, i.e., $2d_1 < d + 2(1 - g)$, the sheaf $\mathcal{E} \otimes \mathcal{L}^{-1}$ will have a section, and we have an inclusion $\mathcal{L} \subseteq \mathcal{E}$. If the quotient $\mathcal{E}/\mathcal{L}$ has torsion, we pull it back to $\mathcal{L}$, so by increasing $\mathcal{L}$ we may assume that the quotient $\mathcal{M} = \mathcal{E}/\mathcal{L}$ is another line bundle. Then there is an exact sequence

$$0 \to \mathcal{L} \to \mathcal{E} \to \mathcal{M} \to 0.$$

This determines an element $\xi \in \mathrm{Ext}^1(\mathcal{M}, \mathcal{L}) = H^1(\mathcal{L} \otimes \mathcal{M}^{-1})$, which must be nonzero, since $\mathcal{E}$ is indecomposable. Therefore $\deg(\mathcal{L} \otimes \mathcal{M}^{-1}) = 2d_1 - d$ must be $\leq 2g - 2$, so $2d_1 \leq d + 2g - 2$.

Thus we see that if we now let $\mathcal{L}$ be a sub-line bundle of maximum degree of $\mathcal{E}$, then its degree d_1 must satisfy

$$d - 2g \leq 2d_1 \leq d + 2g - 2.$$

Thus there is only a finite number of possible values for d_1 and hence also a finite number of possible values for $\deg \mathcal{M} = d - d_1$. By the induction step, the families of such $\mathcal{L}$ and $\mathcal{M}$ are bounded, so there is a universal m_0 making them m_0-regular, and this implies also that $\mathcal{E}$ is m_0-regular, and so the family of $\mathcal{E}$'s is bounded.

We leave to the reader the case of rank $r \geq 3$ (Ex. 28.4).

Corollary 28.3. *The families of stable bundles or simple bundles of given rank and degree are bounded. (For semistable bundles, see (Ex. 28.5).)*

Proof. Follows from (28.1) and (28.2).

Remark 28.3.1. In trying to create a moduli space for vector bundles, the results just proved show that at least the obstacle of boundedness is overcome by restricting to indecomposable bundles. In fact, all that is required is something less than that—some way of limiting the possible degrees of sub-line bundles of maximum degree.

To go further, we need something more. We have seen that the functor of local deformations is pro-representable for simple vector bundles (19.2).

So now we study simple vector bundles on X, and we will show that there is an analogue of the modular families of curves we saw in §§26, 27. One problem is that a simple vector bundle still has automorphisms by scalar multiplication. To eliminate these, as in the case of line bundles, we fix a point $P \in X$, and let $\sigma : S \to X \times S$ be the section sending S to $\{P\} \times S$. Then we consider families of vector bundles $\mathcal{E}$ of rank r and degree d on $X \times S$ together with a given isomorphism $\theta : \sigma^*(\bigwedge^r \mathcal{E}) \xrightarrow{\sim} \mathcal{O}_S$. Note that over the algebraically closed field k, or over an Artin ring over k, the functor of isomorphism classes of pairs $(\mathcal{E}, \theta)$ is equivalent to the functor of isomorphism classes of bundles $\mathcal{E}$. Thus the local pro-representability is the same.

Theorem 28.4. *On the curve X, the problem of classifying families of pairs $(\mathcal{E}, \theta)$ up to isomorphism, where $\mathcal{E}$ is a locally free sheaf on $X \times S$ of rank r and fibers that are simple of degree d, together with an isomorphism $\theta : \sigma^*(\bigwedge^r \mathcal{E}) \xrightarrow{\sim} \mathcal{O}_S$, where $\sigma : S \to P \times S$ is a section, has a modular family (in the sense of §26).*

Proof. Our first step is to construct a family of vector bundles of rank r and degree d, forgetting θ, that contains each isomorphism type a finite number of times. We know that the family is bounded (28.3), so we can find an m_0 such that all our bundles $\mathcal{E}$ are m_0-regular. Then $\mathcal{E}(m_0)$ will be generated by global sections, and $h^0(\mathcal{E}(m_0)) = N$ will be independent of $\mathcal{E}$. Thus each $\mathcal{E}$ in our family can be written as a quotient $\mathcal{O}_X^N \to \mathcal{E}(m_0) \to 0$. We consider the Quot scheme (24.5.2) of all such quotients having the Hilbert polynomial of a rank r degree d vector bundle twisted by m_0. Among these, the locally free quotients form an open set, so the Quot scheme contains all of our bundles. The same bundle will occur many times, depending on the choice of basis of $H^0(\mathcal{E}(m_0))$. The group $GL(N)$ acts on the set of quotients by changing the map $\mathcal{O}_X^N \to \mathcal{E}(m_0) \to 0$, and two quotients $\mathcal{O}^N \to \mathcal{E}_1(m_0) \to 0$ and $\mathcal{O}^N \to \mathcal{E}_2(m_0) \to 0$ correspond to isomorphic bundles $\mathcal{E}_1 \cong \mathcal{E}_2$ if and only if they differ by an element of $GL(N)$. Our Quot scheme is nonsingular, because the obstructions lie in $H^1(\mathcal{H}om(Q, \mathcal{E}(m_0)))$, where $Q = \ker(\mathcal{O}^N \to \mathcal{E}(m_0))$ (7.2), and this vanishes because of an exact sequence in which $H^1(\mathcal{E}(m_0)) = 0$ and $H^2(\mathcal{H}om(\mathcal{E}, \mathcal{E})) = 0$ on the curve X.

Thus we are in a situation similar to the one in the proof of (27.2) for moduli of curves. We have a smooth Quot scheme containing all of our bundles, and the equivalence classes are described by orbits of the group action by $GL(N)$. For any particular point, we take a linear space section of complementary dimension to the orbit in an ambient space and in this way, using a finite number of these, as in the proof of (27.2) we obtain a smooth parameter space S and a universal family $\mathcal{O}_{X \times S}^N \to \mathcal{E}(m_0) \to 0$ on $X \times S$ such that (untwisting by m_0) every isomorphism type of simple bundle $\mathcal{E}$ of rank r and degree d is represented a finite number of times in the family.

To get our modular family, we need to add the isomorphisms θ. If we take the family $\mathcal{E}$ on $X \times S$ constructed above, then $\sigma^*(\bigwedge^r \mathcal{E})$ will be an invertible sheaf on S. Since it is locally free, we can take a cover of S by

Zariski open sets U_i on which it is free, and on each U_i fix an isomorphism $\theta_i : \sigma^*(\bigwedge^r \mathcal{E})|_{U_i} \to \mathcal{O}_{U_i}$. Now replace the earlier S by the disjoint union of the U_i, call it S (by abuse of notation), and then we have the desired family $\mathcal{E}$ on $X \times S$ together with $\theta : \sigma^*(\bigwedge^r \mathcal{E}) \xrightarrow{\sim} \mathcal{O}_S$.

By construction, as in the proof of (27.2) we see that for each point s of S, the induced formal family over $\hat{\mathcal{O}}_{S,s}$ pro-represents the local deformation functor.

It remains to establish the universal property of this family, point (c) in the definition of a modular family in §26, and to do this we need a lemma giving a result analogous to the Isom functor for schemes (24.10.2).

Lemma 28.5. *Let S be a scheme of finite type over k, let $\mathcal{E}_1$ and $\mathcal{E}_2$ be two families of vector bundles of rank r and degree d on $X \times S$, together with isomorphisms $\theta_i : \sigma^*(\bigwedge^r \mathcal{E}_i) \to \mathcal{O}_S$. Then there is a scheme $S' = \mathrm{Isom}_{X \times S}((\mathcal{E}_1, \theta_1), (\mathcal{E}_2, \theta_2))$ of finite type over S and an isomorphism $\varphi : \mathcal{E}_1 \times_S S' \to \mathcal{E}_2 \times_S S'$ on $X \times S'$, compatible with the θ_i and the identity on $\mathcal{O}_{S'}$, and S' is universal with this property. In other words, S' represents the functor of isomorphisms of pairs $(\mathcal{E}, \theta)$.*

Proof. Since S is of finite type, we can choose an m_0 such that the fibers of $\mathcal{E}_1$ and $\mathcal{E}_2$ are all m_0-regular. Then, letting $\pi : X \times S \to S$ be the projection, $\pi_*(\mathcal{E}_1(m_0))$ and $\pi_*(\mathcal{E}_2(m_0))$ will both be locally free of the same rank N on S. Any morphism $f : \mathcal{E}_1 \to \mathcal{E}_2$ determines a map $\pi_* f$ of these sheaves on S, and furthermore, f is uniquely determined by $\pi_* f$, since the $\mathcal{E}_i(m_0)$ are generated by global sections.

Now a morphism of $\pi_*(\mathcal{E}_1(m_0))$ to $\pi_*(\mathcal{E}_2(m_0))$ is locally given on S by an $N \times N$ matrix. Globally, these morphisms are given by an affine scheme of relative dimension N^2 over S, namely

$$\mathbb{V} = \mathbb{V}(\mathrm{Hom}(\pi_*(\mathcal{E}_1(m_0))), \pi_*(\mathcal{E}_2(m_0))).$$

Any map $\varphi : \pi_*(\mathcal{E}_1(m_0)) \to \pi_*(\mathcal{E}_2(m_0))$ determines a map

$$\pi^*\varphi : \pi^*\pi_*(\mathcal{E}_1(m_0)) \to \pi^*\pi_*(\mathcal{E}_2(m_0))$$

and thence by composition to $\mathcal{E}_2(m_0)$. The condition that it should descend to a map $f : \mathcal{E}_1(m_0) \to \mathcal{E}_2(m_0)$ is that it kill the kernel of $\pi^*\pi_*(\mathcal{E}_1(m_0)) \to \mathcal{E}_2(m_0)$. This is an algebraic condition that determines a closed subscheme of $\mathbb{V}$. For a morphism $f : \mathcal{E}_1 \to \mathcal{E}_2$ to be an isomorphism, we need its cokernel to be zero. This is an open condition on f. Compatibility with θ_1, θ_2 is another closed condition. In this way we arrive at the desired Isom scheme.

Proof of (28.4), continued. Let $\mathcal{E}, \theta$ on $X \times S$ be the family constructed above. Let $\mathcal{E}', \theta'$ be any family on $X \times S'$ for some other scheme S'. We imitate the proof of (26.4). We get two families by extending $(\mathcal{E}, \theta)$ and $(\mathcal{E}', \theta')$ to $S \times S'$. Let S'' be the Isom scheme of these two families given by (28.5).

Then for each $s' \in S'$, let $\mathcal{E}'_{s'}$ be the fiber. There are only finitely many points $s \in S$ having an isomorphic fiber, by construction of the family $\mathcal{E}$ on $X \times S$. Furthermore, for each of those, there are only finitely many ways of writing an isomorphism between those bundles compatible with the θ_i, namely, scalar multiples by rth roots of unity. Hence there are only finitely many points of S'' lying over $s' \in S'$. Now as in the proof of (26.4) it follows that $S'' \to S'$ is étale.

Remark 28.5.1. Having now a modular family for simple vector bundles, we can get a coarse moduli space, at least as an algebraic space (27.3.1), or as a complex manifold if $k = \mathbb{C}$ [123, p. 544]. However, it will in general not be separated, as we will see by example (28.7.1). In order to get a quasi-projective moduli space, one needs the condition of stability. We will not explain that proof using geometric invariant theory; we will only show how the condition of stability at least makes the moduli problem separated (§24):

Proposition 28.6. *Let T be a nonsingular curve, $0 \in T$ a point, and let $\mathcal{E}$ and $\mathcal{E}'$ be two families of vector bundles of rank r and degree d on X, parametrized by T, such that for each point $t \neq 0$ in T, the fibers $\mathcal{E}_t$ and $\mathcal{E}'_t$ are isomorphic. Assume that the two fibers $\mathcal{E}_0$ and $\mathcal{E}'_0$ are both stable. Then $\mathcal{E}_0 \cong \mathcal{E}'_0$.*

Proof. Since $\mathcal{E}_t$ and $\mathcal{E}'_t$ are isomorphic for each $t \neq 0$, the function $h^0(\mathcal{H}om(\mathcal{E}_t, \mathcal{E}'_t))$ is nonzero at all points of T different from 0. This function is upper semicontinuous on T [57, III, 12.8], so we conclude that there is a nonzero homomorphism $f : \mathcal{E}_0 \to \mathcal{E}'_0$. If f is not an isomorphism, then since $\mathcal{E}_0$ and $\mathcal{E}'_0$ have the same rank and degree, f must have a kernel. This kernel is locally free, so rank $f < r$. Let $\mathcal{F} = \text{Im } f$. Because $\mathcal{F}$ is a quotient of $\mathcal{E}_0$, its slope $\mu(\mathcal{F})$ is greater than d/r. But $\mathcal{F}$ is also a subsheaf of $\mathcal{E}'_0$, so $\mu(\mathcal{F}) < d/r$. This is a contradiction. (By the way, it would have been enough to assume that one of $\mathcal{E}_0, \mathcal{E}'_0$ is stable and the other semistable.)

If we allow semistable bundles, then the family of stable and semistable bundles is also *complete* (§24). This is a theorem of Langton [91] in general. We give a proof only in the case of rank 2 bundles on a curve, to highlight the ideas more clearly.

Proposition 28.7. *Let T be a nonsingular curve, $0 \in T$ a point, and let $T' = T - \{0\}$. Suppose we are given a family $\mathcal{E}'$ of rank 2 bundles on $X \times T'$ such that for every $t \in T'$, the fiber $\mathcal{E}'_t$ is semistable. Then there exists a family $\mathcal{E}$ on $X \times T$ with $\mathcal{E}_0$ semistable and $\mathcal{E}|_{T'} \cong \mathcal{E}'$.*

Proof. For any rank 2 bundle $\mathcal{E}$ on X we introduce the *stability degree* $\delta(\mathcal{E})$ as follows. Let $\mathcal{L} \subseteq \mathcal{E}$ be a sub-line bundle of maximum degree, let $\mathcal{M} = \mathcal{E}/\mathcal{L}$, and let $\delta(\mathcal{E}) = \deg \mathcal{M} - \deg \mathcal{L}$. Then by definition, $\delta(\mathcal{E}) > 0$ if and only if $\mathcal{E}$ is stable, and $\delta(\mathcal{E}) \geq 0$ if and only if $\mathcal{E}$ is semistable.

Returning to the situation of the proposition, suppose $\mathcal{E}'$ is given on $X \times T'$ with $\mathcal{E}'_t$ semistable for all $t \in T'$. One knows that $\mathcal{E}'$ can be extended (in many ways) to a coherent sheaf $\mathcal{E}$ on $X \times T$. Furthermore, taking double duals, we may assume that $\mathcal{E}$ is locally free, hence is a family of vector bundles over T. Among all such locally free extensions $\mathcal{E}$ of $\mathcal{E}'$, choose one for which the stability degree $\delta(\mathcal{E}_0)$ of the fiber at 0 is maximum. (Such a maximum exists, because in any case $\delta(\mathcal{E}_0) \leq 2g$; cf. proof of (28.2).) If $\delta(\mathcal{E}_0) \geq 0$, we are done.

If not, $\delta(\mathcal{E}_0) < 0$, and we proceed by contradiction. We then have an exact sequence

$$0 \to \mathcal{L}_0 \to \mathcal{E}_0 \to \mathcal{M}_0 \to 0, \tag{10}$$

where $\mathcal{L}_0$ is a sub-line bundle of maximum degree, and $\deg \mathcal{L}_0 > \deg \mathcal{M}_0$. Consider the composed map $\mathcal{E} \to \mathcal{E}_0 \to \mathcal{M}_0$, and let $\mathcal{F}$ be the kernel of $\mathcal{E} \to \mathcal{M}_0$. Then $\mathcal{F}$ is another locally free extension of $\mathcal{E}'$ to $X \times T$, and one sees easily that its closed fiber $\mathcal{F}_0$ belongs to an exact sequence

$$0 \to \mathcal{M}_0 \to \mathcal{F}_0 \to \mathcal{L}_0 \to 0. \tag{11}$$

By choice of $\mathcal{E}$, we know that $\delta(\mathcal{F}_0) \leq \delta(\mathcal{E}_0) < 0$, so $\mathcal{F}_0$ must be *unstable*. Therefore $\mathcal{F}_0$ has a sub-bundle $\mathcal{N}_0$ of rank 1 with $\deg \mathcal{N}_0 \geq \deg \mathcal{L}_0 > \deg \mathcal{M}_0$. Hence the composed map $\mathcal{N}_0 \to \mathcal{L}_0$ is nonzero, and must be an isomorphism by reason of degree. Thus the exact sequence (11) splits.

This sequence (11) is determined by an element $\xi \in \mathrm{Ext}^1(\mathcal{L}_0, \mathcal{M}_0)$, and comparison with the proof of (7.2) shows that this ξ is precisely the obstruction to lifting the sheaf $\mathcal{M}_0$ as a quotient of $\mathcal{E}_0$ to the restriction of $\mathcal{E}$ to the dual numbers $\mathrm{Spec}\, k[t]/t^2$ along T. Since the sequence splits, $\xi = 0$, and we can lift $\mathcal{M}_0$ to a quotient $\mathcal{M}_1$ of $\mathcal{E}_1 = \mathcal{E} \otimes k[t]/t^2$.

Suppose we have lifted $\mathcal{M}_0$ to a quotient $\mathcal{M}_n$ of $\mathcal{E}_n = \mathcal{E} \otimes k[t]/t^{n+1}$. Let $\mathcal{F}$ (a new one) be the kernel of the composed map $\mathcal{E} \to \mathcal{E}_n \to \mathcal{M}_n$. Then $\mathcal{F}$ is another extension of $\mathcal{E}'$ to $X \times T$, and as above, its closed fiber $\mathcal{F}_0$ belongs to an exact sequence as in (11). As before, $\mathcal{F}_0$ must be unstable, so $\xi \in \mathrm{Ext}^1(\mathcal{L}_0, \mathcal{M}_0)$ is zero. This is the obstruction to extending the quotient $\mathcal{E}_n \to \mathcal{M}_n$ one more step to get $\mathcal{E}_{n+1} \to \mathcal{M}_{n+1}$.

Thus we see that the quotient $\mathcal{E}_0 \to \mathcal{M}_0$ from (10) lifts to all infinitesimal neighborhoods of $0 \in T$. Since the Quot scheme represents the functor, and is a scheme of finite type over the base (24.5.2), it follows that there is a component of this Quot scheme that dominates T. Thus the general $\mathcal{E}_t$ has a quotient $\mathcal{M}_t$ of the same degree as $\mathcal{M}_0$, contradicting the semistability of $\mathcal{E}_t$ for $t \neq 0$. So by contradiction, we have shown that there must have been some extension of $\mathcal{E}'$ to an $\mathcal{E}$ on $X \times T$ with $\mathcal{E}_0$ semistable.

Example 28.7.1. We will now consider in some detail an example of a moduli space including simple but unstable bundles, suggested by [123, 12.3].

Let X be a nonsingular projective curve of genus 3. We consider simple vector bundles of rank 2 and degree 1. For these we have a moduli space (28.5.1) that is at least a complex manifold M when $k = \mathbb{C}$. I do not know whether it has a structure of a scheme. It is smooth and of dimension 9 [19.2.1]. We proceed in several steps.

(a) First we show that there exist simple unstable bundles in M. Let $\mathcal{L}, \mathcal{M}$ be line bundles of degrees 1, 0, respectively on X, with the property that $h^0(\mathcal{M}^\vee \otimes \mathcal{L}) = 0$. This is possible because $\mathcal{M}^\vee \otimes \mathcal{L}$ has degree 1, and a general divisor of degree 1 on X is not effective. Then $H^1(\mathcal{M}^\vee \otimes \mathcal{L})$ has dimension 1, and a nonzero element of this group, considered as $\text{Ext}^1(\mathcal{M}, \mathcal{L})$, defines a nonsplit extension

$$0 \to \mathcal{L} \to \mathcal{E} \to \mathcal{M} \to 0. \tag{12}$$

Its stability degree is -1; hence $\mathcal{E}$ is unstable. Now suppose $\mathcal{E}$ were not simple. Then there would be a map $\varphi : \mathcal{E} \to \mathcal{E}$ of rank 1, as in the proof of (28.1). If $\deg \text{Im}\, \varphi \geq 1$, then $\text{Im}\, \varphi = \mathcal{L}$, because the maximal sub-line bundle of an unstable sheaf is unique. But this means (12) splits. So $\deg \text{Im}\, \varphi \leq 0$; hence $\ker \varphi$ has degree ≥ 1, so $\ker \varphi = \mathcal{L}$. In that case $\text{Im}\, \varphi = \mathcal{M}$, so $\mathcal{M}$ must map to $\mathcal{L}$, contradicting the hypothesis that $h^0(\mathcal{M}^\vee \otimes \mathcal{L}) = 0$. Thus $\mathcal{E}$ is simple.

Note that the extension $\mathcal{E}$ is determined up to scalar multiple, so that $\mathcal{E}$ is determined up to isomorphism by $\mathcal{L}$ and $\mathcal{M}$. These line bundles are chosen in $\text{Pic}^1 X$ and $\text{Pic}^0 X$, respectively, of dimension 3 each, subject to the open condition $h^0(\mathcal{M}^\vee \otimes \mathcal{L}) = 0$, so there is an irreducible 6-dimensional family of these simple unstable bundles with stability degree -1. We denote the corresponding subset of the moduli space by $M_{-1} \subseteq M$. It is easy to see that there are no simple bundles with $\delta < -1$.

(b) Next we consider stable bundles with stability degree $\delta = 1$. We can construct these as follows. Let $\mathcal{L}, \mathcal{M}$ be line bundles of degrees $0, 1$, respectively, and consider extensions

$$0 \to \mathcal{L} \to \mathcal{E} \to \mathcal{M} \to 0 \tag{13}$$

given by an element $\xi \in H^1(\mathcal{M}^\vee \otimes \mathcal{L})$, which has dimension 3. If $\xi \neq 0$ it is easy to see that $\mathcal{E}$ is stable. The choices of $\mathcal{L}$ and $\mathcal{M}$ require three dimensions each, and ξ up to scalars, two dimensions, so in this way we obtain a family of dimension 8 of stable bundles.

When are two of these isomorphic? Or in other words, when can the same $\mathcal{E}$ be represented in another way as such an extension? This can happen only if there is another line bundle $\mathcal{N}$ of degree 0 such that $h^0(\mathcal{E} \otimes \mathcal{N}^\vee) \neq 0$. So we examine the sequence

$$0 \to H^0(\mathcal{L} \otimes \mathcal{N}^\vee) \to H^0(\mathcal{E} \otimes \mathcal{N}^\vee) \to H^0(\mathcal{M} \otimes \mathcal{N}^\vee) \xrightarrow{\alpha} H^1(\mathcal{L} \otimes \mathcal{N}^\vee)$$
$$\to \cdots .$$

If the term on the left is nonzero, then $\mathcal{N} \cong \mathcal{L}$ and we have nothing new. So, assuming $H^0(\mathcal{L} \otimes \mathcal{N}^\vee) = 0$, in order for $H^0(\mathcal{E} \otimes \mathcal{N}^\vee) \neq 0$, we must have $H^0(\mathcal{M} \otimes \mathcal{N}^\vee) \neq 0$, and the map α from there to $H^1(\mathcal{L} \otimes \mathcal{N}^\vee)$ must be zero. Since $\deg \mathcal{M} \otimes \mathcal{N}^\vee = 1$, this can happen only if $\mathcal{N} \cong \mathcal{M}(-P)$ for some point $P \in X$. Comparing with the sequence for $\mathcal{E} \otimes \mathcal{M}^\vee$ we have a commutative diagram

$$H^0(\mathcal{O}(P)) \xrightarrow{\ \alpha\ } H^1(\mathcal{M}^\vee \otimes \mathcal{L}(P))$$

$$\uparrow \qquad\qquad\qquad\qquad \uparrow$$

$$H^0(\mathcal{O}) \xrightarrow{\ \alpha\ } H^1(\mathcal{M}^\vee \otimes \mathcal{L})$$

Now $1 \in H^0(\mathcal{O})$ goes by α to $\xi \in H^1(\mathcal{M}^\vee \otimes \mathcal{L})$, which is 3-dimensional. The first vertical arrow takes 1 to the unique section of $H^0(\mathcal{O}(P))$ and our requirement that this go to 0 by α says that ξ must be in the kernel of the second vertical map $H^1(\mathcal{M}^\vee \otimes \mathcal{L}) \to H^1(\mathcal{M}^\vee \otimes \mathcal{L}(P))$. This map is surjective, and $h^1(\mathcal{M}^\vee \otimes \mathcal{L}(P)) = 2$, so by reason of dimension the kernel has dimension 1.

Thus for each $P \in X$ there is a unique ξ (up to scalar) for which the corresponding $\mathcal{E}$ has another representation as

$$0 \to \mathcal{M}(-P) \to \mathcal{E} \to \mathcal{L}(P) \to 0.$$

So for given $\mathcal{L}, \mathcal{M}$, there is a $\mathbb{P}^2$ of possible extensions (13). There is a map of the curve X to $\mathbb{P}^2$ such that for each $P \in X$, the ξ that is its image defines an $\mathcal{E}$ with a second representation.

So we can describe the set M_1 of stable bundles with $\delta = 1$ as a $\mathbb{P}^2$-bundle over $\mathrm{Pic}^0 X \times \mathrm{Pic}^1 X$, in which a certain divisor has been identified in a finite-to-one manner. In particular, $\dim M_1 = 8$.

(c) A theorem of Nagata [122] implies that every rank 2 bundle on X has stability degree $\delta \le 3$. Since the space M has dimension 9, and the bundles with $\delta = -1, 1$ form families of dimension 6, 8, respectively, we conclude that there are bundles with $\delta = 3$ and that these form an open dense subset of M. We denote the set of all these by M_3. Thus M is the union of the subsets M_{-1}, M_1, and M_3.

(d) The whole space M is irreducible. Since we know that it is smooth of dimension 9, and M_{-1}, M_1 have lower dimension, it is enough to show that M_3 is irreducible. This is clear, because we can form an irreducible family of extensions as in (13) with $\deg \mathcal{L} = -1$, $\deg \mathcal{M} = 2$, containing an open set of stable bundles, and which maps surjectively to M_3.

(e) Let $M' \subseteq M$ be the open subset consisting of points corresponding to stable bundles only, i.e., M_1 together with M_3. Since the rank and degree are coprime, there are no semistable bundles. Thus by (28.6) and (28.7), the space M' is separated and complete. In the case $k = \mathbb{C}$, this means that it is a compact Hausdorf space. In the algebraic case, one knows from geometric invariant theory that it is a nonsingular projective variety.

(f) The puzzle is, where do the points of M_{-1} fit in this moduli space? We have to imagine the separated complete space M', which however is dense in M, so the points of M_{-1} are hovering over the space M' in some nonseparated way.

To understand this better, let $\mathcal{E}_0$ be an unstable simple bundle in M_{-1}, given by an extension

$$0 \to \mathcal{L}_0 \to \mathcal{E}_0 \to \mathcal{M}_0 \to 0 \qquad (14)$$

with $\deg \mathcal{L}_0 = 1$, $\deg \mathcal{M}_0 = 0$. We know that it is in the closure of M', so take a family $\mathcal{E}$ on $X \times T$ for a curve T, with $\mathcal{E}_t$ stable for $t \neq 0$, and having $\mathcal{E}_0$ as its fiber at 0. As in the proof of (28.7), let $\mathcal{F} = \ker(\mathcal{E} \to \mathcal{M}_0)$ be a new family. Then $\mathcal{F}_t \cong \mathcal{E}_t$ for $t \neq 0$ and $\mathcal{F}_0$ belongs to an exact sequence

$$0 \to \mathcal{M}_0 \to \mathcal{F}_0 \to \mathcal{L}_0 \to 0. \qquad (15)$$

This extension is defined by an element $\xi \in \mathrm{Ext}^1(\mathcal{L}_0, \mathcal{M}_0)$ that is determined by the family $\mathcal{E}$ we chose having limit $\mathcal{E}_0$. The tangent space to M at $\mathcal{E}_0$ is $\mathrm{Ext}^1(\mathcal{E}_0, \mathcal{E}_0)$, and ξ is the image of the tangent vector $\tau \in \mathrm{Ext}^1(\mathcal{E}_0, \mathcal{E}_0)$ corresponding to the family $\mathcal{E}$. The map $\beta : \mathrm{Ext}^1(\mathcal{E}_0, \mathcal{E}_0) \to \mathrm{Ext}^1(\mathcal{L}_0, \mathcal{M}_0)$ is given by the maps $\mathcal{L}_0 \to \mathcal{E}_0$ and $\mathcal{E}_0 \to \mathcal{M}_0$ from (14), and since X is a curve, this map β is surjective. In other words, by choosing all different approaches to $\mathcal{E}_0$ in the family M, we obtain all possible extensions (15), hence all those elements of M_1 that can be written as extensions in (15).

Summing up, the whole smooth connected moduli space M contains a separated and complete subspace $M' = M_1 \cup M_3$, which is dense in M. A point x of M_{-1} is (in many ways) a limit of a family of points in M', and each such family determines a point $x' \in M'$. Thus we have a correspondence between M_{-1} and M' that explains the nonseparated way in which they are attached. The graph of this correspondence is a $\mathbb{P}^2$-bundle over M_{-1}, and its map to M' collapses the divisor mentioned in (b) and has image a dense open subset of M_1. Not all points of M_1 have a correspondent in M_{-1}, because of the condition $h^0(\mathcal{M}^\vee \otimes \mathcal{L}) = 0$ of (a).

References for this section. Mumford [113] defined the notion of stability for vector bundles on a curve, and announced the existence of a quasi-projective coarse moduli space for stable vector bundles on curves. About the same time, Narasimhan and Seshadri [123] showed that stable bundles on curves over $\mathbb{C}$ correspond to irreducible unitary representations of the fundamental group of the curve. Since then, the moduli of vector bundles on curves have been studied thoroughly. The notion of stability was extended to torsion-free sheaves on varieties of higher dimension by Gieseker [36] and Maruyama [101], who proved the existence of a (coarse) moduli space locally of finite type. The finiteness, and hence projectivity, of this space was at first established only in case of characteristic zero, or rank 2, or dimension of the base space 2. The boundedness of the family of semistable bundles in all ranks and all characteristics was proved only recently [90]. See the appendix to Chapter 5 of [119] for a survey of these results up to 1981, and [70] for further developments and references, too many to include here. For stable vector bundles of rank 2 on $\mathbb{P}^3$, see [58].

Exercises.

28.1. Let $\mathcal{E}$ be a vector bundle on the curve X, let $\mathcal{L}$ be an invertible sheaf, and let $\mathcal{E}' = \mathcal{E} \otimes \mathcal{L}$. Show that $\mathcal{E}'$ is stable (resp., semistable, simple, indecomposable) if and only if $\mathcal{E}$ is.

28.2. Show that $\mathcal{E}$ on X is stable if and only if for every quotient bundle $\mathcal{E} \to \mathcal{G} \to 0$ of $\mathcal{E}$, with $0 < \operatorname{rank} \mathcal{G} < \operatorname{rank} \mathcal{E}$, we have

$$\mu(\mathcal{G}) > \mu(\mathcal{E}).$$

28.3.

(a) Give examples of bundles on curves to show that each of the implications in (28.1) is strict.
(b) Show also that there are bundles that are
  (1) semistable but not indecomposable;
  (2) simple but not semistable.

28.4. Complete the proof of (28.2) for ranks $r \geq 3$. Be careful, because even though $\mathcal{E}$ is assumed to be indecomposable, when you take a sub-line bundle $\mathcal{L}$ of maximum degree, the quotient $\mathcal{E}/\mathcal{L}$ may not be indecomposable. Show, however, that if it decomposes, the possible degrees and ranks of the factors are finite in number.

28.5. Show that the family of semistable bundles of given rank and degree on a curve X is bounded.

28.6.

(a) Let S be an integral scheme of finite type over k, and let $\mathcal{E}$ and $\mathcal{E}'$ be two locally free sheaves on $X \times S$ such that for every closed point $s \in S$, the fibers $\mathcal{E}_s$ and $\mathcal{E}'_s$ are simple and isomorphic to each other. Let $\mathcal{F} = \mathcal{H}om(\mathcal{E}, \mathcal{E}')$, and show that $\pi_* \mathcal{F}$ is locally free of rank 1 on S, where $\pi : X \times S \to S$ is the projection. Conclude that S can be covered by open sets U such that $\mathcal{E}_U$ and $\mathcal{E}'_U$ are isomorphic on $X \times U$ for each U.
(b) Show that a fiberwise trivial family of simple vector bundles on an integral scheme S of finite type over k is locally trivial in the Zariski topology.

28.7. Use (28.4) to show that there can be no jump phenomena for simple vector bundles on a curve X.

28.8. Let $\mathcal{E}$ be a simple rank 2 bundle on a curve X of genus 2. Show that

(a) if $\deg \mathcal{E}$ is even, then $\mathcal{E}$ is semistable;
(b) if $\deg \mathcal{E}$ is odd, then $\mathcal{E}$ is stable.

28.9. Stable rank 2 vector bundles on $\mathbb{P}^2$.

(a) **Chern classes.** Using the theory of Chern classes [57, Appendix A, §3] verify the following. A rank 2 vector bundle $\mathcal{E}$ on $\mathbb{P}^2$ has two Chern classes $c_1, c_2 \in \mathbb{Z}$. Show that
  (1) $\bigwedge^2 \mathcal{E} \cong \mathcal{O}(c_1)$.
  (2) The Euler characteristic $\chi(\mathcal{E}) = \frac{1}{2}(c_1^2 + 3c_1) - c_2 + 2$.
  (3) For any n, $c_1(\mathcal{E}(n)) = c_2(\mathcal{E}) + c_1(\mathcal{E}) \cdot n + n^2$.
  (4) If $\mathcal{E}$ has a global section s vanishing only at points, then there is an exact sequence

  $$0 \to \mathcal{O} \xrightarrow{s} \mathcal{E} \to \mathcal{I}_Z(a) \to 0,$$

  where $a = c_1(\mathcal{E})$ and Z is a zero-scheme of length $c_2(\mathcal{E})$.
  (5) $\chi(\mathcal{E}nd\,\mathcal{E}) = c_1^2 - 4c_2 + 4$.

(b) **Stable bundles.**   We say that a rank 2 bundle $\mathcal{E}$ on $\mathbb{P}^2$ is *stable* if for every line bundle $\mathcal{L} \subseteq \mathcal{E}$ we have $c_1(\mathcal{L}) < \frac{1}{2} c_1(\mathcal{E})$. If $\mathcal{E}$ is a rank 2 bundle with $c_1 = 0$, show that:

(1) $\mathcal{E}$ is simple if and only if $\mathcal{E}$ is stable, if and only if $h^0(\mathcal{E}) = 0$.

(2) If $\mathcal{E}$ is stable, then $c_2(\mathcal{E}) \geq 2$.

(3) For $\mathcal{E}$ stable, the functor of infinitesimal deformations is pro-representable.

(4) For $\mathcal{E}$ stable, $h^2(\mathcal{E}nd\,\mathcal{E}) = 0$.

(5) The pro-representing family is smooth of dimension $4c_2 - 3$.

(c) **Boundedness.**   Show that the family of stable rank 2 bundles on $\mathbb{P}^2$ with Chern classes $c_1 = 0$ and fixed c_2 is bounded, as follows:

(1) Show that there is an $n \leq \sqrt{c_2}$ such that $h^0(\mathcal{E}(n)) \neq 0$.

(2) If n is the least integer for which $h^0(\mathcal{E}(n)) \neq 0$, then a section $s \in H^0(\mathcal{E}(n))$ can vanish only at points.

(3) Use exact sequences of the form

$$0 \to \mathcal{O} \xrightarrow{s} \mathcal{E}(n) \to \mathcal{I}_Z(2n) \to 0,$$

where Z is a zero-scheme of length $c_2 + n^2$, to show that $\mathcal{E}$ is determined by the choice of Z and an element $\xi \in \mathrm{Ext}^1(\mathcal{I}_Z(2n), \mathcal{O})$. Thus show that the family is bounded.

(d) **Existence.**   Show the existence of rank 2 stable bundles on $\mathbb{P}^2$ with $c_1 = 0$ as follows.

(1) For any $c_2 \geq 2$, let Z be a general set of $c_2 + 1$ points in $\mathbb{P}^2$. Show that for a general element $\xi \in \mathrm{Ext}^1(\mathcal{I}_Z(2), \mathcal{O})$, the extension

$$0 \to \mathcal{O} \to \mathcal{E}(1) \to \mathcal{I}_Z(2) \to 0$$

defines a stable bundle $\mathcal{E}$ with $c_1 = 0$ and the given c_2.

(2) If $c_2 = 2, 3, 4, 5$, show that every stable bundle $\mathcal{E}$ with $c_1 = 0$ and the given c_2 can be obtained as in (1).

(3) If $c_2 \geq 6$, show that there exist stable bundles $\mathcal{E}$ with $c_1 = 0$, the given c_2, and $h^0(\mathcal{E}(1)) = 0$.

(e) **Modular families.**   Imitating the methods of this section show that the family of stable rank 2 bundles $\mathcal{E}$ on $\mathbb{P}^2$ with $c_1 = 0$ and given $c_2 \geq 2$ has a modular family that is smooth, separated, of dimension $4c_2 - 3$, and irreducible. (To show that it is irreducible, show that there is a uniform n_0 depending on c_2 such that for every bundle $\mathcal{E}$ in the family, $\mathcal{E}(n_0)$ is generated by global sections. Show then that a general section of $\mathcal{E}(n_0)$ vanishes only at points and use extensions as in (c) (3) above.)

29. Smoothing Singularities

In this section we study the question of simplifying the singularities of a scheme by deformation. At one extreme we have smoothable singularities, which can be deformed to nonsingular varieties. At the other extreme we have rigid sigularities, which have no deformations at all. In between there are singularities that can be deformed but not smoothed. The principal technical difficulty we encounter is how to recognize at the level of infinitesimal deformations when a global deformation will be nonsingular. To do this we introduce the notion of a formally smoothable scheme, and investigate its properties. This gives a framework for passage from local to global.

We give conditions under which a projective variety with locally smoothable singularities is globally smoothable. Applying these techniques to curves, we show that a curve is globally smoothable if and only if its singularities are locally smoothable. We give various examples of curves that are smoothable or not smoothable in $\mathbb{P}^3$. We give Mumford's example of a nonsmoothable abstract curve singularity.

In higher dimensions, we study cones over projective varieties, giving rise to examples of rigid singularities, generic singularities, and Pinkham's examples of nonsmoothable normal surface singularities.

As a final application, we use a smoothing argument to show that for any genus g, the subset of d-gonal curves in the variety of moduli $\mathcal{M}_g$ is in the closure of the set of $(d+1)$-gonal curves, for any d.

Let X_0 be a scheme of finite type over an algebraically closed field k. We say that X_0 is *smoothable* if there exists a flat family X/T of schemes over an integral scheme of finite type T and a point $0 \in T$ such that the fiber over 0 is X_0, and there are fibers X_p for $p \neq 0$ that are nonsingular (hence smooth over k). This definition comes in different flavors, depending on whether X_0 is affine or projective, and whether we seek to smooth X_0 as an abstract scheme (as above), or, when X_0 is a closed subscheme of a smooth scheme P, we seek to smooth X_0 as a subscheme of P.

For any scheme T of finite type over k and a nonisolated point $0 \in T$, we can find a nonsingular curve T' over k (not necessarily complete), a point $0' \in T'$, and a morphism $T' \to T$ sending $0'$ to 0 and $p' \in T'$, $p' \neq 0'$ to a point $p \in T$, $p \neq 0$. Thus X_0 is smoothable if and only if there is a family as in the definition above, where T is required to be a nonsingular curve over k.

Example 29.0.1. A hypersurface in $\mathbb{P}^n$, or more generally, a complete intersection scheme in $\mathbb{P}^n$, is smoothable. This follows from repeated applications of Bertini's theorem [57, II, 8.18], since the space of complete intersections of given degrees is irreducible.

Extending this argument, we can show that a complete intersection X_0 in a smooth affine scheme Z, that is, defined by r elements $f_1, \ldots, f_r$ in the affine ring of Z, where r is the codimension of X_0, is smoothable in Z. Indeed, let $\bar{Z}$ be the projective closure of Z in some projective space $\mathbb{P}^n$, and let $\bar{f}_1, \ldots, \bar{f}_r$, be homogeneous polynomials on $\mathbb{P}^n$ that reduce to $f_1, \ldots, f_r$ on Z. Then Bertini's theorem, applied to the nonsingular part Z of $\bar{Z}$, gives a family of smooth complete intersections smoothing X_0 in Z.

Example 29.0.2. On the other hand, we have seen some examples of nonsmoothable schemes. There are zero-dimensional schemes in $\mathbb{A}^3$ and $\mathbb{A}^4$ that are not smoothable (Ex. 5.8), (Ex. 8.10). Some nonreduced projective curves are not smoothable (Ex. 5.10). We will see later in this section examples of nonsmoothable integral curves (29.10.3) and normal surfaces (29.12).

Example 29.0.3. Recall that a scheme is *rigid* if it has no nontrivial deformations over the dual numbers (5.3.1). We have seen examples of normal rigid singularities in dimension ≥ 3 (5.5.1), (Ex. 5.4), of reduced rigid singularities in dimension 2 (Ex. 5.5), and of nonreduced rigid projective curves (Ex. 5.10). It seems to be an open question whether there are any affine rigid singularities in dimension 0 or 1. Although common sense would suggest that a rigid scheme is not smoothable, the proof is not obvious (29.6). The trouble is, even though all infinitesimal deformations are trivial (Ex. 10.3), there may be nontrivial global deformations (Ex. 4.9).

The main technical difficulty in studying smoothing questions is the relationship between infinitesimal deformations and global families containing the given scheme. Since this point seems not to have been sufficiently addressed in the literature, we will take some time to examine it carefully.

Proposition 29.1. *Let X_0 be an affine scheme of finite type over the algebraically closed field k. Let X/T be a flat family over a nonsingular curve T of finite type over k with X, T both affine, and let $0 \in T$ be a point such that the fiber of X over 0 is X_0. We let $T = \operatorname{Spec} A$, $X = \operatorname{Spec} B$, and let F be the functor $T^1(B/A, \cdot)$ on B-modules. Let $t \in A$ generate the maximal ideal of 0 on T. Then the following conditions are equivalent:*

(i) *X_p is smooth for all $p \in T$, $p \neq 0$, so the family X/T smooths X_0.*
(ii) *There is an integer n_0 such that $t^{n_0} \cdot F(M) = 0$ for all finitely generated B-modules M.*

Proof. We know from (4.11) that a morphism is smooth if and only if the relative T^1 functor vanishes for all modules. Thus condition (i) says that the functor F is zero over $T \setminus \{0\}$. In other words, for every B-module M of finite type, $F(M)$ has support along the closed fiber X_0 defined by t, and so $t^n \cdot F(M) = 0$ for some n (depending on M). Before completing the proof, we recall the definition of a coherent functor [9], [64] and prove a lemma.

Definition. Let B be a noetherian ring, let $\operatorname{Mod}(B)$ be the category of finitely generated B-modules, and let F be a covariant functor from $\operatorname{Mod}(B)$ to itself. We say that the functor F is *coherent* if there are modules $P, Q \in \operatorname{Mod}(B)$ and a homomorphism $f : P \to Q$ such that F appears as the cokernel functor

$$\operatorname{Hom}(Q, \cdot) \to \operatorname{Hom}(P, \cdot) \to F(\cdot) \to 0.$$

Lemma 29.2. *Let B be a noetherian ring, let F be a coherent functor on $\operatorname{Mod}(B)$, and let $t \in B$ be a non-zero-divisor. Suppose that for every $M \in \operatorname{Mod}(B)$ there exists some $n > 0$ such that $t^n \cdot F(M) = 0$. Then there is a uniform $n_0 > 0$ such that for every $M \in \operatorname{Mod}(B)$, we have $t^{n_0} \cdot F(M) = 0$.*

Proof. Our hypothesis implies that the extension of F to the category of modules over the localized ring B_t is identically zero. Hence for any B_t-module N, the map

$$\operatorname{Hom}_{B_t}(Q_t, N) \to \operatorname{Hom}_{B_t}(P_t, N)$$

is surjective. Taking $N = P_t$ and lifting the identity map, we obtain a map $\sigma : Q_t \to P_t$ such that $\sigma \circ f = \operatorname{id}_{P_t}$. Looking at a finite set of generators of Q and their images in P_t, we can find a common denominator, so that $t^{n_0}\sigma$ sends Q to P. It follows that for any $M \in \operatorname{Mod}(B)$, all elements of $t^{n_0} \operatorname{Hom}(P, M)$ are in the image of $\operatorname{Hom}(Q, M)$, and hence $t^{n_0} \cdot F(M) = 0$.

Proof of (29.1), continued. By (3.10) the functor $F = T^1(B/A, \cdot)$ is coherent. Condition (i) is equivalent to the hypothesis of the lemma, so from the lemma we obtain (ii). The converse is immediate.

Definition. Let $X_0 = \operatorname{Spec} B_0$ be an affine scheme of finite type over k. We say that X_0 is *formally smoothable* if there exists a formal family over $A = k[[t]]$, that is, a compatible collection of deformations $X_n = \operatorname{Spec} B_n$ of X_0/k over $A_n = k[t]/t^{n+1}$ for each n, and there exists a uniform $n_0 > 0$ such that for each n, and for each B_n-module of finite type M_n, we have $t^{n_0} \cdot T^1(B_n/A_n, M_n) = 0$.

Remark 29.2.1. With this definition, the implication (i) $\Rightarrow$ (ii) of (29.1) simply says that if X_0 is smoothable, then it is formally smoothable. Indeed, with X/T as in (29.1), for each n let $B_n = B \otimes_A A/t^{n+1}$. Then the collection of $X_n = \operatorname{Spec} B_n$ makes a formal family, and it satisfies the condition of the definition because by base change properties of the T^i functors (Ex. 3.8), for any B_n-module M_n we have $T^1(B/A, M_n) = T^1(B_n/A_n, M_n)$.

Remark 29.2.2. If X_0 is not affine, we can define *formally smoothable* in the same way, using a formal family of deformations X_n/A_n, and requiring that for every coherent sheaf $\mathcal{F}_n$ on X_n, $t^{n_0} \cdot T^1(X_n/A_n, \mathcal{F}_n) = 0$, using the T^1 sheaves. By applying (29.1) to an open affine cover of X_0 we see similarly that if X_0 is smoothable, then it is formally smoothable.

Remark 29.2.3. If X_0 is a closed subscheme of a smooth scheme Z, we say that X_0 is a *formally smoothable subscheme* if in the above definition all the X_n are closed subschemes of Z.

Next, we give some elementary properties of formally smoothable schemes.

Proposition 29.3. *Let X_0 be an affine scheme with isolated singularities. Then X_0 is formally smoothable if and only if each singular point has a formally smoothable affine neighborhood.*

Proof. One direction is obvious. For the converse, let U_0 be the disjoint union of formally smoothable open affine subsets containing all the singular points (each one once only). Then by hypothesis there is a formal family $\{U_n\}$ of deformations of U_0 and an integer n_0 such that $t^{n_0} \cdot T^1(U_n/A_n, M_n) = 0$ for every U_n-module M_n. We need to construct a formal family of deformations X_n of X_0 with the same property.

Suppose inductively we have X_n whose restriction to U_0 is isomorphic to U_n. The obstruction to finding an extension X_{n+1} of X_n over A_{n+1} lies in a certain T^2 group (10.1), which is supported at the singular points, and this obstruction is zero because an extension U_{n+1} of U_n exists. Hence an extension X_{n+1} exists, and the set of such is a torsor under $T^1(X_0/k, \mathcal{O}_{X_0})$. But this T^1 is supported at the singular points, so it is isomorphic to the corresponding T^1 for deformation of U_0. Therefore we can find X_{n+1} whose restriction to U_0 is isomorphic to U_{n+1}. Then the vanishing of $t^{n_0} \cdot T^1(U_n/A_n, M_n)$ implies the same for X_n.

Remark 29.3.1. Because of this proposition, for an isolated singularity, it makes sense to say that it is *locally formally smoothable* if it has a formally smoothable affine neighborhood. This property is independent of the affine neighborhood chosen.

Remark 29.3.2. In the proof of (29.3) each deformation of X_0 induces a deformation of U_0, so that we have a morphism of deformation functors $\mathrm{Def}(X_0) \to \mathrm{Def}(U_0)$. The proof then shows that this morphism is strongly surjective (§15). In fact it is an isomorphism.

Proposition 29.4. Let X_0 and X_0' be affine schemes each having a single isolated singularity at points P, P', and assume that these singular points are analytically isomorphic. Then X_0 is formally smoothable if and only if X_0' is formally smoothable.

Proof. There is an isomorphism of the deformation functors $\mathrm{Def}(X_0)$ and $\mathrm{Def}(X_0')$, so that a formal family X_n of deformations of X_0 corresponds to a formal family X_n' of deformations of X_0', and furthermore, the singularities of X_n and X_n' at P and P' are analytically isomorphic (Ex. 18.6). Since in this case the corresponding T^1 functors are also isomorphic (Ex. 4.4), the condition to be formally smoothable carries over from one to the other.

Proposition 29.5. Let X_0 be a formally smoothable closed subscheme of a nonsingular projective scheme Z over k. Then X_0 is smoothable as a subscheme of Z.

Proof. We make use of the fact that deformations of closed subschemes of a projective scheme are represented by the Hilbert scheme H. Let X_0 correspond to a point $x_0 \in H$. Then a formal family X_n of deformations of X_0 in Z corresponds to a series of compatible maps of $\mathrm{Spec}\, A_n \to H$ landing at the point x_0. Taking their limit we obtain a map of $T = \mathrm{Spec}\, k[[t]]$ to H. Let X/T be the pullback of the universal family over H. I claim that the generic fiber X_τ over the generic point $\tau \in T$ is smooth.

Covering X_0 with open affines and considering the corresponding open affines of X over T, we reduce to the case $X = \mathrm{Spec}\, B$ affine, where now B is a finitely generated $A = k[[t]]$-algebra. We apply the argument of (29.1) to show that X_τ is smooth by showing that $t^{n_0} \cdot F(M) = 0$ for every finitely generated B-module, where n_0 is the integer occurring in our hypothesis that X_0 is formally smoothable, and F is the functor $T^1(B/A, \cdot)$. It is enough to verify for every closed point $x \in X_0$ that $t^{n_0} \cdot F(M)_x^\wedge = 0$, meaning the completion of $F(M)$ over the complete local ring $B_x^\wedge$. Since F is a coherent functor, $F(M)_x^\wedge = \varprojlim F(M/\mathfrak{m}_x^{n+1}M)$. This follows from the definition of a coherent functor and the Mittag-Leffler condition for the finite-length modules $F(M/\mathfrak{m}_x^{n+1}M)$ [64, 3.5]. Now for each n, $M_n = M/\mathfrak{m}_x^{n+1}M$ is a module of finite type over the ring $B_n = B \otimes_A A_n$, so the n_0 that occurs in our hypothesis that X_0 is formally smoothable will do. (Here again we use base extension (Ex. 3.9) for the T^1-functor.)

Thus we have shown that the generic fiber X_τ is smooth. Let ξ be the image of τ in H. This is not a closed point of H. But since the set of points $x \in H$ for which the corresponding closed subscheme is smooth is an open set, the presence of ξ, whose closure contains x_0, shows that x_0 is in the closure of the open subset of H corresponding to smooth subschemes. Now we need only take a morphism of a nonsingular curve over k to H whose image contains x_0 and meets this open set to obtain a smoothing of X_0 in Z.

Now comes our first application.

Proposition 29.6. *A singular rigid scheme is not formally smoothable, and hence not smoothable.*

Proof. If X_0 is singular, then there is some X_0-module M_0 for which $T^1(X_0/k, M_0) \neq 0$. If X_0 is rigid, and X_n is a deformation of X_0 over $A_n = k[t]/t^{n+1}$, then $X_n \cong X_0 \times \operatorname{Spec} A_n$ (Ex. 10.3). Let $M_n = M_0 \otimes_k A_n$. Then $T^1(X_n/k, M_n) = T^1(X_0/k, M_0) \otimes A_n$ is not annihilated by t^n. Taking n sufficiently large, we see that X_0 cannot be formally smoothable (29.2.1). $\quad\blacksquare$

For further applications, we need to compare deformations of a projective scheme with deformations of its affine open subsets. We give a criterion when the restriction map from global to local deformations is strongly surjective.

Theorem 29.7. *Let X_0 be a closed subscheme of the nonsingular projective scheme Z, and assume that X_0 has isolated singularities. Let U_0 be the disjoint union of open affine subsets of X_0 containing all the singular points of X_0, each one once. Suppose that*

(a) $H^1(X_0, \mathcal{N}_{X_0/Z}) = 0$ *and*
(b) $H^0(X_0, \mathcal{N}_{X_0/Z}) \to T^1_{X_0} \to 0$ *is surjective.*

Then the restriction map of functors $\operatorname{Hilb}(X_0, Z) \to \operatorname{Def}(U_0)$ *from embedded deformations of X_0 in Z to abstract deformations of U_0 is strongly surjective.*

Proof. Hypothesis (b) tells us that the map of functors is surjective on tangent spaces. To show strong surjectivity, suppose we are given a small extension $C' \to C$ of Artin rings and a deformation X of X_0 in Z over C, restricting to a deformation U of U_0, and suppose furthermore that we are given an extension U' of U over C'. We must show that there is an extension X' of X over C' restricting to U'. First of all, consider the obstruction to extending X over C'. The existence of an extension U' of U tells us that there are no local obstructions to extending X, since U_0 contains all the singular points of X_0. Therefore (10.9) the obstruction to extending X lies in $H^1(X_0, \mathcal{N}_{X_0/Z} \otimes J)$, which is zero because of hypothesis (a). Thus some extension X' exists. The set of choices for X' is a torsor under $H^0(X_0, \mathcal{N}_{X_0/Z})$, and since this maps surjectively to $T^1_{X_0} = T^1_{U_0}$, we can adjust the choice of X' so as to restrict to U'.

Corollary 29.8. *Suppose that X_0 is a closed subscheme of the nonsingular projective scheme Z, having isolated singularities and satisfying conditions (a), (b) of (29.7). Suppose furthermore that each singular point P_i is locally formally smoothable. Then X_0 is smoothable in Z.*

Proof. By hypothesis we can find a formally smoothable open affine subset U_i of X_0 containing P_i for each i. Let U_0 be the disjoint union of the U_i. Then by (29.7) the restriction map of functors $\mathrm{Hilb}(X_0, Z) \to \mathrm{Def}(U_0)$ is strongly surjective. This means we can lift a family U_n of compatible deformations of U_0 to a family of compatible deformations X_n of X_0 in Z, and we conclude that X_0 is formally smoothable in Z. Then from (29.5) it follows that X_0 is smoothable in Z.

Now we will apply these results to smoothing curve singularities.

Proposition 29.9. *A reduced curve Y in $\mathbb{P}^n$ with locally smoothable singularities and $H^1(Y, \mathcal{O}_Y(1)) = 0$ is smoothable. In particular, this applies if Y has locally complete intersection singularities.*

Proof. Since Y is reduced, its singularities are isolated. Since the singularities are locally smoothable, the obstructions to deforming Y in $\mathbb{P}^n$ will lie in $H^1(Y, \mathcal{N}_{Y/\mathbb{P}^n})$.

The defining sequence for the module T^1_Y, supported at the singular points of Y is (3.10)

$$0 \to \mathcal{T}_Y \to \mathcal{T}_{\mathbb{P}^n} \otimes \mathcal{O}_Y \to \mathcal{N}_{Y/\mathbb{P}^n} \to T^1_Y \to 0.$$

Splitting this sequence with a coherent sheaf $\mathcal{R}$, we have

$$0 \to \mathcal{T}_Y \to \mathcal{T}_{\mathbb{P}^n} \otimes \mathcal{O}_Y \to \mathcal{R} \to 0,$$
$$0 \to \mathcal{R} \to \mathcal{N}_{Y/\mathbb{P}^n} \to T^1_Y \to 0.$$

From the Euler sequence on $\mathbb{P}^n$ we have

$$0 \to \mathcal{O}_Y \to \mathcal{O}_Y(1)^{n+1} \to \mathcal{T}_{\mathbb{P}^n} \otimes \mathcal{O}_Y \to 0.$$

Now since Y is a curve, the hypothesis $H^1(\mathcal{O}_Y(1)) = 0$ implies $H^1(\mathcal{T}_{\mathbb{P}^n} \otimes \mathcal{O}_Y) = 0$ and therefore also $H^1(\mathcal{R}) = 0$. And this in turn, since T_Y^1 is supported at points, implies $H^1(\mathcal{N}_{Y/\mathbb{P}^n}) = 0$ and $H^0(\mathcal{N}_{Y/\mathbb{P}^n}) \to T_Y^1 \to 0$ is surjective. Since smoothable implies formally smoothable (29.2.1), the conditions of (29.8) are satisfied, and Y is smoothable in $\mathbb{P}^n$.

Remark 29.9.1. In $\mathbb{P}^3$ we can strengthen this result to say that any reduced curve Y in $\mathbb{P}^3$ with $H^1(Y, \mathcal{O}_Y(1)) = 0$ is smoothable. The reason is that a curve in $\mathbb{P}^3$ is a Cohen–Macaulay scheme of codimension 2. Schaps [144] shows that a Cohen–Macaulay subscheme of codimension 2 and dimension ≤ 3 in an affine space is smoothable. This is done by adding extra variables and comparing it to a generic Cohen–Macaulay subscheme defined by the $r \times r$ minors of an $r \times (r + 1)$ matrix of indeterminates. One knows that the singular locus of this one is defined by the $(r - 1) \times (r - 1)$ minors of the same matrix and has codimension 6 in the ambient space, hence codimension 4 in the subvariety. Thus for a variety of dimension ≤ 3 one can avoid the singularities by deformation. This result is sharp, because for example the cone in $\mathbb{A}^6$ over the Segre embedding of $\mathbb{P}^1 \times \mathbb{P}^2$ in $\mathbb{P}^5$ is rigid and hence not smoothable (Ex. 5.4), (29.6).

Example 29.9.2. The hypothesis $H^1(Y, \mathcal{O}_Y(1)) = 0$ is necessary in (29.9). Let Y be the union of a plane quartic curve with a line meeting it at one point and not lying in the same plane. This curve is not smoothable in $\mathbb{P}^3$ for the simple reason that there are no nonsingular curves of degree 5 and genus 3 in $\mathbb{P}^3$. Even though this curve is not smoothable in $\mathbb{P}^3$, it is smoothable as an abstract curve because of (29.10) below.

Example 29.9.3. For an example of an irreducible reduced curve with a single node in $\mathbb{P}^3$ that is not smoothable in $\mathbb{P}^3$, see [60].

Example 29.9.4. Ein [23] shows that any integral curve in $\mathbb{P}^3$ with $d \geq p_a + 2$ is smoothable, by carefully counting the dimension of the family of singular curves and showing that it is less than $4d$ (cf. (12.1)).

Corollary 29.10. *An abstract reduced curve Y with locally formally smoothable singularities is smoothable.*

Proof. First embed Y in a complete curve $\bar{Y}$ and normalize at points of $\bar{Y} \setminus Y$ so as to introduce no new singularities. Thus we reduce to the case Y complete. Taking a Cartier divisor of sufficiently high degree on each irreducible component of Y, we can embed Y as a curve in $\mathbb{P}^n$ with $H^1(Y, \mathcal{O}_Y(1)) = 0$. Then by (29.9) Y is smoothable in $\mathbb{P}^n$, and a fortiori is smoothable as an abstract curve.

Example 29.10.1. Any stable curve (27.4) is smoothable.

Example 29.10.2. A connected, reduced curve Y in $\mathbb{P}^n$ with $p_a(Y) = 0$ is smoothable. If Y is irreducible, being reduced with $p_a = 0$, it is already isomorphic to $\mathbb{P}^1$, hence smooth. If it is reducible and connected, then we can find an irreducible component C such that the union of the remaining irreducible components, D, is still connected. Let s be the length of the intersection scheme $S = C \cap D$. Then $p_a(Y) = p_a(C) + p_a(D) + s - 1$. Since the arithmetic genus of any reduced connected curve is ≥ 0, we conclude that $p_a(C) = p_a(D) = 0$ and $s = 1$. By induction on the number of irreducible components, D is smoothable. Then smoothing D while keeping C fixed and always meeting at one point, we reduce to the case D irreducible and smooth. Since C is irreducible with $p_a = 0$, it is already smooth. Now Y consists of the two nonsingular irreducible components C, D meeting with multiplicity one. Thus their intersection is a node. To show that Y is smoothable it is now sufficient, by (29.9), to show that $H^1(\mathcal{O}_Y(1)) = 0$.

Consider the exact sequence

$$0 \to \mathcal{O}_Y(1) \to \mathcal{O}_C(1) \otimes \mathcal{O}_D(1) \to k_S \to 0.$$

Since C and D are rational curves, $H^1(\mathcal{O}_C(1)) = H^1(\mathcal{O}_D(1)) = 0$. Also clearly $H^0(\mathcal{O}_C(1)) \to k_S$ is surjective. Hence $H^1(\mathcal{O}_Y(1)) = 0$, and Y is smoothable.

Example 29.10.3. A nonsmoothable integral curve. This example is due to Mumford [118]. Let C be a nonsingular projective curve of genus $g \geq 3$ with no automorphisms. Let $P \in C$ be a point, and inside the local ring $\mathcal{O}_{C,P}$ with maximal ideal $\mathfrak{m}$, let V be a sub-k-vector space of $\mathcal{O}_{C,P}$ with $\mathfrak{m}^{2n} \subseteq V \subseteq \mathfrak{m}^n$ for some even integer n. We define a new curve $C(V)$ by "pinching" C at P using V: the curve $C(V)$ is the union of $C \setminus \{P\}$ with a new point P' whose local ring on $C(V)$ is $k \oplus V$. The new curve $C(V)$ is integral, with arithmetic genus $p_a(C(V)) = g + \text{length}(\mathcal{O}_{C,P}/V)$.

If V_1 and V_2 are two such vector spaces, then $C(V_1)$ and $C(V_2)$ are isomorphic if and only if $V_1 = V_2$. Indeed their normalizations are both equal to C, which has no automorphisms, by hypothesis, so an isomorphism would have to carry V_1 to V_2.

If we fix n and fix the dimension of $V/\mathfrak{m}^{2n}$, then the family of such curves can be made into a flat family parametrized by the Grassmann variety of $V/\mathfrak{m}^{2n}$ inside $\mathfrak{m}^n/\mathfrak{m}^{2n}$. If we take $\dim V/\mathfrak{m}^{2n} = \frac{1}{2}n$, then the dimension of this family is $\frac{1}{4}n^2$. Meanwhile, the arithmetic genus of these curves is $p_a = g + \left(\frac{3}{2}n - 1\right)$. Thus for n sufficiently large, the dimension of the family exceeds $3p_a - 3$. We will show in this case that the general curve of the family is not smoothable.

Changing notation, let X/S be an irreducible flat family of singular integral projective curves of arithmetic genus p_a. Assume that $\dim S \geq 3p_a - 3$ and for each $s \in S$ the set of $s' \in S$ for which $X_s \cong X_{s'}$ is finite. Then the general fiber X_s is not smoothable.

We proceed by contradiction. If a fiber X_s is smoothable, then there exists a flat family X'/T whose general fiber X_t' is a smooth curve of genus $g = $

$p_a(X_s)$ and whose special fiber X_0' is X_s. Choose a projective embedding $X_s \hookrightarrow \mathbb{P}^n$ with $h^1(\mathcal{O}_{X_s}(1)) = 0$. Then by (Ex. 21.5), after an étale base extension $T' \to T$, one can embed the family X'/T in $\mathbb{P}_T^n$ restricting to the given embedding of X_s. Hence $X_s \subseteq \mathbb{P}^n$ is in the closure of the set of non-singular curves of the same degree and genus in $\mathbb{P}^n$. Assuming this holds for all $s \in S$, and that the embedding $X_s \hookrightarrow \mathbb{P}^n$ was chosen uniformly for all $s \in S$, we conclude that there is an irreducible component of the Hilbert scheme of nonsingular curves of genus g in $\mathbb{P}^n$ whose closure contains a dense subset of the family S. Cutting this with hypersurfaces, we can find a flat family X'/S' where $\dim S' = \dim S + 1$ and an embedding $S \subseteq S'$ such that $X'|_S = X$. Now for general $s' \in S'$, the fiber $X_{s'}'$ is nonsingular, so it is still true that for general $s \in S$ the set of $s' \in S'$ with $X_s \cong X_{s'}'$ is finite.

To arrive at a contradiction, we consider the two families $X' \times S'$ and $S' \times X'$ over $S' \times S'$, and let I be the Isom scheme $I = \mathrm{Isom}_{S' \times S'}(X' \times S', S' \times X')$ (24.10.2). Let $\pi : I \to S'$ be the projection on the first factor. Then for a point $s' \in S'$, the fiber of π over s' consists of triples (s', s'', φ) where $\varphi : X_{s'} \to X_{s''}$ is an isomorphism. Since $\dim S' = \dim S + 1 > 3g - 3$, for a general point $s' \in S'$ the fiber $\pi^{-1}(s')$ will have dimension ≥ 1. Think of a modular family of curves of genus g (§27), which has dimension $3g - 3$, and the corresponding morphism of S' (after an étale base extension) to that modular family.

On the other hand, for a general point $s \in S$, the number of s' for which $X_s \cong X_{s'}$ is finite, and since the normalization is a curve of genus $g \geq 3$, with no automorphisms, the fiber $\pi^{-1}(s)$ is of dimension zero. This contradicts the semicontinuity of dimensions of fibers of a morphism [57, II, Ex. 3.22].

We conclude that such a family X'/S' cannot exist and hence that the general curve $C(V)$ in our family is not smoothable.

Because of (29.10), an open affine piece of this curve containing the singular point is also not smoothable.

Furthermore, because of (29.4), any singularity analytically isomorphic to one in this example is not smoothable. In particular, we could have made the same construction on an affine line.

We now apply the methods of this section to deformations of cones.

Theorem 29.11 (Pinkham). *Let Y be a nonsingular, projectively normal subvariety of dimension ≥ 1 in $\mathbb{P}_k^n$. Let X be the affine cone over Y in $\mathbb{A}_k^{n+1}$, and let $\bar{X}$ be its projective closure in $\mathbb{P}_k^{n+1}$. Assume that $H^1(\mathcal{O}_Y(\nu)) = 0$ and $H^1(\mathcal{T}_Y(\nu)) = 0$ for all $\nu > 0$. Then every (abstract) infinitesimal deformation of the affine cone X is induced by an (embedded) infinitesimal deformation of the projective cone $\bar{X}$ in $\mathbb{P}_k^{n+1}$. More precisely, the morphism of functors of Artin rings $\mathrm{Hilb}(\bar{X}, \mathbb{P}^{n+1}) \to \mathrm{Def}(X)$ is strongly surjective.*

Proof. We recall the notation of §5. Let x be the origin in $\mathbb{A}_k^{n+1}$, and let $U = X - \{x\}$ be the punctured cone. We denote by $\bar{U}$ the punctured projective cone $\bar{X} - \{x\}$. Also let $V = \mathbb{A}_k^{n+1} - \{x\}$ and $\bar{V} = \mathbb{P}_k^{n+1} - \{x\}$.

Step 1. As we saw in (Ex. 5.3), there is an exact sequence

$$0 \to \mathcal{O}_U \to \mathcal{T}_U \to \pi^* \mathcal{T}_Y \to 0,$$

where $\pi : U \to Y$ is the projection. Comparing this with the analogous sequence on V, restricted to U, we find that $\mathcal{N}_{U/V} \cong \pi^* \mathcal{N}_{Y/\mathbb{P}^n}$. Furthermore, since U is an $\mathbb{A}^1 - \{0\}$ bundle over Y, we have

$$\pi^* \mathcal{T}_Y \cong \bigoplus_{\nu \in \mathbb{Z}} \mathcal{T}_Y(\nu)$$

and

$$\mathcal{N}_{U/V} = \pi^* \mathcal{N}_Y \cong \bigoplus_{\nu \in \mathbb{Z}} \mathcal{N}_Y(\nu).$$

By the same reasoning on $\bar{U}$, which is an $\mathbb{A}^1$-bundle over Y, we find that

$$\mathcal{N}_{\bar{U}/\bar{V}} \cong \bigoplus_{\nu \leq 0} \mathcal{N}_Y(\nu).$$

Step 2. From the definition of the T^1-module we have

$$0 \to T_X \to T_{\mathbb{A}^{n+1}}|_X \to N_{X/\mathbb{A}^{n+1}} \to T_X^1 \to 0.$$

Since $T_{\mathbb{A}^{n+1}}|_X$ and N_X are reflexive modules, and $\mathrm{depth}_x X \geq 2$, we can recover these from their sections over U, so that there is an exact sequence

$$H^0(\mathcal{T}_V|_U) \xrightarrow{\alpha} H^0(\mathcal{N}_{U/V}) \to T_X^1 \to 0.$$

The map α factors through $H^0(\pi^* \mathcal{T}_{\mathbb{P}^n}|_U)$, by Step 1, so with the identifications given there we have

$$\bigoplus_{\nu \in \mathbb{Z}} H^0(\mathcal{T}_{\mathbb{P}^n}|_Y(\nu)) \to \bigoplus_{\nu \in \mathbb{Z}} H^0(\mathcal{N}_Y(\nu)) \to T_X^1 \to 0.$$

Thus we can regard T_X^1 as a graded module over $R = k[x_0, \ldots, x_n]$.

Step 3. From (5.4) we have an inclusion

$$0 \to T_X^1 \to H^1(U, \mathcal{T}_U),$$

and from the sequence of Step 1 we can write an exact sequence

$$H^1(U, \mathcal{O}_U) \to H^1(U, \mathcal{T}_U) \to H^1(U, \pi^* \mathcal{T}_Y).$$

Now our hypotheses $H^1(Y, \mathcal{O}_Y(\nu)) = 0$ and $H^1(Y, \mathcal{T}_Y(\nu)) = 0$ for all $\nu > 0$ imply that $H^1(U, \mathcal{O}_U)$ and $H^1(U, \pi^* \mathcal{T}_Y)$ are zero in positive degrees, hence $H^1(U, \mathcal{T}_U)$ also, and we conclude that T_X^1 is zero in positive degrees. This is what Pinkham calls the case of *negative grading*.

Step 4. Recall from Step 2 that T_X^1 is a quotient of $\bigoplus_{\nu \in \mathbb{Z}} H^0(\mathcal{N}_Y(\nu))$. Since T_X^1 has negative grading, it is in fact a quotient of the submodule $\bigoplus_{\nu \leq 0} H^0(\mathcal{N}_Y(\nu))$. But this is just $H^0(\mathcal{N}_{\bar{U}/\bar{V}})$ by Step 1. And again using the fact that $\mathcal{N}_{\bar{X}}$ is reflexive and depth$_x$ $\bar{X} \geq 2$, we find that $H^0(\mathcal{N}_{\bar{U}/\bar{V}}) = H^0(\mathcal{N}_{\bar{X}/\mathbb{P}^{n+1}})$. Thus we have shown that

$$H^0(\mathcal{N}_{\bar{X}/\mathbb{P}^{n+1}}) \to T_X^1 \to 0$$

is surjective. In other words, every first-order infinitesimal abstract deformation of X comes from a first-order embedded deformation of $\bar{X}$.

Step 5. Now we wish to show that the morphism of functors $\text{Hilb}(\bar{X}) \to \text{Def}(X)$ is strongly surjective. This is similar to the proof of (29.8), except that in this case we do not have $H^1(\mathcal{N}_{\bar{X}}) = 0$. Suppose we are given a small extension of Artin rings $C' \to C$ and a deformation ξ of $\bar{X}$ over C, inducing a deformation η of X, and suppose η lifts to a deformation η' of X over C'. We must show that there is a ξ' restricting to ξ and η'.

Since any abstract affine deformation is embeddable (Ex. 10.1), we can regard η' as an embedded deformation η'' of X. In particular, the restriction of η to U extends over C' as an embedded deformation, and so the corresponding obstruction in $H^1(U, \mathcal{N}_U)$ must be zero.

Now the obstruction to extending ξ over C', since we know it is locally extendable around x, and $\bar{X}$ is nonsingular elsewhere, lies by (10.4) in $H^1(\bar{X}, \mathcal{N}_{\bar{X}})$. Because of the depth hypothesis once more, there is an injective map of this group to $H^1(\bar{U}, \mathcal{N}_{\bar{U}})$. Finally, since we have seen in Step 1 that $\mathcal{N}_{\bar{U}} = \bigoplus_{\nu \leq 0} \mathcal{N}_Y(\nu)$ and $\mathcal{N}_U = \bigoplus_{\nu \in \mathbb{Z}} \mathcal{N}_Y(\nu)$, it follows that $H^1(\bar{U}, \mathcal{N}_{\bar{U}})$ injects into $H^1(U, \mathcal{N}_U)$. Hence the obstruction to extending ξ reduces to the obstruction to extending η, which is zero, and so ξ extends to a deformation ξ' of $\bar{X}$ over C'.

It remains to show that we can modify ξ' so as to restrict to η'. This follows from the fact that the map on tangent spaces of $\text{Hilb}(\bar{X})$ to $\text{Def}(X)$ is surjective (Step 4), and the choices of extensions in each case are torsors under the action of those tangent spaces.

Example 29.11.1. Rigid singularities.   We have seen many examples of rigid singularities, and we have shown that rigid singularities are not smoothable (29.6). Now we can show, for the particular class of rigid cone singularities described in (5.5), that the associated projective cone is also rigid.

So let $Y \subseteq \mathbb{P}_k^n$ be a nonsingular projectively normal variety with $H^1(\mathcal{O}_Y(\nu)) = H^1(\mathcal{T}_Y(\nu)) = 0$ for all $\nu \in \mathbb{Z}$. Let $\bar{X}_0$ be the projective cone over Y in $\mathbb{P}^{n+1}$. The tangent space to the functor of abstract deformations of $\bar{X}_0$ fits in an exact sequence (Ex. 5.7)

$$0 \to H^1(\mathcal{T}_{\bar{X}_0}) \to \text{Def}(\bar{X}_0) \to H^0(\mathcal{T}_{\bar{X}_0}^1).$$

Now $\bar{X}_0$ is nonsingular except at the vertex, so $T_{\bar{X}_0}^1$ is concentrated there, hence is the same as for the affine cone, which is zero (5.5). On the other hand, using the depth condition as in the proof of (29.11), Step 5, we see that $H^1(\mathcal{T}_{\bar{X}_0})$ injects into $H^1(\bar{V}, \mathcal{T}_{\bar{V}}) = \bigoplus_{\nu \leq 0} H^1(Y, \mathcal{T}_Y(\nu))$, which is zero. Hence $\bar{X}_0$ is rigid. Furthermore, in any global family of projective deformations of $\bar{X}_0$ the nearby fibers are also isomorphic to $\bar{X}_0$ (Ex. 24.7e).

Example 29.11.2. Generic singularities. These curious birds are singularities that are neither rigid nor smoothable. They have nontrivial deformations, but these all resemble the original singularity. We have already seen some (Ex. 5.8). To make more, we start with a projectively normal nonsingular variety $Y \subseteq \mathbb{P}^n_k$ satisfying $H^1(\mathcal{O}_Y(\nu)) = 0$ for all ν and $H^1(\mathcal{T}_Y(\nu)) = 0$ for all $\nu \neq 0$. Let X_0 be the affine cone over Y. Then from the proof of (29.11), we see that $T^1_{X_0}$ is included in $H^1(Y, \mathcal{T}_Y)$. On the other hand, any deformation of Y in $\mathbb{P}^n_k$ gives a deformation of the cone, so we see that $T^1_{X_0}$ is just the subspace of $H^1(Y, \mathcal{T}_Y)$ induced by embedded deformations. In particular, any infinitesimal deformation of X_0 is again a cone. Assuming that Y has nontrivial abstract deformations embedded in $\mathbb{P}^n$, X_0 is a *generic singularity* in the sense that it has nontrivial infinitesimal deformations, but they are all cones over deformations of Y.

To make examples, let Y_0 be a nonsingular surface of degree ≥ 5 in $\mathbb{P}^3$. Then Y_0 has deformations in $\mathbb{P}^3$ that are nontrivial as abstract deformations of Y_0, and every abstract deformation of Y_0 is realizable inside $\mathbb{P}^3$ (20.2.2). Let $Y \subseteq \mathbb{P}^n$ be the d-uple embedding of Y_0, for d sufficiently large. Then Y will satisfy the conditions above.

Theorem 29.12 (Pinkham). *Let $Y \subseteq \mathbb{P}^n$ be a nonsingular projectively normal curve of genus $g \geq 1$ and degree $d > 4g + 5$ if $g = 1$ or $d > 4g + 4$ if $g \geq 2$. Then the affine cone $X_0 \subseteq \mathbb{A}^{n+1}$ is not smoothable. In fact, it is not even locally formally smoothable.*

Proof. By (29.11), if $\bar{X}_0$ is the closure of X_0 in $\mathbb{P}^{n+1}$, then the morphism of functors $\varphi : \mathrm{Hilb}(\bar{X}_0, \mathbb{P}^{n+1}) \to \mathrm{Def}(X_0)$ is strongly surjective. Therefore, as in the proof of (29.8), if X_0 is smoothable as an abstract scheme, then $\bar{X}_0$ is smoothable in $\mathbb{P}^{n+1}$. In that case there are smooth surfaces X in $\mathbb{P}^{n+1}$ having the same Hilbert polynomial as $\bar{X}_0$, and having smooth hyperplane sections that are deformations of Y. We will show that this is not possible.

First we compute the arithmetic genus $p_a(\bar{X}_0)$. By taking cohomology of the sequence

$$0 \to \mathcal{O}_{\bar{X}_0}(-1) \to \mathcal{O}_{\bar{X}_0} \to \mathcal{O}_Y \to 0$$

and its twists and summing, we find for the Euler characteristics that

$$\chi(\mathcal{O}_{\bar{X}_0}(m)) = \chi(\mathcal{O}_{\bar{X}_0}) + \sum_{i=1}^{m} \chi(\mathcal{O}_Y(i)).$$

On the other hand, for m large, $\chi(\mathcal{O}_{\bar{X}_0}(m)) = h^0(\mathcal{O}_{\bar{X}_0}(m))$, and this, because $\bar{X}_0$ is a cone, is simply

$$\chi(\mathcal{O}_{\bar{X}_0}(m)) = \sum_{i=0}^{m} h^0(\mathcal{O}_Y(i)).$$

Since $h^1(\mathcal{O}_Y(i)) = 0$ for $i > 0$, by comparing these two equalities we find that $\chi(\mathcal{O}_{\bar{X}_0}) = 1$, and hence $p_a(\bar{X}_0) = 0$. Therefore also $p_a(X) = 0$, since the arithmetic genus is constant in a flat family.

Now consider the hyperplane section H of X, which is a deformation of Y in $\mathbb{P}^n$. Its normal bundle is $\mathcal{O}_H(1)$, which has degree d. By the adjunction formula, $2g - 2 = H \cdot (H + K)$, where K is the canonical divisor on X. It follows that $H.K = 2g - 2 - d$ is negative. In particular, no multiple of K is effective, and we obtain the second plurigenus $P_2 = 0$. By Castelnuovo's criterion [57, V, 6.2] therefore, X is a rational surface. Now H is a curve of genus g on X with self-intersection $H^2 = d > 4g + 5$ if $g = 1$, and $d > 4g + 4$ if $g \geq 2$. This contradicts a theorem of Hartshorne [54], and thus completes the proof.

Remark 29.12.1. By a similar method, Pinkham studies the cone over a set of points in general position in $\mathbb{P}^n$, and shows that there are bundles of lines through a point in $\mathbb{P}^{n+1}$ that form a nonsmoothable embedded singular curve. His first example is a union of 13 general lines through a point in $\mathbb{P}^7$.

As another application of our smoothing theorems, we study d-gonal curves of genus g and the corresponding subsets of the moduli space. A nonsingular curve C is called d-gonal if it admits a base-point-free linear system g_d^1 of dimension 1 and degree d (27.2.1). This corresponds to a morphism of C to $\mathbb{P}^1$ of degree d. Given a curve C, it is always d-gonal for sufficiently large d, but those that are d-gonal for smaller d are special in the variety of moduli. Thus we showed (27.3) that a general curve of genus $g > 2d - 2$ does not have a g_d^1, while it is known that for $g \leq 2d - 2$, every curve of genus g has one. In showing that the variety of moduli $\mathcal{M}_3$ is irreducible (Ex. 27.2) we saw that the hyperelliptic curves of genus 3 are in the closure of the set of trigonal curves, by reason of counting dimensions. A similar argument works for $\mathcal{M}_4$, since a nonhyperelliptic curve of genus 4 is also trigonal. We will show now that for any genus, the set of d-gonal curves is in the closure of the set of $(d + 1)$-gonal curves. The first nontrivial case is genus 5, where the general curve is not trigonal, so counting dimensions is not sufficient to show that hyperelliptic curves are limits of trigonal curves.

Theorem 29.13. *If C_0 is a d-gonal curve of genus g, then there exists a family X/T of projective nonsingular curves over a nonsingular curve T, and a point $0 \in T$ whose fiber X_0 is C_0, such that for all $p \in T$, $p \neq 0$, the fiber X_p is $(d + 1)$-gonal.*

Proof. Let $\pi_0 : C_0 \to \mathbb{P}^1$ be the map of degree d determined by the g_d^1 on C_0. Let L be another copy of $\mathbb{P}^1$, mapping isomorphically to $\mathbb{P}^1$, and let $X_0 = C_0 \cup L$, meeting at a single point P. Then X_0 is a projective curve of (arithmetic) genus g, with a projection to $\mathbb{P}^1$ of degree $d+1$. Our goal is to smooth X_0 along with the projection $\pi : X_0 \to \mathbb{P}^1$.

Choose an embedding $i : X_0 \hookrightarrow \mathbb{P}^n$ such that $H^1(\mathcal{O}_{X_0}(1)) = 0$, and consider the resulting diagonal map $X_0 \to \mathbb{P}^1 \times \mathbb{P}^n$. We will smooth X_0 as an embedded subscheme of $Z = \mathbb{P}^1 \times \mathbb{P}^n$. For this purpose we consider the associated sequence of tangent and normal sheaves

$$0 \to \mathcal{T}_{X_0} \to \mathcal{T}_Z|_{X_0} \to \mathcal{N}_{X_0} \to T_P^1 \to 0.$$

Here we write T_P^1 as a module, since the sheaf $T_{X_0}^1$ is concentrated at its unique singular point P. We wish to show that $H^1(\mathcal{N}_{X_0}) = 0$ and $H^0(\mathcal{N}_{X_0}) \to T_P^1 \to 0$ is surjective. Then, by (29.8), X_0 will be smoothable in Z and a general path in the Hilbert scheme will smooth the singularity at P.

Since Z is a direct product, $\mathcal{T}_Z = p_1^* \mathcal{T}_{\mathbb{P}^1} \oplus p_2^* \mathcal{T}_{\mathbb{P}^n}$, where p_1, p_2 are the projections. Note that $p_2^* \mathcal{T}_{\mathbb{P}^n}|_{X_0}$ is just $i^* \mathcal{T}_{\mathbb{P}^n}$. Since we assumed $H^1(\mathcal{O}_{X_0}(1)) = 0$, it follows from the usual Euler sequence that $H^1(i^* \mathcal{T}_{\mathbb{P}^n}) = 0$. On the other hand, the sequence of $\mathcal{T}^i$-sheaves for the map $\pi : X_0 \to \mathbb{P}^1$ gives

$$0 \to \mathcal{T}_{X_0} \to \pi^* \mathcal{T}_{\mathbb{P}^1} \to T_{X/\mathbb{P}^1}^1 \to T_X^1 \to 0.$$

These last two T^1-modules are concentrated at points, so $H^1(\mathcal{T}_{X_0}) \to H^1(\pi^* \mathcal{T}_{\mathbb{P}^1})$ is surjective. It follows that $H^1(\mathcal{N}_{X_0}) = 0$ and $H^0(\mathcal{N}_{X_0}) \to T_P^1$ is surjective from the sequence above.

Thus the Hilbert scheme is smooth, and by (29.8) smooths the singularity at P. Taking an open set of a nonsingular curve T passing through that point of the Hilbert scheme, we obtain a flat family of projective curves X/T whose fiber at $0 \in T$ is X_0 and whose general fiber X_t is a nonsingular curve of the same genus g, whose first projection $p_1 : X_t \to \mathbb{P}^1$ gives a g_{d+1}^1 on the curve. Note also that since the miniversal deformation space of a node, defined by $xy - t = 0$, has smooth total family, it follows also that the total space X of our family is nonsingular.

Now consider the special fiber $X_0 = C_0 \cup L \subseteq X$. Being a fiber of the projection to T, its self-intersection is zero; in fact, its normal bundle $\mathcal{N}_{X_0/X}$ is trivial. But then $\mathcal{N}_{L/X} = \mathcal{N}_{X_0/X}|_L(-P)$, so $L^2 = -1$. The curve L is isomorphic to $\mathbb{P}^1$, so it is an exceptional curve of the first kind on the nonsingular surface X, and we can blow it down [57, V, 5.7], getting another nonsingular surface X'/T. This is still a flat family, since it is irreducible and T is a nonsingular curve, and now the special fiber X_0' is just C_0. This is the required family.

Corollary 29.14. *There exist d-gonal curves of genus g for all $d \geq 2$.*

Proof. Indeed, since there exist hyperelliptic curves of every genus (for example curves of bidegree $(2, g+1)$ on nonsingular quadric surface in $\mathbb{P}^3$), the theorem provides us with d-gonal curves for every $d \geq 2$.

References for this section. My main references for this section have been Schlessinger's work on rigid singularities [147], Pinkham's thesis [135], Artin's lecture notes explaining these results [8], and my paper with Hirschowitz [65]. Further references mentioned along the way are [117], [118], [144], [60], and [23]. See also [56] for some other criteria for smoothing singularities, not using deformation theory. The definition and properties of formally smoothable schemes are new. With them we could justify a method already used in [135] and [65]. The book of Stevens [160] has many more computations of smoothable and nonsmoothable cones over curves and over sets of points in $\mathbb{P}^n$.

As for the last result on d-gonal curves, many experts assured me that this was "obvious" or "well known" or "follows from the general theory," but since I could not understand their arguments, I gave a proof here using only the techniques of this book.

Exercises.

29.1.

(a) Generalize the proof of (29.10.2) to show that any reduced connected curve Y in $\mathbb{P}^n$ having $p_a(Y) = 1$ or 2 is smoothable.

(b) Show that the corresponding statement is false for $p_a(Y) \geq 3$.

29.2.   Show that an integral curve Y of degree $d > 2p_a - 2$ in $\mathbb{P}^3$ is smoothable in $\mathbb{P}^3$. This is weaker but much more elementary than Ein's theorem (29.9.4).

29.3. (cf. [65, 6.1.2])

(a) Let C be a nonsingular plane quartic curve, and let D be a nonsingular rational curve of degree d in $\mathbb{P}^3$ meeting C, transversal to the plane of C, in a single point P. Let $Y = C \cup D$. Show that Y is not smoothable in $\mathbb{P}^3$, even though its only singularity is a node, and its degree $d + 4$ is much larger than its arithmetic genus $p_a = 3$. *Hint:* Use the fact that D is transversal to the plane of C to argue that $H^1(\mathcal{N}_Y) = 0$. Then show that the Hilbert scheme near Y is irreducible and non singular, and that by reason of dimension, every nearby curve is again of the form $Y' = C' \cup D'$ for nearby curves C' and D'.

(b) Show, however, that the union of a plane quartic curve with two skew lines, each meeting it at one point, is smoothable in $\mathbb{P}^3$ [65, 4.3.1].

29.4. Smoothing cones over curves.

(a) Let X be a nonsingular surface in $\mathbb{P}^{n+1}$. Assume that X contains the point $P = (1, 0, \ldots, 0)$, and that the hyperplane section Y defined by $x_0 = 0$ is projectively normal in $\mathbb{P}^n$. For any $t \neq 0$ consider the automorphism of $\mathbb{P}^{n+1}$ that sends the point $(1, a_1, \ldots, a_{n+1})$ to $(t, a_1, \ldots, a_{n+1})$, and let X_t be the image of X. Show that the limit as t approaches zero is the projective cone X_0 over Y, and that this is a flat family smoothing X_0.

(b) For any $r = 0, 1, \ldots, 6$, let X be the image in $\mathbb{P}^{9-r}$ of $\mathbb{P}^2$ by the linear system of cubic curves passing through r general fixed points of $\mathbb{P}^2$. This is the *Del Pezzo surface* of degree $9 - r$. Using the family of (a), these smooth the cone over an elliptic curve Y of degree $9 - r$ in $\mathbb{P}^{8-r}$. Thus for curves of genus 1, (29.12) is sharp.

(c) For $r = 0, 1, \ldots, 7$, let X be the image in $\mathbb{P}^{n+1}$ of $\mathbb{P}^2$ given by the linear system of quartic curves through $2P_0, P_1, \ldots, P_r$, where P_i are general points of $\mathbb{P}^2$, and $n + 1 = 11 - r$. For $r = 7$ this is the *Castelnuovo surface* of degree 5 in $\mathbb{P}^4$. Again using (a) show that these surfaces provide smoothing of the cone over a curve of genus 2 and degree $12 - r$ in $\mathbb{P}^{10-r}$ for $r = 0, 1, \ldots, 7$. Thus (29.12) is sharp for $g = 2$.

(d) Show also that (29.12) is sharp for $g = 3$ using embeddings of $\mathbb{P}^2$ by quartic curves through $r = 0, 1, \ldots, 10$ general points. For $r = 10$, this is the *Bordiga surface* in $\mathbb{P}^4$.

29.5. In (29.13) it would be nice to know that a d-gonal curve is a limit of $(d + 1)$-gonal curves that are not also d-gonal. And in (29.17) it would be nice to know that there exist d-gonal curves that are not $(d - 1)$-gonal.

(a) Show that if a curve C is simultaneously d-gonal and $(d + 1)$-gonal for some $d \geq 2$, then there is a birational morphism from C to a curve C' of bidegree $(d, d + 1)$ in $Q = \mathbb{P}^1 \times \mathbb{P}^1$, and therefore $g \leq d(d - 1)$.

(b) Conclude that for every $g \geq 3$, the hyperelliptic curves are limits of trigonal nonhyperelliptic curves, and there exist trigonal curves that are not hyperelliptic.

29.6. Cones over the rational quartic curves in $\mathbb{P}^4$. These were studied by Pinkham [135], and such cones provide examples of embedded and abstract obstructed deformations of Cohen–Macaulay schemes in codimension 3.

Following the notation of (29.11), let Y be a rational normal curve of degree 4 in $\mathbb{P}^4$, let X be the affine cone over Y in $\mathbb{A}^5$, and let $\bar{X}$ be its projective closure in $\mathbb{P}^5$.

(a) Show that the hypotheses of (29.11) are satisfied, and so the restriction map of functors $\mathrm{Hilb}(\bar{X}, \mathbb{P}^5) \to \mathrm{Def}(X)$ is strongly surjective.

(b) Use the analysis of the proof of (29.11) to compute that $h^0(\mathcal{N}_{\bar{X}/\mathbb{P}^5}) = 30$ and $\dim T_X^1 = 4$, so that if R is the complete local ring pro-representing the functor $\mathrm{Def}(X)$, and S is the completion of the local ring of the Hilbert scheme at the point corresponding to $\bar{X}$, then S is isomorphic to a power series ring in 26 variables over R (Ex. 15.7).

(c) One knows [57, V, 2.19.1, 2.19.2], [162] that any integral surface of degree 4 in $\mathbb{P}^5$ is one of the following:

    (i) a rational scroll with $e = 0$ embedded by $C_0 + 2f$,
    (ii) a rational scroll with $e = 2$ embedded by $C_0 + 3f$,
    (iii) the Veronese surface: $\mathbb{P}^2$ embedded by $\mathcal{O}(2)$,
    (iv) a cone over a rational quartic curve in $\mathbb{P}^4$.

Show that any two in the same family differ by an automorphism of $\mathbb{P}^5$, and compute the dimensions of the families: 29, 28, 27, and 26 respectively. Furthermore, all four types have the same Hilbert polynomial, and are ACM schemes in $\mathbb{P}^5$.

(d) To see the structure of the Hilbert scheme of these integral surfaces in $\mathbb{P}^5$, show that:

    (1) The family (ii) is in the closure of the family (i): cf. (Ex. 2.3).
    (2) Using the stretching method of (Ex. 29.4) show that the family (iv) is contained in the closures of each of the families (i), (ii), and (iii).

(3) The closures of families (i) and (iii) form two distinct irreducible components of the Hilbert scheme.

Thus the Hilbert scheme has two irreducible components of dimensions 29 and 27, meeting along a subvariety of dimension 26, and having embedding dimension 30 along the intersection.

(e) Conclude finally that the formal deformation space $\operatorname{Spec} R$ of the functor of abstract deformations of the affine cone X has two irreducible components of dimensions 3 and 1 meeting at a point, and has embedding dimension 4. Note that since Y is projectively normal, X is a Cohen–Macaulay scheme.

(f) These calculations also show that X is smoothable in two essentially different ways.

29.7. Nonlicci schemes.   Recall (Ex. 9.4) that a scheme is called licci if it can be linked in a finite number of steps to a complete intersection.

(a) Using (Ex. 9.4), show that the local ring of the vertex of the cone over a rational quartic curve in $\mathbb{P}^4$ (Ex. 29.6) is not licci.

(b) A subscheme of $\mathbb{P}^n_k$ is called (globally) licci if it can be linked by complete intersection schemes in a finite number of steps to a complete intersection in $\mathbb{P}^n$ (Ex. 8.4). Show that as a consequence of (a) above, the rational quartic curve in $\mathbb{P}^4$ is not licci.

(c) Let Y be a set of four points in general position in $\mathbb{P}^3$. By an argument analogous to (Ex. 29.6) show that Y is not licci.

29.8.   Show that the example of (29.11.2) still works if we allow $H^1(\mathcal{O}_Y)$ to be nonzero. Show also that in a global family of projective deformations of the projective cone $\bar{X}_0$, the nearby fibers are also cones over deformations of Y.

References

1. Altman, A., and Kleiman, S., *Introduction to Grothendieck duality theory*, Springer LNM **146**, 1970.
2. Arbarello, E., Cornalba, M., Griffiths, P. A., and Harris, J., *Geometry of Algebraic Curves, Vol. I*, Springer, 1985.
3. Artin, M., The implicit function theorem in algebraic geometry, *Proc. Bombay Colloq.*, Bombay–Oxford, 1968, 13–34.
4. Artin, M., Algebraization of formal moduli, I, in *Global Analysis*, ed. Spencer and Iyanaga, Princeton, 1969.
5. Artin, M., Algebraic approximation of structures over complete local rings, *Publ. Math. IHES* **36** (1969), 23–58.
6. Artin, M., Algebraization of formal moduli: II. Existence of modifications, *Annals of Math.* **91** (1970), 88–135.
7. Artin, M., Versal deformations and algebraic stacks, *Invent. Math.* **27** (1974), 165–189.
8. Artin, M., *Lectures on Deformations of Singularities*, Tata Inst. Fund. Res. Bombay (1976) (notes by C. S. Seshadri and Alan Tannenbaum).
9. Auslander, M., Coherent functors, in *Proc. Conf. Categorical Algebra*, La Jolla 1965, Springer (1966), 189–231.
10. Bloch, S., Semi-regularity and de Rham cohomology, *Invent. Math.* **17** (1972), 51–66.
11. Buchsbaum, D., Complexes associated with the minors of a matrix, *Inst. Naz. Alta Mat.*, Symposia Mat. IV (1970), 255–281.
12. Buchsbaum, D. A., and Eisenbud, D., Algebra structures for finite free resolutions, and some structure theorems for ideals of codimension 3, *Amer. J. Math.* **99** (1977), 447–485.
13. Buchweitz, R. O., *Contributions à la théorie des singularités*, thèse, Univ. Paris VII (1981).
14. Burch, L., On ideals of finite homological dimension in local rings, *Proc. Camb. Phil. Soc.* **64** (1968), 941–948.
15. Burns, D. M., Jr., and Wahl, J. M., Local contributions to global deformations of surfaces, *Invent. Math.* **26** (1974), 67–88.
16. Carlson, J., Green, M., Griffiths, P., and Harris, J., Infinitesimal variation of Hodge structure, I., *Compos. Math.* **50** (1983), 109–205.

R. Hartshorne, *Deformation Theory*, Graduate Texts in Mathematics 257,
DOI 10.1007/978-1-4419-1596-2, © Robin Hartshorne 2010

17. Curtin, D. J., Obstructions to deforming a space curve, *Trans. AMS* **267** (1981), 83–94.

18. Deligne, P., Le théorème de Noether, in: SGA7, II, by P. Deligne and N. Katz, *SLN* **340** (1973), exposé XIX, 328–340.

19. Deligne, P., *Relèvement des surfaces K3 en caractéristique nulle* (prepared for publication by L. Illusie), Springer LNM **868** (1981), 58–79.

20. Deligne, P., and Illusie, L., Relèvements modulo p^2 et décomposition du complexe de de Rham, *Invent. Math.* **89** (1987), 247–270.

21. Deligne, P., and Mumford, D., The irreducibility of the space of curves of given genus, *Publ. Math. IHES* **36** (1969), 75–109.

22. Ein, L., An analogue of Max Noether's theorem, *Duke Math. J.* **52** (1985), 895–907.

23. Ein, L., Hilbert scheme of smooth space curves, *Ann. Sc. ENS* **19** (1986), 469–478.

24. Eisenbud, D., *Commutative Algebra with a View Toward Algebraic Geometry*, Springer, GTM **150** (1995).

25. Ekedahl, T., On non-liftable Calabi-Yau threefolds, preprint math.AG.0306435 (2003).

26. Ellia, Ph., and Hartshorne, R., Smooth specializations of space curves: questions and examples, in *Commutative Algebra and Algebraic Geometry*, ed. F. van Oystaeyen, Dekker, Lecture Notes **206** (1999), 53–79.

27. Ellingsrud, G., Sur le schéma de Hilbert des variétés de codimension 2 dans $\mathbb{P}^e$ à cône de Cohen–Macaulay, *Ann. Sci. ENS* **8** (1975), 423–432.

28. Ellingsrud, G., and Peskine, C., Equivalence numérique pour les surfaces génériques d'une famille lisse de surfaces projectives, in: Problems in the theory of surfaces and their classification (Cortona 1988), *Sympos. Math.* XXXII, Academic Press (1991), 99–109.

29. Ellingsrud, G., and Peskine, C., Anneau de Gorenstein associé à un fibré inversible sur une surface de l'espace et lieu de Noether–Lefschetz, in: *Proc. Indo-French Conf. on Geometry*, Bombay 1989, Hindustan Book Agency (1993), 29–42.

30. Fano, G., Sulle varietà algebriche che sono intersezioni complete di più forme, *Atti. R. Accad. Sci. Torino* **44** (1909), 633–648.

31. Fantechi, B., and Manetti, M., On the T^1-lifting theorem, *J. Alg. Geom.* **8** (1999), 31–39.

32. Fantechi, B., Stacks for everybody, in: *European Congress of Mathematics*, Barcelona 2000, Birkhäuser (2001) Vol. I, 349–359.

33. Fogarty, J., Algebraic families on an algebraic surface, *Amer. J. Math.* **90** (1968), 511–521.

34. Franchetta, A., Sulle curve appartenenti a una superficie generale d'ordine $n \geq 4$ dell' S_3, *Atti Accad. Naz. Lincei* (8)**3** (1947), 71–78.

35. Fulton, W., On the irreducibility of the moduli space of curves, Appendix to the paper of Harris and Mumford, *Invent. Math.* **67** (1982), 87–88.

36. Gieseker, D., On the moduli of vector bundles on an algebraic surface, *Annals of Math.* **106** (1977), 45–60.

37. Gomez, T. L., Algebraic stacks, *Proc. Indian Acad. Sci. Math. Sci.* **111** (2001), 1–31.

38. Grauert, H., Über die Deformation isolierter Singularitäten analytischer Mengen, *Invent. Math.* **15** (1972), 171–198.

39. Grauert, H., and Kerner, H., Deformationen von Singularitäten Komplexer Räume, *Math. Ann.* **153** (1964), 236–260.

40. Green, M., A new proof of the explicit Noether–Lefschetz theorem, *J. Diff. Geom.* **27** (1988), 155–159.

41. Green, M., Components of maximal dimension in the Noether–Lefschetz locus, *J. Diff. Geom.* **29** (1989), 295–302.

42. Greuel, G.-M., Lossen, C., and Shustin, E., *Introduction to Singularities and Deformations*, Springer, 2007.

43. Griffiths, P., and Harris, J., On the Noether–Lefschetz theorem and some remarks on codimension-two cycles, *Math. Ann.* **271** (1985), 31–51.

44. Grothendieck, A., Géométrie formelle et géométrie algébrique, *Sem. Bourbaki* **182**, May (1959) (also reprinted in [45] q.v.).

45. Grothendieck, A., Fondements de la géométrie algébrique [Extraits du Séminaire Bourbaki 1957–1962], *Secr. Math.* 11, rue Pierre Curie, Paris (1962).

46. Grothendieck, A., Le groupe de Brauer, I, *Séminaire Bourbaki* **290** (1964/65).

47. Grothendieck, A., *Revêtements étales et groupe fondamental*, Springer LNM **224**, 1971.

48. Grothendieck, A., and Dieudonné, J., Elements de géométrie algébrique, *Publ. Math. IHES* **4**, **8**, **11**, **17**, **20**, **24**, **28**, **32** (1960–1967).

49. Gruson, L., and Peskine, C., Genre des courbes de l'espace projectif, *Springer LNM* **687**, 1977, 31–59.

50. Gruson, L., and Peskine, C., Genre des courbes de l'espace projectif, II, *Ann. Sci. ENS* (4) **15** (1981), 401–418.

51. Harris, J. (with the collaboration of D. Eisenbud), *Curves in projective space*, Sem. Math. Sup., Univ. Montreal (1982).

52. Harris, J., and Morrison, I., *Moduli of Curves*, Springer, 1998.

53. Hartshorne, R., Connectedness of the Hilbert scheme, *Publ. Math. IHES* **29** (1966), 261–304.

54. Hartshorne, R., Curves with high self-intersection on algebraic surfaces, *Publ. Math. I.H.E.S.* **36** (1969), 111–125.

55. Hartshorne, R., *Ample subvarieties of algebraic varieties*, Springer Lecture Notes in Math. **156**, 1970.

56. Hartshorne, R., Topological conditions for smoothing algebraic singularities, *Topology* **13** (1974), 241–253.

57. Hartshorne, R., *Algebraic Geometry*, Springer, 1977.

58. Hartshorne, R., Stable vector bundles of rank 2 on $\mathbb{P}^3$, *Math. Ann.* **238** (1978), 229–280.

59. Hartshorne, R., On the classification of algebraic space curves, in *Vector Bundles and Differential Equations* (Nice 1979), Birkhäuser, Boston (1980), 83–112.

60. Hartshorne, R., Une courbe irréductible non lissifiable dans $\mathbb{P}^3$, *C. R. Acad. Sci. Paris* **299** (1984), 133–136.

61. Hartshorne, R., On the classification of algebraic space curves, II, *Proc. Symp. Pure Math.*, AMS **46** (1987), 145–164.

62. Hartshorne, R., Genre des courbes dans l'espace projectif (d'après L. Gruson et C. Peskine), *Sém. Bourbaki* **592** (1981/82).

63. Hartshorne, R., Generalized divisors on Gorenstein schemes, *K-theory* 8 (1994), 287–339.

64. Hartshorne, R., Coherent functors, *Advances in Math.* **140** (1998), 44–94.

65. Hartshorne, R., and Hirschowitz, A., Smoothing algebraic space curves, in: *Algebraic Geometry*, Sitges 1983, Springer Lecture Notes in Math. **1124**, 1985, 98–131.

66. Hartshorne, R., Martin-Deschamps, M., and Perrin, D., Triades et familles de courbes gauches, *Math. Ann.* **315** (1999), 397–468.

67. Hilbert, D., Über die Theorie der algebraischen Formen, *Math. Ann.* **36** (1890), 473–534.

68. Hirokado, M., A non-liftable Calabi–Yau threefold in characteristic 3, *Tohoku Math. J.* **51** (1999), 479–487.

69. Huneke, C., and Ulrich, B., The structure of linkage, *Ann. of Math.* **126** (1987), 277–334.

70. Huybrechts, D., and Lehn, M., *The geometry of moduli spaces of sheaves*, Vieweg, 1997.

71. Iarrobino, A., Reducibility of the families of 0-dimensional schemes on a variety, *Invent. Math.* **15** (1972), 72–77.

72. Iarrobino, A., and Emsalem, J., Some zero-dimensional generic singularities; finite algebras having small tangent space, *Compositio Math.* **36** (1978), 145–188.

73. Illusie, L., Complexe cotangent et déformations, I, II, Springer LNM **239**, 1971; **283**, 1972.

74. Illusie, L., Grothendieck's existence theorem in formal geometry, in *Fundamental Algebraic Geometry*, by B. Fantechi et al., Amer. Math. Soc., 2005, 179–233.

75. Joshi, K., A Noether–Lefschetz theorem and applications, *J. Alg. Geom.* **4** (1995), 105–135.

76. Kaplansky, I., *Commutative Rings*, Allyn and Bacon, Boston, 1970.

77. Kas., A., On obstructions to deformations of complex analytic surfaces, *Proc. Nat. Acad. Sci.* **58** (1967), 402–404.

78. Kas, A., and Schlessinger, M., On the versal deformation of a complex space with an isolated singularity, *Math. Ann.* **196** (1972), 23–29.

79. Kleiman, S., The Picard scheme, in *Fundamental Algebraic Geometry*, by B. Fantechi et al., Amer. Math. Soc., 2005, 235–321.

80. Kleiman, S., and Piene, R., Enumerating singular curves on surfaces, *Contemp. Math.* **241** (1999), 209–238.

81. Kleppe, J. O., *The Hilbert-flag scheme, its properties and its connection with the Hilbert scheme, Applications to curves in 3-space*, preprint (part of thesis) March (1981), Univ. Oslo.

82. Kleppe, J. O., Non-reduced components of the Hilbert scheme of smooth space curves, in *Rocca di Papa* 1985, Springer LNM **1266**, 1987, 181–207.

83. Kleppe, J. O., Liaison of families of subschemes in $\mathbb{P}^n$, in *Algebraic Curves and Projective Geometry*, Springer, Lec. Notes in Math., **1389**, 1989, 128–173.

84. Kleppe, J. O., The Hilbert scheme of space curves of small Rao modules with an appendix on nonreduced components (preprint).

85. Knutson, D., *Algebraic Spaces*, Lecture Notes in Math., Springer **203**, 1971.

86. Kodaira, K., *Complex Manifolds and Deformation of Complex Structures*, Springer, 1986.

87. Kodaira, K., and Spencer, D., On deformations of complex analytic structures, I, II, *Annals of Math.* **67** (1958), 328–466.

88. Kollár, J., *Rational Curves on Algebraic Varieties*, Springer, Ergebnisse **32**, 1996.

89. Lang, W. E., Examples of surfaces of general type with vector fields, in *Arithmetic and Geometry*, Vol II, dedicated to I. R. Shafarevich, Birkhäuser, 1983, 167–173.

90. Langer, A., Semistable sheaves in positive characteristic, *Ann. Math.* **159** (2004), 251–276.

91. Langton, S. G., Valuative criteria for families of vector bundles on algebraic varieties, *Annals of Math.* **101** (1975), 88–110.

92. Laudal, O. A., *Formal moduli of algebraic structures*, Springer LNM **754**, 1979.

93. Laumon, G., and Moret-Bailly, L., *Champs algébriques*, Springer, Ergebnisse **39**, 2000.

94. Lefschetz, S., On certain numerical invariants of algebraic varieties with applications to Abelian varieties, *Trans. Amer. Math. Soc.* **22** (1921), 327–406; 407–482.

95. Lefschetz, S., *L'analysis situs et la Géométrie Algébrique*, Paris (1924), Gauthier–Villars.

96. Lichtenbaum, S., and Schlessinger, M., The cotangent complex of a morphism, *Trans. AMS* **128** (1967), 41–70.

97. Looijenga, E., Smooth Deligne–Mumford compactifications by means of Prym level structures, *J. Alg. Geom.* **3** (1994), 283–293.

98. Lopez, A. F., *Noether–Lefschetz theory and the Picard group of projective surfaces*, Amer. Math. Soc. Memoirs **438**, 1991.

99. Martin-Deschamps, M., and Perrin, D., Sur la classification des courbes gauches, *Astérisque* **184–185** (1990).

100. Martin-Deschamps, M., and Perrin, D., Le schéma de Hilbert de courbes localement de Cohen–Macaulay n'est (presque) jamais réduit, *Ann. Sci. ENS* **29** (1996), 757–785.

101. Maruyama, M., Stable vector bundles on an algebraic surface, *Nagoya Math. J.* **58** (1975), 25–68.

102. Maruyama, M., Construction of moduli spaces of stable sheaves via Simpson's idea, in *Moduli of Vector Bundles*, Dekker, Lect. Notes Math. **179**, 1996, 147–187.

103. Matsumura, H., *Commutative Algebra*, W. A. Benjamin Co., New York, 1970.

104. Matsumura, H., *Commutative Ring Theory*, Cambridge Univ. Press, 1986.

105. Migliore, J. C., *Introduction to Liaison Theory and Deficiency Modules*, Birkhäuser, Boston, 1998.

106. Miró-Roig, R. M., Non-obstructedness of Gorenstein subschemes of codimension 3 in $\mathbb{P}^n$, *Beiträge z. Alg. u. Geom.* **33** (1992), 131–138.

107. Mohan Kumar, N., and Srinivas, V., The Noether–Lefschetz theorem (preprint).

108. Moishezon, B., On algebraic cohomology classes on algebraic varieties, *Math. USSR-Izvestia* **1** (1967), 209–251.

109. Mori, S., Projective manifolds with ample tangent bundles, *Annals of Math.* **110** (1979), 593–606.

110. Mori, S., On degrees and genera of curves on smooth quartic surfaces in $\mathbb{P}^3$, *Nagoya Math. J.* **96** (1984), 127–132.

111. Morrow, J., and Kodaira, K., *Complex Manifolds*, Holt, Rinehart and Winston, 1971.

112. Mumford, D., Further pathologies in algebraic geometry, *Amer. J. Math.* **84** (1962), 642–648.

113. Mumford, D., Projective invariants of projective structures and applications, *Proc. Int. Cong. Math. Stockholm* (1962), 526–530.
114. Mumford, D., Picard groups of moduli problems, in *Arithmetical Algebraic Geometry (Purdue)*, Harper and Row, 1965, 33–81.
115. Mumford, D., *Lectures on curves on an algebraic surface*, Annals of Math. Studies **59**, Princeton (1966).
116. Mumford, D., Biextensions of formal groups, in *Algebraic Geometry*, Bombay 1968, Oxford Univ. Press, London (1969), 307–322.
117. Mumford, D., A remark on the paper of M. Schlessinger, *Rice University Studies* **59** (1973), 113–117.
118. Mumford, D., Pathologies IV, *Amer. J. Math.* **97** (1975), 847–849.
119. Mumford, D., *Geometric Invariant Theory*, Springer (1965); second enlarged edition, with J. Fogarty, Springer, 1982.
120. Mumford, D., and Oort, F., Deformations and liftings of finite, commutative group schemes, *Invent. Math.* **5** (1968), 317–334.
121. Mumford, D., and Suominen, K., Introduction to the theory of moduli, in *Algebraic Geometry*, Oslo 1970, Wolters–Noordhoff, Groningen, 1972.
122. Nagata, M., On self-intersection number of a section on a ruled surface, *Nagoya Math. J.* **37** (1970), 191–196.
123. Narasimhan, M. S., and Seshadri, C. S., Stable and unitary vector bundles on a compact Riemann surface, *Ann. Math.* **82** (1965), 540–567.
124. Nitsure, N., Construction of Hilbert and Quot schemes, in *Fundamental Algebraic Geometry*, by B. Fantechi, et al., Amer. Math. Soc., 2005, 105–137.
125. Noether, M., Zur Grundlegung der Theorie der algebraischen Raumcurven, *Abh. Kön, Preuss. Akad. Wiss.*, Berlin (1882).
126. Ogus, A., and Bergman, G., Nakayama's lemma for half-exact functors, *Proc. AMS* **31** (1972), 67–74.
127. Ogus, A., Supersingular K3 crystals, *Astérisque* **64** (1979), 3–86.
128. Oh, K., and Rao, A. P., Dimension of the Hilbert scheme of curves in $\mathbb{P}^3$, *Amer. J. Math.* **118** (1996), 363–375.
129. Oort, F., Finite group schemes, local moduli for Abelian varieties, and lifting problems, in *Algebraic Geometry*, Oslo 1970, Wolters–Noordhoff, Groningen, 1972, 223–254.
130. Oort, F., Lifting algebraic curves, Abelian varieties, and their endomorphisms to characteristic zero, *Proc. Symp. Pure Math.* **46** AMS, Providence (1987), Vol II, 165–195.
131. Perrin, D., Courbes passant par m points généraux de $\mathbb{P}^3$, *Bull. Soc. Math. France*, Mémoire **28/29** (1987).
132. Peskine, C., and Szpiro, L., Dimension projective finie et cohomologie locale, *Publ. Math. IHES* **42** (1973), 47–119.
133. Peskine, C., and Szpiro, L., Liaison des variétés algébriques, *Invent. Math.* **26** (1974), 271–302.
134. Piene, R., and Schlessinger, M., On the Hilbert scheme compactification of the space of twisted cubics, *Amer. J. Math.* **107** (1985), 761–774.
135. Pinkham, H. C., Deformations of algebraic varieties with $\mathbb{G}_m$-action, *Astérisque* **20** Soc. Math. Fr. (1974).
136. Pinkham, H. C., Deformations of cones with negative grading, *J. Algebra* **30** (1974), 92–102.
137. Popp, H., *Moduli theory and classification theory of algebraic varieties*, Springer, Lecture Notes in Math. **620**, 1977.

138. Quillen, D., On the (co-)homology of commutative rings, in *Applications of Categorical Algebra*, Proc. Symp. Pure Math., Amer. Math. Soc. (1970), 65–87.

139. Rauch, H. E., The singularities of the modulus space, *Bull. AMS* **68** (1962), 390–394.

140. Raynaud, M., Contre-exemple au "Vanishing theorem" en caractéristique $p > 0$, in *C. P. Ramanujan—A Tribute*, Springer (1978), 273–278.

141. Rim, D. S., *Formal deformation theory*, exposé VI of SGA7,I, Springer LNM **288**, 1972.

142. Rohn, K., Die Raumcurven auf den Flächen IV^{ter} Ordnung, *Leipziger Bericht* **49** (1897), 631–663.

143. Rohn, K., and Berzolari, L., Algebraische Raumkurven und abwickelbare Flächen, *Enz. Math. Wiss.* $III_2.9$ (1928), 1229–1436.

144. Schaps, M., Deformations of Cohen–Macaulay schemes of codimension 2 and nonsingular deformations of space curves, *Amer. J. Math.* **99** (1977), 669–685.

145. Schlessinger, M., Functors of Artin rings, *Trans. AMS* **130** (1968), 208–222.

146. Schlessinger, M., Rigidity of quotient singularities, *Invent. Math.* **14** (1971), 17–26.

147. Schlessinger, M., On rigid singularities, *Rice Univ. Studies* **59**, 1973, 147–162.

148. Schlessinger, M., and Stasheff, J., The Lie algebra structure of tangent cohomology and deformation theory, *J. Pure Appl. Algebra* **38** (1985), 313–322.

149. Schröer, S., Some Calabi–Yau threefolds with obstructed deformations over the Witt vectors, *Compos. Math.* **140** (2004), 1579–1592.

150. Sernesi, E., Un esempio di curva ostruita in $\mathbb{P}^3$, *Sem. Variabili Complesse*, Univ. Bologna (1981), 223–231.

151. Sernesi, E., *Topics on families of projective schemes*, Queen's University, Kingston, Ont (1986).

152. Sernesi, E., *Deformations of Algebraic Schemes*, Springer, Grundlehren **334**, 2006.

153. Serre, J.-P., Géométrie algébrique et géométrie analytique, *Ann. Inst. Fourier* **6** (1956), 1–42.

154. Serre, J.-P., Exemples de variétés projectives en caractéristique p non relevable en caractéristique zéro, *Proc. Nat. Acad. Sci.* **47** (1961), 108–109.

155. Serre, J.-P., *Corps locaux*, Hermann, 1962.

156. Serre, J.-P., *Algèbre locale-multiplicités*, Springer LNM **11**, 1965.

157. Seshadri, C. S., Theory of moduli, *Proc. Symp. Pure Math.* (AMS) **29** (1975), 263–304.

158. Severi, F., Una proprietà delle forme algebriche prive di punti multipli, *Rend. Acc. Lincei* **15** (1906), 691.

159. Stevens, J., Computing versal deformations, *Exper. Math.* **4** (1994), 129–144.

160. Stevens, J., *Deformations of Singularities*, Springer LNM **1811**, 2003.

161. Strømme, S. A., Elementary introduction to representable functors and Hilbert schemes, in: *Parameter Spaces*, Banach Center Publications **36**, 1996, 179–198.

162. Swinnerton-Dyer, H. P. F., An enumeration of all varieties of degree 4, *Amer. J. Math.* **95** (1973), 403–418.

163. Tannenbaum, A., Families of algebraic curves with nodes, *Compos. Math.* **41** (1980), 107–126.

164. Vakil, R., Murphy's law in algebraic geometry: badly behaved deformation spaces, *Invent. Math.* **164** (2006), 569–590.

165. Vistoli, A., Intersection theory on algebraic stacks and on their moduli spaces, *Invent. Math.* **97** (1989), 613–670.

166. Vistoli, A., The deformation theory of local complete intersections, arXiv, 12.Apr (1999).
167. Vistoli, A., Grothendieck topologies, fibered categories and descent theory, in: *Fundamental Algebraic Geometry*, by B. Fantechi et al., Amer. Math. Soc., 2005, 1–104.
168. Voisin, C., Une précision concernant le théorème de Noether, *Math. Ann.* **208** (1988), 605–611.
169. Voisin, C., Composantes de petite codimension du lieu de Noether–Lefschetz, *Comm. Math. Helv.* **64** (1989), 515–526.
170. Voisin, C., Sur le lieu de Noether–Lefschetz en degrés 6 et 7, *Compos. Math.* **75** (1990), 47–68.
171. Wahl, J., Deformations of plane curves with nodes and cusps, *Amer. J. Math.* **96** (1974), 529–577.
172. Wahl, J., Equisingular deformations of plane algebroid curves, *Trans. AMS* **193** (1974), 143–170.
173. Watanabe, J., A note on Gorenstein rings of embedding codimension three, *Nagoya Math. J.* **50** (1973), 227–232.
174. Zariski, O., Introduction to the problem of minimal models in the theory of algebraic surfaces, *Pub. Math. Soc. Japan* **4** (1958).
175. Zariski, O., *Algebraic Surfaces*, Springer, Ergebnisse (1935); second supplemented edition, Springer, 1971.
176. Zariski, O., *The moduli problem for plane branches*, with an appendix by Bernard Teissier, Amer. Math. Soc., University Lecture Series **39** (2006) [translated from original French edition of 1973].
177. Zariski, O., and Samuel, P., *Commutative Algebra*, van Nostrand, Princeton, 1958, 1960.

Index